Ecological Silvicultural Systems

Ecological Silvicultural Systems

Exemplary Models for Sustainable Forest Management

Edited by

Brian J. Palik
USDA Forest Service
Northern Research Station
Grand Rapids, MN, USA

Anthony W. D'Amato
Rubenstein School of Environment and Natural Resources
University of Vermont
Burlington, VT, USA

WILEY

This edition first published 2024

Registered Office(s)
John Wiley & Sons, Inc., 111 River Street, Hoboken, NJ 07030, USA
John Wiley & Sons Ltd, The Atrium, Southern Gate, Chichester, West Sussex, PO19 8SQ, UK

For details of our global editorial offices, customer services, and more information about Wiley products visit us at www.wiley.com.

Wiley also publishes its books in a variety of electronic formats and by print-on-demand. Some content that appears in standard print versions of this book may not be available in other formats.

Library of Congress Cataloging-in-Publication Data
Names: Palik, Brian, editor. | D'Amato, Anthony W., editor.
Title: Ecological silvicultural systems : exemplary models for sustainable forest management / Brian J. Palik, Anthony W. D'Amato.
Other titles: Exemplary models for sustainable forest management
Description: First edition. | Hoboken, NJ : Wiley, 2024 | Includes index.
Identifiers: LCCN 2023023190 (print) | LCCN 2023023191 (ebook) | ISBN 9781119890904 (paperback) | ISBN 9781119890928 (adobe pdf) | ISBN 9781119890935 (epub)
Subjects: LCSH: Sustainable forestry. | Forest ecology. | Forest management–Environmental aspects.
Classification: LCC SD387.S87 E24 2024 (print) | LCC SD387.S87 (ebook) | DDC 634.9/2–dc23/eng/20230825
LC record available at https://lccn.loc.gov/2023023190
LC ebook record available at https://lccn.loc.gov/2023023191

Cover Design: Wiley
Cover Image: Courtesy of Lukas Kopacki; Brian J. Palik; Douglas Kastendick; Anthony W. D'Amato

Set in 9.5/12.5pt STIXTwoText by Straive, Pondicherry, India

Contents

List of Contributors

William J. Beese
Vancouver Island University (Retired)
Nanaimo, BC, Canada

John-Pascal Berrill
Cal Poly Humboldt, Arcata, CA, USA

Derek J. Churchill
Forest Resilience Division,
Washington Department of Natural
Resources, Olympia, WA, USA

Philip G. Comeau
Department of Renewable Resources
University of Alberta, Edmonton, AB, Canada

Justin S. Crotteau
USDA Forest Service Rocky Mountain
Research Station, Missoula, MT, USA

Miranda T. Curzon
Department of Natural Resource Ecology
and Management, Iowa State University
Ames, IA, USA

Anthony W. D'Amato
Rubenstein School of Environment and
Natural Resources, University of Vermont
Burlington, VT, USA

Robert L. Deal
USDA Forest Service Pacific Northwest
Research Station (Retired), Portland, OR, USA

Daniel C. Dey
Northern Research Station, USDA Forest
Service, Columbia, MO, USA

Pablo J. Donoso
Facultad de Ciencias Forestales y
Recursos Naturales, Instituto de Bosques
y Sociedad, Universidad Austral de Chile
Valdivia, Chile

Daniel Dumais
Direction de la recherche forestière
Ministère des Ressources naturelles et des
Forêts du Québec, Canada

Jodi A. Forrester
Department of Forestry and Environmental
Resources, North Carolina State University
Raleigh, NC, USA

Jerry F. Franklin
School of Environmental and Forest Sciences
University of Washington, Seattle, WA, USA

Miguel Montoro Girona
Groupe de Recherche en Écologie de la
MRC Abitbi (GREMA), Institute of Forest
Research, Université du Québec en Abitibi-
Témiscamingue, Amos, Quebec, Canada

J. Davis Goode
Department of Geography and the
Environment, University of Alabama
Tuscaloosa, AL, USA

Constance A. Harrington
Harrington, USDA Forest Service Pacific Northwest Research Station (Emeritus)
Olympia, WA, USA

Justin L. Hart
Department of Geography and the Environment, University of Alabama
Tuscaloosa, AL, USA

Steven B. Jack
Boggy Slough Conservation Area
T.L.L. Temple Foundation, Lufkin, TX, USA

John M. Kabrick
Northern Research Station, USDA Forest Service, Columbia, MO, USA

Tara L. Keyser
USDA Forest Service, Southern Research Station, Asheville, NC, USA

Benjamin O. Knapp
School of Natural Resources, University of Missouri, Columbia, MO, USA

Randy K. Kolka
USDA Forest Service, Northern Research Station, Grand Rapids, MN, USA

Timo Kuuluvainen
Department of Forest Sciences, University of Helsinki, Helsinki, Finland

S. Ellen MacDonald
Department of Renewable Resources
University of Alberta, Edmonton
AB, Canada

Andrew J. Larson
Department of Forest Management
University of Montana, Missoula, MT, USA

David B. Lindenmayer
Fenner School of Environment and Society
Australian National University, Canberra
ACT, Australia

Maxence Martin
Universidad de Huelva
dr. Cantero Cuadrado 6
Huelva 21004, Spain

R. Kevin McIntyre
Jones Center at Ichauway, Newton, GA, USA

Thomas A. Nagel
Department of Forestry and Renewable Forest Resources, University of Ljubljana, Ljubljana
Slovenia

Kellen N. Nelson
USDA Forest Service Pacific Northwest Research Station, Juneau, AK, USA

Charles A. Nock
Department of Renewable Resources
University of Alberta, Edmonton, AB, Canada

Martin Alcala Pajares
Universidad de Huelva
dr. Cantero Cuadrado 6
Huelva 21004, Spain

Brian J. Palik
USDA Forest Service, Northern Research Station, Grand Rapids, MN, USA

Brad D. Pinno
Department of Renewable Resources
University of Alberta, Edmonton, AB, Canada

Timo Pukkala
Forest Science and Technology Centre of Catalonia, Solsona, Spain

Patricia Raymond
Direction de la recherche forestière, Ministère des Ressources naturelles et des Forêts du Québec, Canada

Laura F. Reuling
Department of Forest Resources, University of Minnesota, Twin Cities MN, USA

Dušan Roženbergar
Department of Forestry and Renewable Forest Resources, University of Ljubljana, Ljubljana Slovenia

David K. Schnake
Department of Forestry and Environmental Resources, North Carolina State University Raleigh, NC, USA

Robert S. Seymour
School of Forest Resources, University of Maine, Orono, ME, USA

Robert A. Slesak
USDA Forest Service, Pacific Northwest Research Station, Olympia, WA, USA

Daniel P. Soto
Department of Natural Sciences and Technology University of Aysen
Coyhaique, Chile

Miroslav Svoboda
Department of Forest Ecology, Faculty of Forestry and Wood Sciences, Czech University of Life Sciences Prague, Prague Czech Republic

Lucie Vítková
Department of Forest Ecology, Faculty of Forestry and Wood Sciences, Czech University of Life Sciences Prague, Prague Czech Republic

Stephanie J. Wessell
Northwest Oregon District Office
USDI Bureau of Land Management
Salem, OR, USA

Abraham Wheeler
OR/WA State Office, USDI Bureau of Land Management, Portland, OR, USA

Marcella A. Windmuller-Campione
Department of Forest Resources, University of Minnesota, Twin Cities MN, USA

Robert A. York
Department of Environmental Science Policy and Management, University of California Berkeley, CA, USA

Preface

In a nutshell, ecological silviculture is the toolbox to manage forests as ecosystems, based on emulation of natural models of disturbance and development. It follows that an ecological silvicultural system is the long-term sequence of treatments for managing composition and structure of a forest, as informed by natural disturbance and development. One could say that ecological silviculture is guilty of putting the forest first, rather than focusing one or a few tree species of commercial value, which can be the focus of timber-focused silviculture. This is not to say that ecological silviculture does not include an objective of managing for timber and commodity output. It does and, in fact, it could not be considered for wide application if there was not a path for profitable forest management as an objective. The distinction, from timber-focused silviculture, is that sustainability of non-commodity goods and services is given high priority as an objective. Moreover, the conceptual framework for devising ecological silvicultural systems is that *nature knows best,* rather than building on agronomic models like classic silviculture. The former involves understanding natural disturbance dynamics for a forest and how that forest develops over time to become a structurally complex and often species rich ecosystem, with a goal of reducing, as much as feasible, the disparity between a natural ecosystem and its managed counterpart. As you will learn, this attempt to better emulate natural forest development is important even in the face of unprecedented global changes, including climate change and invasive pests.

You will learn that not all contributors to this book have the same interpretation of a natural model for forests, nor are their formulations of ecological silvicultural systems the same. In fact, the silviculture presented in this text reflects highly diverse views about how forests work naturally and how silviculture can be used to better emulate natural processes. The diversity reflects geography, with contributors working in forests across the United States, in eastern, central, and western Canada; in Chile and Australia; and several forests in Europe. It also reflects differing views about the degree to which a natural model can be used in silviculture. Embrace this diversity, as it captures the most innovative thinking on the topic of ecological silviculture from the world's most innovative thinkers in the field. Importantly, all contributors do share one central goal, that of presenting silvicultural systems for forests that better sustain ecosystems broadly, so as to provide the full array of services people expect from forests.

Brian J. Palik and Anthony W. D'Amato

Acknowledgments

Many influential thinkers have shaped our ideas about forest ecology and silviculture, through their writings and mentorship, but more often through thoughtful discussions. In fact, several of these people have contributed to this book, while the influence of others is reflected in the chapters we authored. Topping our list is Dr. Jerry Franklin, whose thinking on the topic of managing forests ecologically predated most, if not all western scientists. Without his mentorship and inspiration, we would have very different careers. We thank others, in no particular order, including Robert (Bob) Mitchell, Bob Seymour, Mac Hunter, Kurt Pregitzer, Shawn Fraver, John Bradford, Craig Lorimer, Ellen Macdonald, Jim Long, Patricia Raymond, Dan Kneeshaw, Linda Nagel, David Foster, Klaus Puettmann, Steve Jack, Bill Leak, Steve Bédard, Kevin Evans, Paul Catanzaro, David Lindenmayer, Mariko Yamasaki, Mike Snyder, and David Kittredge. We also acknowledge the generations of Indigenous peoples and the historically overlooked models of ecological silviculture they advanced and continue to practice today. These contributions far predate the formalization of silviculture by Western scientists and foresters and deserve a long overdue acknowledgment for providing critical models to emulate as we grapple with how best to steward forests into the future. Finally, our home institutions provided the time and financial support needed to complete this project including the USDA Forests Service-Northern Research Station (Palik) and The University of Vermont-Rubenstein School of Environment and Natural Resources (D'Amato). We thank them for this support.

Brian J. Palik and Anthony W. D'Amato

1

The Context of Ecological Silviculture

Brian J. Palik[1] *and Anthony W. D'Amato*[2]

[1] *USDA Forest Service, Northern Research Station, Grand Rapids, MN, USA*
[2] *Rubenstein School of Environment and Natural Resources, University of Vermont, Burlington, VT, USA*

1.1 What Is Ecological Silviculture?

In forestry, the tools to carry out on-the-ground management are found in the toolbox of silviculture, that is, the part of forestry that deals with the development and care of forests. This development and care can be all encompassing, including establishment, growth, composition, health, and quality of trees and stands to meet the diverse needs of society. The planned program of all these activities, spanning the entire life of a forest stand, is termed a silvicultural system.

Ostensibly, such a definition for silviculture would seem to recognize most forests, outside of plantations, as natural ecosystems and that the tools to manage these ecosystems would be based on ecological principles. In reality, classical silviculture is largely based on principles of production agriculture, more so than the ecology of natural forests [1]. The outcome of the long-term application of an agricultural model to managed forest is near global-wide simplification of structure and composition [2], as well as a reduced ability to adapt to unexpected and novel threats [3].

This is where ecological silviculture comes in, which we define as an approach for managing forests, including trees, associated organisms, and ecological functions, based on emulation of natural models of development [2]. There are other closely aligned approaches that also are ecological that have similar aims, such as close-to-nature forestry (e.g. [4]). An explicit goal of ecological silviculture and allied approaches is to manage forests in ways that reduce the disparity between natural forests and their managed counterparts in terms of composition, structure, and function. Recognizing that this disparity cannot be eliminated in managed forests, particularly those where timber resources are highly valued, ecological silviculture still rests on the premise that managers can get closer to the natural model than classical silvicultural has allowed. Let us look at why this might be, by contrasting some of the main differences between the two approaches.

Ecological Silvicultural Systems: Exemplary Models for Sustainable Forest Management,
First Edition. Edited by Brian J. Palik and Anthony W. D'Amato.

1.2 How Does Ecological Silviculture Differ from Classical Commodity-Focused Silviculture?

The central emphasis of ecological silviculture is on maintaining the full array of structures, functions, and species found in a healthy natural forest ecosystem [2]. To achieve this goal, ecological silviculture builds from an understanding of natural disturbances and forest development to craft silvicultural systems that generate and maintain complexity and heterogeneity in ecosystem attributes. This often includes the application of regeneration harvests patterned after the prevailing natural disturbance regime for an ecosystem, including their scale, severity, and frequency. In addition, the legacies of these disturbances, namely surviving trees, and coarse woody material, are accounted for by placing an equal emphasis on what is left behind relative to what is removed at each silvicultural intervention over the course of the silvicultural system. Although economic objectives are still a priority with ecological silviculture, those associated with ecosystem diversity and resilience are given high priority in the design and implementation of ecological silvicultural systems.

In contrast, commodity-focused silviculture emphasizes the "forest crop" and is correspondingly concerned with silvicultural regimes that generate forest conditions to optimize growth and development of economically desirable species and tree forms. Silviculture treatments under this approach are largely based on agronomic models with a corresponding emphasis on simplifying and homogenizing forest-wide structure (e.g. spacing) and composition (i.e. primarily commercial tree species) to maximize the area occupied by economically desirable trees and to enhance the efficiency of harvest operations. Not surprisingly, the timing, frequency, and severity of silvicultural treatments are largely governed by economic criteria, including maximizing net present value, and avoiding risk of crop-tree loss to damaging agents [5]. Forests managed under this approach can satisfy noneconomic objectives to a degree [6]; however, achievement of economic objectives is the central priority under timber-focused regimes.

1.3 Why Is Ecological Silviculture Needed?

To appreciate the need for ecological silviculture, consider how the forces shaping forest management have changed over the last 70 years. First, globally, it is estimated that there are 4.0 bha of forest, including 0.3 bha of plantations and 3.7 bha of natural-origin forests; moreover, of the latter, 0.65 bha are legally protected reserves [7]. Generally, plantations are managed intensively for wood, such that ecological benefits are marginalized, while reserves usually exclude harvesting. This leaves about 3 bha of natural-origin forests where multiple environmental, economic, and social objectives are pursued [8] and, for the portion of this forest that is managed, ecological silviculture may be the appropriate tool for meeting diverse objectives.

In fact, the drivers of policy and practice across the global forest have changed fundamentally to be more ecological [2]. For example, conservation stewards, such as The Nature Conservancy, have emerged as key players in the forest management world, with their goals being overtly ecological. Also, the objectives of traditional forest stewards, e.g. United States National Forests, have evolved to focus on sustaining whole ecosystems. Additionally, third-party certification, as well as best management practices (BMP's), have led forest stewards to adopt ecological approaches for silviculture. Many largely unprecedented health threats facing forests, including climate change, catastrophic wildfires, and invasive species, require silviculture that restores the structural and functional integrity of forests to better position them to adapt to uncertainty [9]. Finally, the vast increase in our understanding of forests – how they are structured, how they function, and how they are connected across the landscape – argues for a different kind of silviculture, one that focuses on forests as ecosystems, not agricultural ecosystems.

1.4 What Are the Foundational Concepts of Ecological Silviculture?

While the definition of ecological silviculture may vary a bit among stakeholders, all are based on an understanding and modeling of natural disturbance and forest development. For silvicultural application, such a model begins with canopy disturbance and proceeds through stages that reflect structural maturation. The model may include a *preforest* stage, followed by *young*, *mature*, and *old forest* stages (Box 1.1). The model may apply to whole-stands or patches within stands. As you will learn in the chapters that follow, ecological *silvicultural systems* often emulate this developmental model and are catered to the ecological and socioeconomic contexts of the ecosystem being managed.

Box 1.1 Some Concepts and Terms Associated with Ecological Silviculture

Biological legacies: biological legacies are healthy trees, decadent trees, large deadwood, and derivatives of dead trees, e.g. tip-up mounds, and pre-disturbance vegetation (advance regeneration, understory plants) that survive from the previous forest into the new forest after disturbance.

Deadwood: dead trees, including snags and downed wood, and tree components (branches, roots).

Forest developmental stages

Disturbance and legacy creation: a disturbance that kills trees and leaves a legacy of organisms and structures from the pre-disturbance forest.

Preforest: a period when herbs and shrubs dominate growing space and tree abundance is not enough to dominate the site.

Young forest: establishment of new cohort trees of relatively uniform size, along with large legacy trees; density-dependent mortality is a dominant process in this stage.

Mature forest: new cohort trees dominate; mortality shifts to density-independent agents; advance regeneration of shade-tolerant trees are released and species of lesser tolerance may establish in canopy gaps; spatial heterogeneity of structure increases; new deadwood accumulates.

Old forest: a diverse mix of live and dead trees of varying size and condition; high levels of spatial heterogeneity; increasing decline and mortality of trees from exogenous agents; significant inputs of large deadwood.

Recovery period: the time between regeneration harvests of sufficient duration to allow for development of mature forest structures.

Retention tree: a tree retained during a regeneration harvest to live out its natural lifespan so as to provide mature forest structural elements and associated functions.

Structural complexity: abundance, size, conditions, and degree of heterogeneity in living and dead components in a forest.

Variable retention harvesting: retention of living trees and deadwood at harvest in a range of spatial patterns *(dispersed and aggregated) and* abundances; includes retention of species or functional groups, e.g. conifers or hardwoods, and can occur stand-wide or at gap-scales depending on natural developmental model being emulated.

Variable-density thinning: silvicultural strategy that varies the density of removal across a stand, including gaps, standard thinning, and no removal; accelerates the development of complexity and heterogeneity. Although termed a "thinning," this approach includes deliberate consideration for regenerating new cohorts in gaps, so it can also be considered as a regeneration method.

Another commonality of ecological silvicultural systems is that they address several foundational principles derived from a natural model. We have discussed these principles in detail elsewhere [2, 10], which include *continuity*, *complexity/diversity*, *timing*, and *context*. Here we provide a brief review, along with some definitions of terms associated with the principles (Box 1.1).

Continuity is the provisioning of structure, biota, and function from a pre- to post-disturbance forest. The currency of continuity are *biological legacies*, especially *retention trees* and large *deadwood*, and the principle is often addressed using gap-scale or forest-wide *variable retention harvesting* (see Box 1.2) during regeneration harvests.

Box 1.2 Clarifying and Advancing Ecological Silviculture Terminology: Variable Retention Harvests and Variable-Density Thinning

> *When the important task of inventing the solution has proceeded far enough, the less important task of attaching a name to it can be taken. Standard terminology should be used to the extent of its limited capacity for providing information in terms meaningful to all foresters and then supplemented with additional, detailed information that is usually necessary.*
>
> —David Smith [11]/John Wiley & Sons.

The use of a common and standard terminology has been a central aspect of the discipline of silviculture with the core intent of minimizing confusion over intended prescription outcomes and maintaining professionalism and consistency of communication in the field. The silvicultural terms and practices used in North America and many other regions generally owe their origin to European silviculture methods, which were primarily designed to address the predominant socioecological conditions in Europe during the nineteenth and twentieth centuries [1]. These terms and practices were codified into North American professional standards, texts, and curricula in the early twentieth century [12, 13] and have served as the cornerstone of silviculture ever since. It is noteworthy that Indigenous models of forest stewardship actually predated the formalization and organization of the American forestry profession [14], yet these models were largely ignored by early, European-trained forest scientists and foresters and have only recently been recognized by non-tribal institutions as powerful models for informing ecological silviculture (e.g. [15, 16]).

Given the general adherence of North American silviculturists to European terminology and approaches, a long-standing, albeit unfounded, criticism of ecological silviculture has been that it largely reinvented terms for practices that already have names in the traditional silviculture vernacular. This is particularly true for two of the most commonly applied elements of ecological silvicultural systems, *variable retention harvests (VRH) and variable density thinning (VDT)*. For both of these practices, criticism has stemmed from their similarity to existing regeneration methods that include "reserves" or that apply multi-cohort systems that work from and maintain spatial heterogeneity in structure and composition in a forest over time (e.g. group selection and some irregular shelterwood systems). The outcomes of these existing methods may resemble those following VRH or VDT at different stages of implementation; however, there are key differences in the intent of historical silvicultural methods that prevent making them synonymous with certain ecological silviculture terms and approaches.

In an attempt to move terminology forward, particularly given how widespread the use of VRH and VDT has become (as evidenced in the chapters that follow), it is worth defining and clarifying the intent of these ecological silvicultural methods. Our hope is to introduce consistency in communicating the intent and application of these practices and to move beyond forcing the use of traditional terms in cases where they do not make sense nor convey silvicultural intent. Importantly, the use of standard and historical terminology still is a key aspect of any silvicultural application, whether it is ecological or not. Nevertheless, we should strive as a profession to allow for evolution of historical and new terms to allow for operationalization of strategies that match new and emerging objectives in a twenty-first century context.

Variable Retention Harvests

VRH is a two-aged regeneration method that emphasizes the permanent retention of mature canopy trees, as well as other important structural legacies, such as snags and downed logs at the time of regeneration harvest. VRH was first popularized in the Pacific Northwest region of the United States during the 1990s in response to social and ecological concerns surrounding the widespread use of clear-cutting methods [17], with their use expanding globally over the past two decades into many regions where clear-cutting-based methods historically prevailed [18]. In practice, this method encapsulates the ecological silviculture principle of continuity by ensuring structural, functional, and compositional elements from the preharvest forest are carried forward into the new, regenerating stand.

The use of the term "variable" to describe these systems reflects the need to vary the amount and distribution of retained trees depending on management objectives and ecological conditions, with this flexibility contributing to the wide range of global ecosystems and ownerships now applying this approach. In regions where there is a longer history and greater experience with VRH, such as British Columbia, Canada, fairly specific guidelines exist surrounding levels of retention, and spatial patterns of retention when applying VRH as an integrated silvicultural system for achieving ecological objectives [19]. For other regions and forest types, less guidance may exist and the flexibility of VRH should be utilized to determine the retention levels and patterns necessary to achieve ecological goals over time. To this end, although originally termed a "harvest system," VRH is better thought of as a regeneration method, so the seedbed conditions and resource environments required by the desired species on a site are a critical consideration in the design of VRH, with differing patterns of retention often used to accommodate species with varying tolerance [20]. Unfortunately, applications of VRH have led to a common misconception that solely retaining trees at the time of harvest is all that ecological silviculture entails [10]; however, true ecological silvicultural systems often use VRH as the regeneration method occurring at the Disturbance and Legacy Creation stage of a given system. As with other silvicultural systems, subsequent intermediate treatments, such as release treatments and thinning, are applied over time to enhance ecological conditions and future economic outputs and achieve long-term desired future ecological conditions (Figure 1.1).

Existing two-aged regeneration methods, such as clear-cutting, seed-tree, and shelterwood, with reserves certainly have similarities to VRH given these methods include the retention of some mature trees beyond 20% of the rotation age of the new regenerating cohort (the definition of *with reserves* systems). Nevertheless, a key and important distinction between VRH and these older methods is intent. In particular, the historical intent of retaining "reserve trees" largely revolved around allowing canopy trees to grow to larger, more economically valuable sizes before their ultimate harvest [13]. Modern silviculture texts certainly acknowledge that we now also retain "reserves" for many ecological and cultural values [21]; however, future

Figure 1.1 Variable retention harvest in red pine-dominated forests in north-central Minnesota, USA. Left photo represents disturbance/legacy creation stage with right hand photo taken from same site during early portion of the young forest stage. Release treatments were applied to site during young forest stage to enhance development of natural and artificial regeneration established on site.

economic returns from these reserve trees remains an objective linked to this terminology, making it incongruous with the intent of VRH.

Of note, permanent retention of legacy trees is certainly not restricted to VRH, as gap-level retention of legacy trees is a common feature of ecological silviculture systems for ecosystems dominated by gap and mesoscale-disturbance regimes. As with VRH, simply modifying the regeneration methods used for these forests to include "reserves" does not fully convey silvicultural intent given the primary objective for retention is for those trees to live out their natural lifespan. As such, adding modifiers such as "with permanent legacy retention" to regeneration methods, such as group selection or expanding gap irregular shelterwoods, maybe a more effective way to communicate these methods going forward to avoid confusion with traditional strategies developed during a time (i.e. eighteenth and nineteenth century) where permanent retention was not acceptable from a socioeconomic standpoint.

Variable-Density Thinning

VDT is a harvesting method in which the removal of trees is varied spatially across a stand or management unit to increase heterogeneity in horizontal and vertical structure. As with VRH, VDT developed in the Pacific Northwest region of the United States, with the intent of restoring the spatial variability in canopy tree densities and age classes that are a common attribute of natural forests around the globe [22]. This approach reflects the ecological silviculture principle of complexity and is generally applied to forests in the late young forest or mature forest stages of stand development to accelerate heterogeneity in forest conditions. To this end, VDT is a common component of ecological silvicultural systems applied to second-growth forests homogenized by past management practices (e.g., plantations). In its application, VDT includes the creation of a combination of canopy gaps and thinned areas, as well as designation of permanent, unharvested areas as patch reserves (i.e. groups of legacy trees) (Figure 1.2). This pattern of harvesting and retention creates and reinforces structural and compositional conditions found in the old forest stage of development, including canopy gaps, lower density areas

Figure 1.2 Variable density thinning (VDT) within 6-ha harvest unit on the Nulhegan Basin of the Silvio Conte National Wildlife Refuge in Brunswick, VT, USA. Overarching goal of VDT is to meet objectives surrounding increasing ecolcgical complexity in spruce-fir forests simplified by past land use to enhance habitat conditions for priority bird species. (a) Second-growth spruce-fir forest homogenized by historic, intensive management; (b) 0.1 ha gaps for recruiting new age-class and increasing compositional diversity; (c) thinned matrix conditions for accelerating large-tree development; and (d) patch reserves (i.e. skips) to protect unique ecological features and maintain areas of high stem density and canopy cover.

with larger diameter trees, and dense vegetation, all-natural outcomes of a long-term history of gap-phase disturbance and tree maturation.

A primary reason for the widespread appeal of VDT is the relative simplicity with which it operationalizes and achieves desired ecological outcomes (e.g. [23]). In particular, tree marking focuses on identifying logical areas for gap creation (e.g. pockets of advance regeneration) and patch reserves or "skips" (e.g. rare plant communities, areas with high accumulations of deadwood) with thinning applied in matrix areas between these gap and anti-gap elements to accelerate tree growth. Adaptations of this approach have been extended to frequent-fire mixed conifer systems where widely spaced trees, canopy openings, and clumps of trees are created using the Individuals, Clumps, and Openings (ICO) approach [24]. As with VDT, the varying tree densities created by the ICO approach are intended to restore natural patterns of tree density, which in the case of frequent-fire forests are often mosaics of large-diameter individual trees, gaps, and tree clumps of different sizes.

An unfortunate aspect of VDT is its name, as this approach includes provisions for regeneration in the gap elements created. Thinning treatments in a traditional sense are focused on improving the growth and quality of existing trees on a site with no intent to regenerate new cohorts. The use of "thinning" in describing VDT has created confusion over its intent and where it fits with regard to existing silvicultural treatments (i.e. is it an intermediate treatment or regeneration method?). Given the inclusion of regeneration as part of VDT, it is best classified as a regeneration method with aspects similar to hybrid single-tree and group selection and expanding gap and continuous cover irregular shelterwoods. A key difference is that VDT is meant to restore a spatial structure that includes stand areas that will never be harvested (skips). In addition, VDT is best viewed in many applications as a transformative silviculture treatment moving simplified, even-aged stands into more complex, multi-aged conditions [25].

As such, future entries following the application of VDT may take the form of the aforementioned selection and irregular shelterwood methods by either expanding gaps created by VDT or applying single-tree and group removals in thinned matrix areas to further emulate natural developmental models for a given forest. In short, it is an unfortunate name but has been widely demonstrated as a highly valuable tool for conveying and achieving ecological silviculture objectives, so we are sticking with it.

Complexity/diversity addresses the need to create and maintain *structural complexity* and species diversity in managed stands. Often, application of this principle takes the form of restoration, particularly in stands lacking complexity and tree diversity. Some of the tools for addressing complexity/diversity include *variable-density thinning (VDT)* (see Box 1.2), decadence and deadwood creation, and tree enrichment.

Timing is the application of silvicultural interventions at ecologically appropriate intervals, especially the time necessary for *recovery* and development of structure after a major disturbance. It also reflects the need to allow appropriate time intervals between intermediate treatments, such as thinning.

Finally, *context* underscores the importance of implementing silviculture with the landscape perspective in mind. Context recognizes that stand-level actions accumulate to influence the structure and function of landscapes.

1.5 What to Expect from the Chapters that Follow?

The chapters that follow present ecological silvicultural systems for various forest types in North American, as well as examples from Europe, South American, and Australia. These systems have been conceived by innovative researchers and forest stewards who often have been studying their forests for decades. The chapters will vary a bit, depending on the forest type, but all will have some commonalities. Each will present (i) the natural history of the forest, (ii) a summary of the forest's cultural, economic, and ecological importance, (iii) the natural dynamics model for the ecosystem, (iv) a review of the traditional silvicultural system, (v) a presentation of an ecological silvicultural system and a reflection on how the latter reduces the disparity between the natural and management models, and finally, (vi) a consideration of ecological silviculture in the context of climate change.

Another common thread is that all the authors recognize that for their forest, a tweaking of classical, often timber-focused, silviculture is not sufficient to meet the needs of forest management broadly in the twenty-first century. Moreover, the contributors recognize that a simple bifurcation of the forest landscape into production and protected forests [26] is not sufficient to sustain natural forest ecosystems broadly; that is, for much of the global forest ecological silvicultural approaches may better sustain the diverse suite of ecosystem services society expects [27], even under changing climate and disturbance regimes [9]. Providing the tools to meet these expectations in diverse forests is what you can expect from this book.

References

1 Puettmann, K.J., Coates, K.D., and Messier, C.C. (2009). *A Critique of Silviculture: Managing for Complexity*. Washington, DC: Island Press.

2 Palik, B.J., D'Amato, A.W., Franklin, J.F., and Johnson, K.N. (2020). *Ecological Silviculture: Foundations and Applications*. Long Grove, IL: Waveland Press.

3 Puettmann, K.J. (2021). Extreme events: managing forests when expecting the unexpected. *Journal of Forestry* 119: 422–431.

4 Schütz, J.P. (1999). Close-to-nature silviculture: is this concept compatible with species diversity? *Forestry* 72: 359–366.

5 Bettinger, P., Boston, K., Siry, J.P., and Grebner, D.L. (2009). *Forest Management and Planning*. Amsterdam, The Netherlands: Academic Press/Elsevier.

6 Brockerhoff, E.G., Jactel, H., Parrotta, J.A. et al. (2008). Plantation forests and biodiversity: oxymoron or opportunity? *Biodiversity and Conservation* 17: 925–951.

7 Food and Agriculture Organization of the United Nations (2015). *Global Forest Resources Assessment 2015: How Are the World's Forests Changing?* Rome, Italy: FAO.

8 Lindenmayer, D.B., Franklin, J.F., Lohmus, A. et al. (2012). A major shift to the retention approach for forestry can help resolve some global forest sustainability issues. *Conservation Letters* 5: 421–431.

9 D'Amato, A.W. and Palik, B.J. (2021). Building on the last "new" thing: exploring the compatibility of ecological and adaptation silviculture. *Canadian Journal of Forest Research* 51: 172–180.

10 Palik, B.J. and D'Amato, A.W. (2017). Ecological forestry: much more than retention harvesting. *Journal of Forestry 115*: 51–53.

11 Smith, D.M. (1962). *The Practice of Silviculture*, 7e. New York, NY USA: Wiley.

12 Fernow, B.E. (1917). Forest terminology—report of committee. *Journal of Forestry* 15: 68–101.

13 Hawley, R.C. (1921). *The Practice of Silviculture*. New York, NY: Wiley.

14 Waller, D.M. and Reo, N.J. (2018). First stewards: ecological outcomes of forest and wildlife stewardship by indigenous peoples of Wisconsin, USA. *Ecology and Society* 23: 45.

15 Dockry, M.J. and Hoagland, S.J. (2017). A special issue of the Journal of Forestry—tribal forest management: innovations for sustainable forest management. *Journal of Forestry* 115: 339–340.

16 Knight, C.A., Anderson, L., Bunting, M.J. et al. (2022). Land management explains major trends in forest structure and composition over the last millennium in California's Klamath Mountains. *Proceedings of the National Academy of Sciences* 119: e2116264119.

17 Franklin, J.F. and Donato, D.C. (2020). Variable retention harvesting in the Douglas-fir region. *Ecological Processes* 9: 8.

18 Gustafsson, L., Baker, S.C., Bauhus, J. et al. (2012). Retention forestry to maintain multifunctional forests: a world perspective. *Bioscience* 62: 633–645.

19 Mitchell, S.J. and Beese, W.J. (2002). The retention system: reconciling variable retention with the principles of silvicultural systems. *The Forestry Chronicle* 78: 397–403.

20 Roberts, M.W., D'Amato, A.W., Kern, C.C., and Palik, B.J. (2017). Effects of variable retention harvesting on natural tree regeneration in *Pinus resinosa* (red pine) forests. *Forest Ecology and Management* 385: 104–115.

21 Ashton, M.S. and Kelty, M.J. (2018). *The Practice of Silviculture: Applied Forest Ecology*, 10e. Hoboken, NJ: Wiley Press.

22 Carey, A.B. (2003). Biocomplexity and restoration of biodiversity in temperate coniferous forest: inducing spatial heterogeneity with variable-density thinning. *Forestry* 76: 127–136.

23 Puettmann, K.J., Ares, A., Burton, J.I., and Dodson, E.K. (2016). Forest restoration using variable density thinning: lessons from Douglas-fir stands in Western Oregon. *Forests* 7: 310.

24 Churchill, D.J., Larson, A.J., Jeronimo, S.M.A. et al. (2014). *The ICO Approach to Quantifying and Restoring Forest Spatial Pattern: Implementation Guide. Version 2.2*. Vashon, WA: Stewardship Forestry. 37 p.

25 O'Hara, K.L. (2001). The silviculture of transformation – a commentary. *Forest Ecology and Management* 151: 81–86.

26 Bennett, B.M. (2015). *Plantations and Protected Areas: A Global History of Forest Management.* Cambridge, MA: MIT Press.

27 Lindenmayer, D.B. and Franklin, J.F. (2002). *Conserving Forest Biodiversity: A Comprehensive Multiscaled Approach.* Washington, DC: Island press.

2

Ecological Silviculture for Great Lakes Red Pine Ecosystems

Brian J. Palik[1] *and Anthony W. D'Amato*[2]

[1] *USDA Forest Service, Northern Research Station, Grand Rapids, MN, USA*
[2] *Rubenstein School of Environment and Natural Resources, University of Vermont, Burlington, VT, USA*

2.1 Introduction

Red pine (*Pinus resinosa* Ait.) ecosystems of the western Great Lakes region of United States and Canada are the focus of this chapter. These forests are found in northern Michigan, Wisconsin, and Minnesota in the United States, and portions of southern Ontario, Canada (Figure 2.1).

Historically, because of high economic value of red pine, this ecosystem has been managed using a timber-focused silvicultural system in both plantations and natural-origin stands, generally based on even-aged regeneration systems followed by planting of red pine. The timber-focused model results in composition and structure that deviate substantially from the natural model, favoring high tree densities, monospecific composition, a narrowly even-aged structure, limited structural complexity (see Chapter 1), and rotations of 50–90 years [1]. As a consequence, approximately 76% of red pine-dominated forests in the United States Lake States are plantations [1].

Interest has grown in silvicultural approaches that better reflect the natural model of disturbance and development for this ecosystem, i.e. ecological silviculture; such approaches can be used to transition plantations and simplified natural forests to structurally complex and diverse conditions that better reflect the native ecosystem condition. There is limited experience with actual ecological silviculture approaches for red pine; this chapter provides a framework for developing these approaches.

2.2 Characteristics of Red Pine Ecosystems

2.2.1 Composition and Structure

There are many variants of red pine ecosystems across the region, but broadly we are considering ecosystems where red pine, often with eastern white pine (*Pinus strobus* L.), is the dominant species. This said, these ecosystems are actually better characterized as having mixed-species composition, potentially containing up to 12 tree species (Table 2.1). The species differ in shade tolerance and life span, but most establish on mineral soil, reflecting adaptation to post-fire environments. The species are present in varying abundance depending on the specific variant of the ecosystem and the stage of development.

Ecological Silvicultural Systems: Exemplary Models for Sustainable Forest Management, First Edition. Edited by Brian J. Palik and Anthony W. D'Amato.

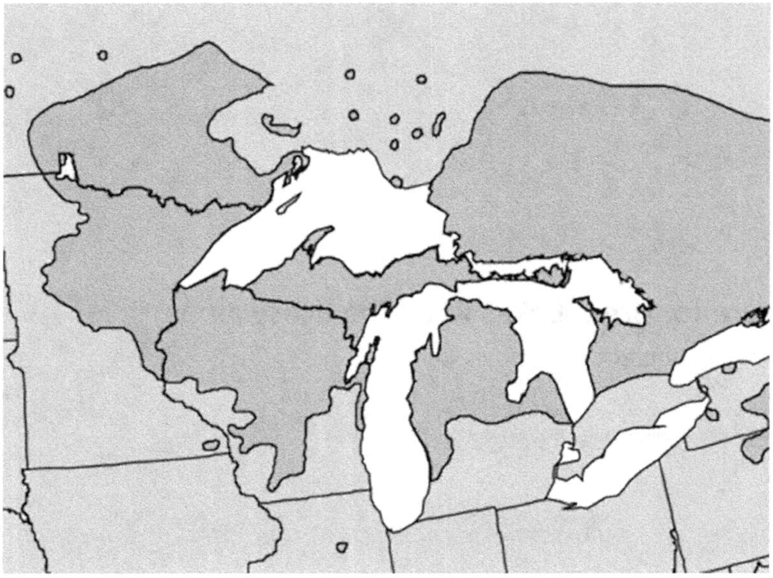

Figure 2.1 Distribution of red pine in the western Great Lakes region of North America. *Source:* commons.wikimedia.org.

Table 2.1 Silvical characteristics of tree species found in Great Lakes red pine ecosystems.

Common name	Species	Shade tolerance	Preferred seed bed	Maximum reported age (years)
Balsam fir	*Abies balsamea* (L.) Mill.	Very tolerant	Moist mineral	245
Bigtooth aspen	*Populus grandidentata* Michx.	Very intolerant	Moist mineral	113
Trembling aspen	*Populus tremuloides* Michx.	Very intolerant	Moist mineral	213
Black spruce	*Picea mariana* (Mill.) Britton, Sterns & Poggenb.	Tolerant	Moist mineral/organic	330
Bur oak	*Quercus macrocarpa* Michx.	Intermediate	Mineral	401
Eastern white pine	*Pinus strobus* L.	Intermediate	Moist mineral	408
Jack pine	*Pinus banksiana* Lamb.	Very intolerant	Moist mineral	246
Northern red oak	*Quercus rubra* L.	Intermediate	Moist mineral with litter	326
Paper birch	*Betula papyrifera* Marshall	Intolerant	Moist mineral	240
Red maple	*Acer rubrum* L.	Tolerant	Litter	300
Red pine	*Pinus resinosa* Ait.	Intolerant	Moist mineral	500
White spruce	*Picea glauca* (Moench) Voss	Tolerant	Moist mineral	262

Sources: Wisconsin Department of Natural Resources [24] and Rocky Mountain Tree-Ring Research, Inc. and The Tree-Ring Laboratory of Lamont-Doherty Earth Observatory and Columbia University [25]. Modified from and reprinted by permission of Waveland Press, Inc. from Palik et al. [1], © 2021 All rights reserved.

Figure 2.2 Old-growth red pine forest: northern Minnesota, USA. Note multiple tree species, dead wood, and diversity of tree sizes. *Source:* With permission from Shawn Fraver.

Age structures of natural origin red pine ecosystems range from broadly single-cohort to two- or three-cohort [2]. Stands are characterized by high levels of structural complexity and heterogeneity (Figure 2.2). Mature and old stands are characterized by woodland structure, with open and variable canopies and low tree densities [3]. Younger developmental stages can be structurally complex, containing substantial legacies of a pre-disturbance forest, including live trees and large deadwood (Chapter 1).

2.2.2 Natural Disturbance Regime

The natural disturbance regime for red pine ecosystems is largely characterized by wildfire. The fire regime is complex, including frequent surface fires punctuated by infrequent crown fires. Historically, the mean combined surface/crown fire return interval has been estimated to range from less than 10 years up to 25 years [4, 5]. The variation in the return interval may reflect burning activity of indigenous peoples, with shorter return intervals associated with higher population areas [6]. Lightning is also a source of ignition [7], but it is unclear if lightning ignitions were/are frequent enough to generate the shortest fire return intervals noted.

Surface fires periodically intensified to kill canopy trees, but mortality was often less-than-stand replacing, as evidenced from reconstructions of tree establishment in old-growth stands [2, 8], as well as many early descriptions of this ecosystem (e.g. [9]). New pine cohorts established in moderate to large openings within stands. True stand-replacement fires occurred [10] but with red pine establishment periods potentially lasting decades. Historically, some red pine ecosystems contained abundant overstory oaks (*Quercus* spp.) [11, 12]. This likely occurred where surface fire intervals were protracted or fires were patchy, allowing these species to avoid fires while of a susceptible size. The natural fire regime is augmented by patch-scale tree mortality from wind and

Armillaria fungi [10, 13]. These disturbances can be important drivers of variation in stand structure and composition, such as hardwoods versus conifers.

2.2.3 Natural Development Model

The natural development model for red pine ecosystems is presented in detail elsewhere [1]; here we present an updated and abbreviated version of the model (Table 2.2). The model begins with a canopy disturbance and proceeds through young, mature, and old forest stages.

2.2.3.1 Disturbance and Legacy Creation

After a canopy disturbance, red pine ecosystems will contain live tree legacies and abundant deadwood structures. Live tree legacies vary from few after severe disturbance to abundant after partial disturbance. Spatial patterns vary from dispersed to aggregated or the pattern may occur as large openings within a less disturbed matrix. Large snag and dead tree legacies will mostly be red and eastern white pines, as these are the dominant species, but other species of deadwood will be present.

2.2.3.2 Preforest Stage

A preforest stage (Chapter 1) is dominated by herbs and shrubs. It may range from a few years, if there is aggressive sprouting of hardwoods (Figure 2.3), or if pines establish after a mast seed year, to decades. Evidence for a several decades-long preforest stage is the often-slow establishment of pines after disturbance (see below).

Table 2.2 A developmental model of red pine ecosystems.

Stage or event	Attribute	Cause or consequence
Disturbance and Legacy Creation	Live trees and deadwood legacies; dispersed to aggregated spatial patterns	Near stand replacing to partial canopy disturbance
Preforest Stage (*Locally absent to one-several decades*)	Herb/shrub dominated; tree regeneration present but not controlling the site	Non-forest organisms and processes dominate; sprouting of hardwoods may shorten length
Young Forest Stage (*Lasting 50–70 years*)	New pine cohort establishes, often in mixtures with other species	Pine establishment may last 40–70 years; densities of regenerating pine may increase slowly; tree species mixtures develop
Mature Forest Stage (*Lasting 50–100 years*)	Red pine and often eastern white pine dominate; shorter-lived species decline; dead and decadent trees accumulate; canopy heterogeneity develops	Patchy fire creates openings for pine establishment; recruitment of fire-sensitive species may occur in fire skips; root-rot gaps are foci for hardwood establishment; small-scale disturbance increases heterogeneity
Old Forest Stage (*Begins after ~150 years*)	Dominated by old emergent pines; broadly even-aged to two+ cohorts; patchy canopy; abundant dead wood; pine regeneration in openings; other canopy species present	Pines accumulate complex morphologies; patchy disturbance creates heterogeneity, fosters tree diversity, and pine establishment

Source: Palik et al. [1]/Reproduced with permission from Waveland Press, Inc.

Figure 2.3 Trembling aspen root suckering in an opening of a mature red pine forest. *Source:* With permission from Christel Kern.

2.2.3.3 Young Forest Stage

In the young forest stage, regenerating trees begin to take control of a site, but this may occur over decades, for example up to 60–70 years [2]. This can occur across whole stands, or it may be focused within larger openings with less-severe disturbance. Tree dominance in the young forest stage may be shared between red pine and other intolerant species, including aspens (*Populus* spp.), paper birch (*Betula papyrifera* Marsh.), and jack pine (*Pinus banksiana* Lamb.).

2.2.3.4 Mature Forest Stage

The mature forest stage is characterized by increasing abundance of eastern white pine, northern red oak (*Quercus rubra* L.), and red maple (*Acer rubrum* L.), along with red pine. These species may be present early after disturbance, but they gain abundance slowly below the red pine canopy. The additional species rarely replace red pine, under a natural fire regime, but historically they often were abundant in older stands [11, 12]. The mature forest stage may also have a spatially heterogeneous canopy structure, developing after patchy disturbances. New cohorts of red pine may establish in larger openings [2].

2.2.3.5 Old Forest Stage

Old forests are structurally complex, with large old trees, diverse tree sizes, and abundant large deadwood. Old pines develop complex morphologies, with large diameters and branch systems, broken tops, and cavities. Snags and downed logs will be equally large. Canopy heterogeneity develops from patchy disturbance. Emergent red pine may co-occur with other species in the main canopy, including eastern white pine, northern red oak, and red maple. Early successional aspens, paper birch, and jack pine can maintain a presence with establishment in larger gap disturbances.

2.2.3.6 The Cycle of Disturbance and Development

Depending on spatial extent of a canopy disturbance, stands in the mature or old-forest stages may continue to develop into yet older conditions, or they may cycle back to the beginning of the development model, i.e. with a new disturbance and legacy creation event. However, given that the latter disturbances are often less than stand-replacing, the structural condition consists of openings, initiating a preforest stage and ultimately a new cohort of trees, as well as residual individuals and clumps of trees from the previous stand, resulting in a multi-cohort age structure.

2.3 An Ecological Silvicultural System for Red Pine Ecosystems

An earlier version of our ecological silvicultural system, based on the natural development model presented above, has been presented elsewhere [1]; here we present a revised version of the approach (Table 2.3). In most cases, actions will be applied at the whole-stand scale, however, as stands develop, some actions might be applied at patch scales within a stand. In general, the system is applicable to both natural origin and plantation stands.

Table 2.3 An ecological silvicultural system for red pine ecosystems.

Stage[a]	Years[b]	Potential actions	Outcomes
Disturbance/ Legacy Creation[c]	0	Variable retention harvest (VRH) Creation of down deadwood and snags Site preparation to reduce slash, expose mineral soil	Continuity of structure; initiate a preforest stage
Preforest	1–20+	Reduce overabundant shrubs Site preparation as needed	Herb/shrub dominated
Young Forest (early)	5–30	Plant, seed, capture mast seed crop Release natural regeneration	Establish a new tree cohort
Young Forest (later)	30–70	Crop-tree release; Variable density thinning (VDT)[d] In VDT gaps: expose mineral soil, reduce shrubs, regenerate trees	Increase growth; enhance canopy heterogeneity; may initiate preforest or new cohort in openings
Mature Forest	70–150	Crop-tree release or VDT in even-aged stands or in younger cohort of two-cohort stands[d] In new VDT gaps: expose mineral soil, reduce shrubs, regenerate trees Existing VDT gaps: expand, site preparation, regenerate, competition control Decadence/deadwood creation Reduce aggressive shrubs and fire-sensitive hardwoods Variable retention harvest	Increased tree growth; enhanced heterogeneity; enriched tree species; may initiate preforest or new cohort in VDT openings or cycle stand back to *Disturbance and Legacy Creation*

Table 2.3 (Continued)

Stage[a]	Years[b]	Potential actions	Outcomes
Old Forest	+150	Decadence/deadwood creation VDT in dense standsd New VDT gaps: expose mineral soil, reduce shrubs, potentially regenerate Existing gaps: expand, site preparation, regenerate, competition control Reduce aggressive shrubs and fire-sensitive hardwoods	Enhanced structural complexity and heterogeneity, enriched species diversity, restored woodland structure

[a] Harvest initiates the developmental sequence, or in existing stands, the table is entered according to current developmental stage, and actions are applied from that point onward.
[b] Timing reflects a general length of time that a particular stage might last.
[c] In the context of a managed forest, legacy creation includes retention of from the pre-disturbance forest and creation of new deadwood.
[d] VDT that includes larger gaps may initiate a disturbance/legacy event in the gap and return to time zero in the context of the gap, but not the whole stand; in the gap, preforest and early young forest conditions and actions may apply.
Source: Palik et al. [1]/Reproduced with permission from Waveland Press, Inc.

2.3.1 Disturbance and Legacy Creation

An initial focus of a regeneration-scale harvest is retention of legacies, including live trees and deadwood, in conjunction with creating resource and seedbed conditions necessary for the eventual establishment of a new cohort of trees. It may also include creation of deadwood if this is lacking. The harvest will also initiate a preforest stage. Natural regeneration may establish early after disturbance, but the goal may not be full site occupancy of regeneration early, as that may undermine the ecological value of a preforest stage.

A variable retention harvest (VRH; Chapter 1) will be used to initiate this stage (Figure 2.4). Live tree retention may range from few to many individuals in dispersed to aggregated patterns. Openings might be the dominant matrix condition, or the pattern may consist of patch cuts within forested matrix that is thinned or unthinned. Existing large deadwood should be retained, and some deadwood may be created to emulate the legacies of disturbance. Note that disturbance and legacy creation may also be a feature of larger openings cut as part of variable-density thinning (VDT; Chapter 1).

2.3.2 Preforest Stage

Most foresters will have little experience managing for a preforest stage, often because they have time-sensitive requirements for regeneration, e.g. fully stocked within five years. With planning, a preforest stage can still be included. For example, artificial regeneration can be deferred for five years, or it can be added in stages over five years, much like natural regeneration accumulates in some settings. When early full-site stocking of regeneration is not a constraint, regeneration can be added over time or natural regeneration can be used exclusively to extend the preforest stage.

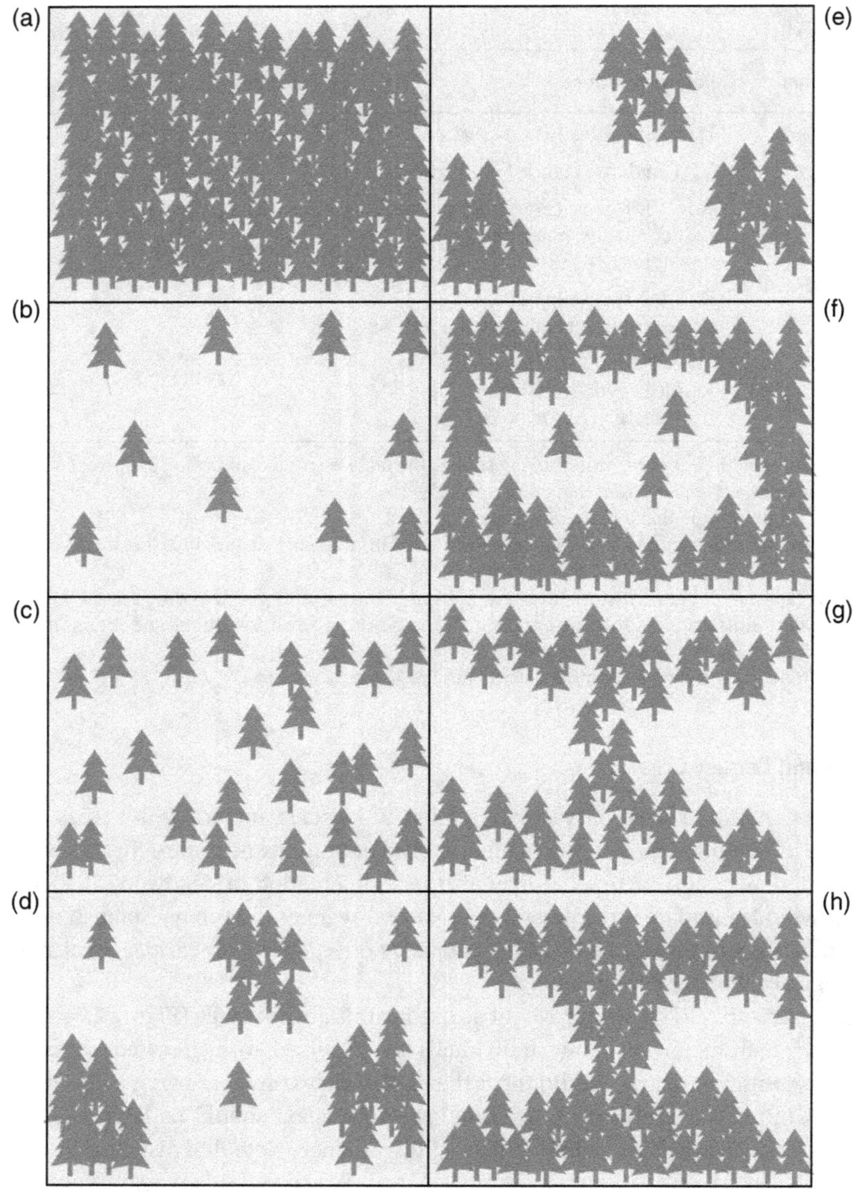

Figure 2.4 Examples of variable retention harvesting in red pine stands: (a) uncut stand, (b, c) dispersed retention, (d, e) patch retention with or without retention between patches, (f) large opening with retained trees, (g, h) large openings with and without thinning in the matrix. *Source:* Palik et al. [1]/Reproduced with permission from Waveland Press, Inc.

Silvicultural actions during a preforest stage may include site preparation and control of aggressive shrubs. While some exposure of mineral soil may occur with harvesting, additional site preparation might be needed to increase establishment from seed, particularly of pines, or to add planting microsites. Competition control may be needed during the preforest stage, particularly to control invasive shrubs, including hazels (*Corylus* spp.) and raspberries (*Rubus* spp.). Site preparation and competition control might be accomplished with prescribed fire or mechanical methods.

Area-wide application of site preparation should be avoided to protect already established tree regeneration, other vegetation, and large deadwood.

2.3.3 Young Forest Stage (Early)

Silviculture in young stage stands focuses on ensuring establishment of a new cohort of trees. This may be natural regeneration that started to develop in the preforest stage, including sprouting hardwoods or species establishing from seed. Natural regeneration may be supplemented with artificial regeneration, but ideally not replaced by it. Ensuring adequate regeneration may require site preparation if it is not sufficient at this point. Planted species likely will include red pine, jack pine, and eastern white pine. In an ecological system, planting densities will be at the lower range of, or below, those typically recommended with a timber-focused approach (typically 1000–2500 trees ha^{-1}). Mechanical release of established regeneration may be needed.

2.3.4 Young Forest Stage (Later)

Competition control around developing regeneration may be needed in this stage. This may include thinning established cohorts of trees using a crop-tree release and/or variable-density thinning (VDT). Thinning might be applied to an even-aged stand, or to a younger cohort in multi-cohort stands if such exists (Figure 2.5). An initial thinning might be pre-commercial, but eventually, it will be applied to merchantable trees. Subsequent thinning entries should be applied to maintain or augment the heterogeneity of density developed in an initial thinning. The length of the young forest stage is such that two or three thinning entries on a

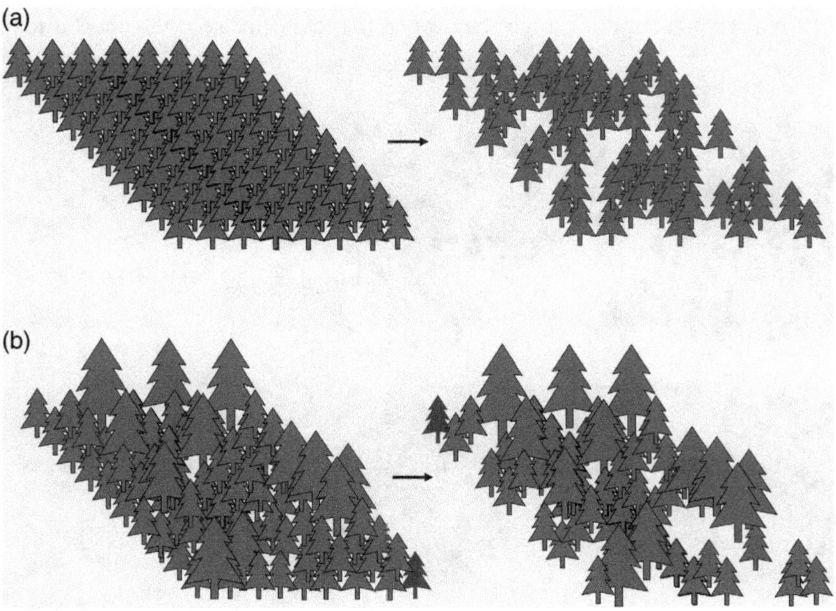

Figure 2.5 An example of variable-density thinning (VDT) in a young forest stage red pine stand. (a) VDT in an even-aged stand. (b) VDT of the younger cohort in stands with two cohorts. *Source:* Palik et al. [1]/ Reproduced with permission from Waveland Press, Inc.

10–15-year interval are possible. If thinning includes cutting larger gaps as part of VDT, then there may be a need for site preparation, regeneration, and competition control in the openings (as above).

2.3.5 Mature Forest Stage

There are two silvicultural pathways to consider in mature-stage stands, including transitioning to an increasingly older condition or cycling back to the start of the development model with a harvest. The pathways are not mutually exclusive; transition treatments may include creation of some larger openings that trigger a preforest stage within the opening.

The first pathway likely includes VDT, either as a first entry or building on an earlier VDT (Figure 2.6). Objectives for a second VDT likely will include releasing established regeneration and crop trees. New gaps might be created in areas with desired advance regeneration or may require site preparation and regeneration (eventually). Prescribed fire or other treatments might be used to reduce aggressive shrubs and fire-sensitive hardwoods in new gaps and the matrix, while avoiding established pine regeneration in older gaps. A VDT in the mature stage presents opportunities to diversify tree species composition. This could focus on VDT gaps, or the matrix for more shade-tolerant species. Deliberate deadwood and decadence creation might be pursued in stands that lack these structures. If repeated VDT entries are used, this pathway emulates the development into complex, multi-cohort, age structures found in some natural stands.

The second silvicultural pathway uses a VRH that retains structural elements, including trees that continue to develop, and deadwood. As above, tree retention may range from low to moderate density in dispersed or aggregate patterns, or large gap/intact matrix retention (Figure 2.4). Deliberate deadwood and decadence creation might be included as part of a VRH.

With the second pathway, actions will be taken eventually to secure regeneration (as above). In doing so, the pathway will emulate the two-cohort age structural development characteristic of many natural stands. Creative prescriptions might be needed to secure this second cohort due to

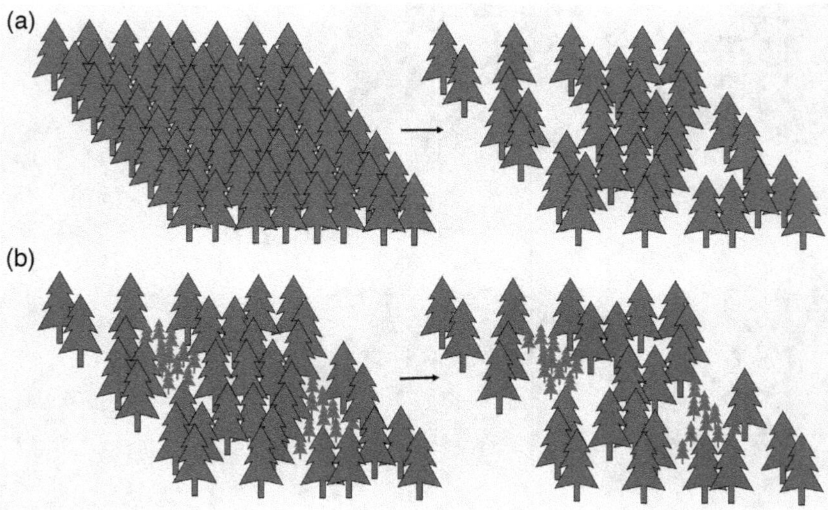

Figure 2.6 Variable-density thinning (VDT) in a mature stage red pine stand. (a) Stand before and after an initial VDT. (b) Thinning that builds on an earlier VDT by expanding gaps, releasing a younger cohort of trees, and thinning the matrix. *Source:* Palik et al. [1]/Reproduced with permission from Waveland Press, Inc.

the impacts of shoot blight diseases including *Sirococcus conigenus* and *Diplodia pinea*. Red pine is susceptible to mortality from these fungal pathogens, particularly when growing near infected overstory trees [14]. However, survival is higher in larger openings [15], as would be present with VRH. Moreover, infection may be minimized with alteration of the dominant pines, so as to prevent red pine dominance in both the overstory and regeneration layers at the same time. For instance, a VRH might retain mostly red pine, followed by regeneration of eastern white pine, jack pine, and oaks. A subsequent VRH might retain mostly eastern white pine while regenerating red pine and other species.

The second pathway is similar to a continuous cover irregular shelterwood [16]. However, when emulating a natural model, it is likely that more trees will be retained to live out their natural lifespans, while also creating spatial patterns of retention that better reflect the outcome of natural disturbance. Moreover, retaining or creating deadwood structures will likely be a bigger focus. Ultimately, providing an opportunity for expression of a preforest stage after a harvest, instead of more immediate site occupancy by regeneration, is another difference in intent from an irregular shelterwood method.

2.3.6 Old Forest Stage

Silviculture in old-stage forests likely will include objectives other than timber production, since old stands are rare on the landscape and may be best managed for ecosystem services other than timber production. Note that mature-stage stands that are managed ecologically will eventually contain old trees and their derivative structures, and these stands may include timber production as an objective, but this is a different situation than management of true old-stage stands.

Silviculture in old stands might focus on restoration of structure and composition, particularly in long-fire excluded stands or those with a history of timber management. Activities might include reducing a dense hardwood and shrub understory and reducing fuels. Prescribed fire or mechanical removal are options to meet these objectives. Enrichment planting or seeding of underrepresented species and creation of deadwood might be considered.

If harvesting is pursued at all in old forests it likely will be VDT, used to reduce density and expedite development of structural heterogeneity. In older stands there may be opportunities to place VDT gaps around individuals of species other than red pine. VDT may initiate a preforest stage in gaps as well as establishment of a new cohort of trees. New gaps may need site preparation. Artificial regeneration and competition control in gaps may also be warranted.

2.3.7 Natural-Origin Stands Versus Plantations

There are some distinctions between natural-origin stands and plantations in application of the silvicultural model. Plantations generally will be dominated by healthy red pines of similar size and will lack large deadwood. Thus, leave trees with VRH in plantations may be exclusively red pine, versus multiple species in natural-origin stands. Also, in natural-origin stands there may be several different species to select for deliberate deadwood creation, unlike in plantations.

In the preforest stage, the diversity of native herbs and shrubs may be limited in harvested plantations, while the abundance of invasive shrubs will vary depending on past management history. The potential for mixed-species natural regeneration in harvested plantations will be limited, as seed sources will be lacking, so greater emphasis on planting or direct seeding might be warranted.

Application of thinning in the young- and mature-stage stands may be different in plantations compared to natural-origin stands, primarily because the spatial pattern of trees in natural stands may be heterogenous to some degree already. Also, natural-origin stands are likely to contain tree species other than red pine that can be released or regenerated naturally in VDT gaps. In mature- and old-stage stands, large-diameter dead and decadent trees will be more abundant in natural-origin stands, so deadwood and decadence creation may be less of a need compared to plantations.

2.4 Climate Change Considerations

Managing red pine ecosystems following a natural development model will be challenging with climate change, which in the western Great Lakes region includes increased growing season droughts, but warmer, wetter winters [17]. Many tree species of the ecosystem potentially may face reduced habitat suitability [18], with red pine subject to reduced growth during drought [19]. Still, there are ways to use a natural development model to adapt to climate change, at least in the near to mid-term [1, 20]. Such approaches use a resistance–resilience–transition framework [21, 22].

A resistance strategy in red pine ecosystems includes thinning to maintain tree basal area at levels of low vulnerability, for instance in the range of 14 and 23 $m^2\ ha^{-1}$, compared to ~32 $m^2\ ha^{-1}$ used in many timber-focused applications [19]. With this, stand-level growth is better maintained with drought because individual trees have greater access to soil moisture. Also, greater tree vigor may increase resistance to insects and diseases. Moreover, lower basal area is consistent with managing the natural woodland structure of this ecosystem.

A resilience strategy in red pine creates heterogeneity of structure and increases abundance of native future-adapted species. The approach might include a VRH or VDT to create heterogeneity, including gaps and uncut forest, with thinning between these bookends. A resilience strategy will encourage natural regeneration, augmented with artificial regeneration, of native future-adapted species, for example, eastern white pine, northern red oak, and bur oak (*Quercus macrocarpa* Michx.).

A transition strategy presents a challenge for use of the natural model. The strategy likely would reduce red pine abundance substantially over time and increase species that are projected to be better adapted to the developing climate, often using assisted migration. For example, transition may include a VRH in the mature stage and, eventually, regeneration of native species predicted to be climate-adapted (as with resilience), but also species not yet part of the ecosystem in any abundance, for example, white oak (*Quercus alba* L.), bitternut hickory (*Carya cordiformis* (Wangenh.) K. Koch), and black cherry (*Prunus serotina* Ehrh.) [23]. Future entries may expand gaps, thin the matrix, and progressively shift composition to future-adapted species.

2.5 Summary

An ecological silviculture system for red pine ecosystems looks different than a timber-focused model, reflecting closer emulation of the natural development model. Key points to keep in mind that follow from the natural development model include:

1) Retention of structural legacies is the initial aim of regeneration harvesting, with legacies being variable in amount and pattern among stands, reflecting the variable nature of natural disturbance.
2) A VRH initiates a preforest stage, where shrub and herbaceous vegetation dominate, rather than early full site occupancy of regenerating trees. Prescribed fire may be used to initiate the stage.

3) In the young forest stage, when tree regeneration is secured, seedling densities may be lower and their accumulation may be slower than with a timber-focused model, reflecting the woodland structure of these ecosystems.
4) Regeneration of all tree species found in the ecosystem should be encouraged, not just red pine. Activities might include site preparation and competition control, direct seeding, or planting.
5) In the young and mature forest stages, thinning can be used to release trees, create heterogeneity of structure, maintain wildlife trees; this may involve crop-tree release or VDT that includes gaps where regeneration is pursued.
6) Decadence and deadwood creation might be considered in the mature and old forest stages, as these features often are lacking in stands managed using a timber-focused model.
7) Longer intervals between regeneration harvests are likely to be the case when using an ecological silvicultural system, leading to development of structures and conditions that depend on large, old trees. Moreover, regeneration harvests may be less than stand-replacing, resulting in multi-cohort age structures.
8) Variability and heterogeneity – of harvesting, legacies, thinning, decadence creation – is the norm rather than the exception of an ecological silvicultural system, with a goal of promoting native biodiversity, greater habitat diversity, and maintaining options for an uncertain climate future.

References

1 Palik, B.J., D'Amato, A.W., Franklin, J.F., and Johnson, K.N. (2021). *Ecological Silviculture: Foundations and Applications*. Long Grove, IL: Waveland Press.

2 Fraver, S. and Palik, B. (2012). Stand and cohort structures of old-growth *Pinus resinosa*-dominated forests of northern Minnesota, USA. *Journal Vegetation Science* 23: 249–259.

3 Aaseng, N., Almendinger, J., Rusterholz, K. et al. (2003). *Field Guide to the Native Plant Communities of Minnesota: The Laurentian Mixed Forest Province*. St. Paul, MN: Minnesota Department of Natural Resources.

4 Stambaugh, M.C., Abadir, E.R., Marschall, J.M. et al. (2021). Red pine (*Pinus resinosa* Ait.) fire history and management implications in the Mississippi River headwaters, Minnesota, USA. *Forest Ecology and Management* 494: 119313.

5 Drobyshev, I., Goebel, P.C., Hix, D.M. et al. (2008). Pre-and post-European settlement fire history of red pine dominated forest ecosystems of Seney National Wildlife Refuge, Upper Michigan. *Canadian Journal of Forest Research* 38: 2497–2514.

6 Johnson, L.B. and Kipfmueller, K.F. (2016). A fire history derived from *Pinus resinosa* Ait. for the Islands of Eastern Lac La Croix, Minnesota, USA. *Ecological Applications* 26: 1030–1046.

7 Loope, W.L. and Anderton, J.B. (1998). Human vs. lightning ignition of presettlement surface fires in coastal pine forests of the upper Great Lakes. *The American Midland Naturalist* 140: 206–218.

8 Bergeron, Y. and Brisson, J. (1990). Fire regime in red pine stands at the northern limit of the species' range. *Ecology* 71: 1352–1364.

9 Eyre, F.H. and Zehngraff, P. (1948). *Red Pine Management in Minnesota*, USDA Circular, vol. 778. USDA. 70 p.

10 Silver, E.J., Fraver, S., D'Amato, A.W. et al. (2013). Long-term mortality rates and spatial patterns in an old-growth *Pinus resinosa* forest. *Canadian Journal of Forest Research* 43: 809–816.

11 Whitney, G.G. (1986). Relation of Michigan's presettlement pine forests to substrate and disturbance history. *Ecology* 67: 1548–1559.

12 Nowacki, G.J., Abrams, M.D., and Lorimer, C.G. (1990). Composition, structure, and historical development of northern red oak stands along an edaphic gradient in north-central Wisconsin. *Forest Science* 36: 276–292.

13 McLaughlin, J.A. (2001). Impact of *Armillaria* root disease on succession in red pine plantations in southern Ontario. *Forestry Chronical* 77: 519–524.

14 Ostry, M.E., Moore, M.J., Kern, C.C. et al. (2012). Multiple diseases impact survival of pine species planted in red pine stands harvested in spatially variable retention patterns. *Forest Ecology and Management* 286: 66–72.

15 Palik, B.J. and D'Amato, A.W. (2019). Variable retention harvesting in Great Lakes mixed-pine forests: emulating a natural model in managed ecosystems. *Ecological Processes* 8: 1–15.

16 Raymond, P., Bédard, S., Roy, V. et al. (2009). The irregular shelterwood system: review, classification, and potential application to forests affected by partial disturbances. *Journal of Forestry* 107: 405–413.

17 Handler, S., Duveneck, M.J., Iverson, L. et al. (2014). *Minnesota Forest Ecosystem Vulnerability Assessment and Synthesis: A Report from the Northwoods Climate Change Response Framework project*, USDA Forest Service General Technical Report NRS-133. Newtown Square, PA: USDA, Forest Service, Northern Research Station. 228 p.

18 Prasad, A.M., Iverson, L.R., Matthews, S., and Peters, M. (2007-Ongoing). A Climate Change Atlas for 134 Forest Tree Species of the Eastern United States. USDA Forest Service, Northern Research Station. https://www.nrs.fs.fed.us/atlas/tree (accessed 12 June 2023).

19 Bottero, A., D'Amato, A.W., Palik, B.J. et al. (2017). Density-dependent vulnerability of forest ecosystems to drought. *Journal of Applied Ecology* 54: 1605–1614.

20 D'Amato, A.W. and Palik, B.J. (2021). Building on the last "new" thing: exploring the compatibility of ecological and adaptation silviculture. *Canadian Journal of Forest Research* 51: 172–180.

21 Millar, C.I., Stephenson, N.L., and Stephens, S.L. (2007). Climate change and forests of the future: managing in the face of uncertainty. *Ecological Applications* 17: 2145–2151.

22 Nagel, L.M., Palik, B.J., Battaglia, M.A. et al. (2017). Adaptive silviculture for climate change: a national experiment in manager-scientist partnerships to apply an adaptation framework. *Journal of Forestry* 115: 167–178.

23 Muller, J.J., Nagel, L.M., and Palik, B.J. (2019). Forest adaptation strategies aimed at climate change: assessing the performance of future climate-adapted tree species in a northern Minnesota pine ecosystem. *Forest Ecology and Management* 451: 117539.

24 Wisconsin Department of Natural Resources (2003). *Wisconsin DNR Silvicultural Handbook*. https://dnr.wi.gov/topic/ForestManagement/silviculture.html (accessed 20 June 2023).

25 Rocky Mountain Tree-Ring Research, Inc. and The Tree-Ring Laboratory of Lamont-Doherty Earth Observatory and Columbia University (2013). *Eastern OldList*. https://www.ldeo.columbia.edu/~adk/oldlisteast (accessed 20 June 2023).

3

Ecological Silviculture for Northern Hardwood Ecosystems of Northeastern U.S.

Anthony W. D'Amato

Rubenstein School of Environment and Natural Resources, University of Vermont, Burlington, VT, USA

3.1 Introduction

This chapter focuses on northern hardwood-dominated forests found across large portions of northeastern North America. These forests are largely defined by the dominance of sugar maple (*Acer saccharum* Marsh.), with American beech (*Fagus grandifolia* Ehrh.) also constituting an important canopy associated throughout much of the range of this forest type (Figure 3.1). These forests are economically and culturally important across this region, with silvicultural systems historically focused primarily on management of sugar maple for wood products and maple syrup production. In some regards, timber-focused models for these forests share consistency with aspects of natural models of disturbance and development given their emphasis on uneven-aged forest structures, particularly tree-size structures. Nevertheless, a historical timber focus on sugar maple dominance and retention of solely "acceptable growing stock" has resulted in lower diversity in species, a narrow range of canopy gap and tree sizes, as well as a lower abundance of dead and decadent trees [1, 2].

Interest in ecological silviculture for northern hardwood forests has increased over the past several decades, catalyzed in part by seminal work in the northern Lake States region in the 1990s [3, 4]. This work led to a wide range of studies and applications to restore complex conditions to the comparatively simple, second-growth forests predominating contemporary landscapes [5–7]. Growth in use of ecological silviculture for these forests partially reflects the economic compatibility of integrating aspects of ecological silviculture with timber-focused models, including the use of multi-aged systems and management for mature structures. In addition, family and public forest ownerships are dominant ownership types across the northern hardwood region, with the rising use of ecological silviculture also reflecting greater compatibility with objectives for these areas (e.g. [8]). This chapter summarizes and builds from this past work to provide an ecological silviculture framework based on natural models of disturbance and development for this ecosystem.

Ecological Silvicultural Systems: Exemplary Models for Sustainable Forest Management, First Edition. Edited by Brian J. Palik and Anthony W. D'Amato.

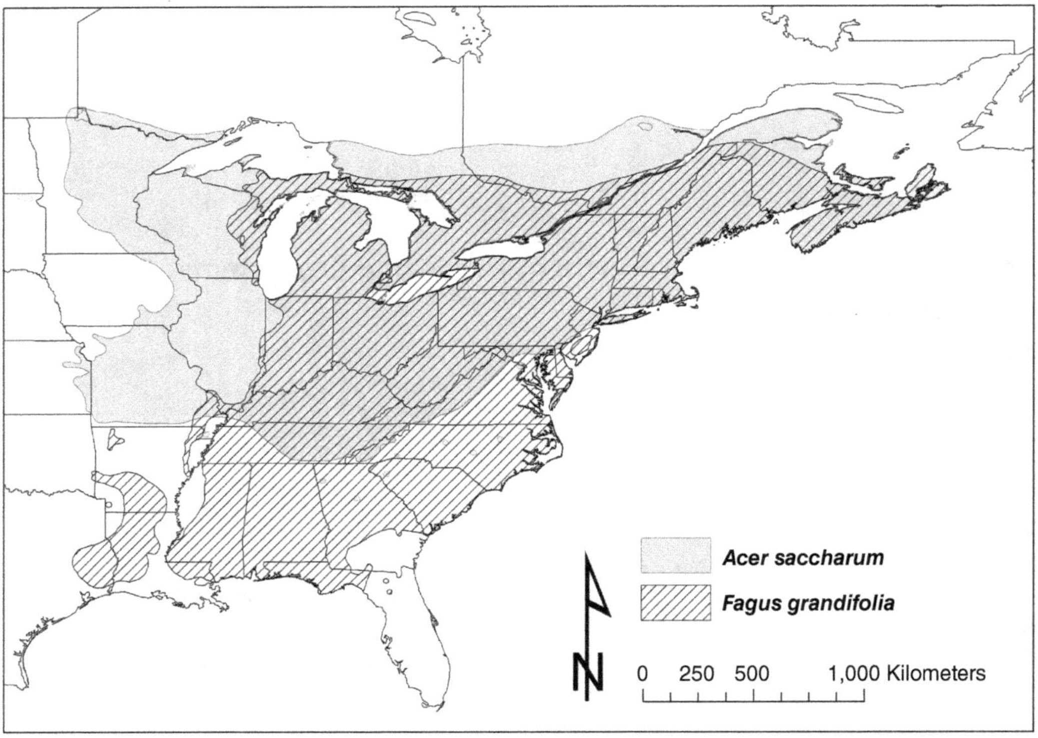

Figure 3.1 Distribution of northern hardwood forests in northeastern North America, as defined by the range of sugar maple (*Acer saccharum*). Distribution of American beech (*Fagus grandifolia*) is also shown to highlight regions where both canopy species are important components of these ecosystems.

3.2 Characteristics of Northern Hardwood Ecosystems

3.2.1 Composition and Structure

There are numerous variants of the northern hardwood forest reflecting differences in site conditions and disturbance regime; however, a general constant is the dominance by shade-tolerant canopy species, namely sugar maple and American beech. Both species successfully germinate on undisturbed leaf litter, highlighting the potential for strong dominance by each in the absence of significant canopy and forest floor disturbance. Nevertheless, these ecosystems are typically quite mixed in composition, reflecting the diversity of light and seedbed conditions necessary for supporting a range of species (Table 3.1).

Stand-replacing disturbance is rare for northern hardwood forests with return intervals two to three times greater than average longevity of canopy tree species. As a result, northern hardwood forests are generally uneven aged with fairly constant recruitment of shade-tolerant canopy species over time. Despite this constancy in recruitment, old stands are often dominated by two to three cohorts reflecting the influence of partial canopy disturbance [9]. Stands often contain a fine-grained mosaic of tree sizes and age and abundance of structural legacies generated by frequent gap-scale disturbances (Figure 3.2).

Table 3.1 Silvical characteristics of canopy tree species commonly found in northern hardwood forests.

Common name	Species	Shade tolerance	Preferred seed bed	Minimum gap size for canopy ascension (ha)	Maximum reported age (years)	Currently impacted by non-native insect or pathogen
Sugar maple	*Acer saccharum* Marsh.	Tolerant	Litter	0.05	315	Yes (Asian longhorned beetle)
American beech	*Fagus grandifolia* Ehrh.	Very tolerant	Litter	0.05	246	Yes (Beech bark disease and beech leaf disease)
Yellow birch	*Betula alleghaniensis* Britton	Intermediate	Humus mix	0.1	387	No
Basswood	*Tilia americana* L.	Intermediate	Litter	0.1	200[a]	No
Paper birch	*Betula papyrifera* Marsh.	Intolerant	Humus mix	0.5	240	No
White ash	*Fraxinus americana* L.	Intermediate	Litter	0.08	198	Yes (emerald ash borer)
Red maple	*Acer rubrum* L.	Tolerant	Litter	0.08	300	Yes (Asian longhorned beetle)
Aspen (bigtooth and trembling)	*Populus grandidentata* Michx. and *P. tremuloides* Michx.	Very intolerant	Moist mineral	>1	113, 213	Yes (spongy moth)
N. red oak	*Quercus rubra* L.	Intermediate	Litter	0.2	326	Yes (oak wilt and spongy moth)
Black cherry	*Prunus serotina* Ehrh.	Intolerant	Litter	0.5	250[a]	No
Red spruce	*Picea rubens* Sarg.	Tolerant	Litter, moist mineral	0.05	445	No
Eastern hemlock	*Tsuga canadensis* (L.) Carrière	Very tolerant	Humus mix	0.05	555	Yes (hemlock woolly adelgid)
Eastern white pine	*Pinus strobus* L.	Intermediate	Moist mineral	0.1	408	No

[a] General estimate; no reported maximum age in Eastern OLDLIST.

Source: From Anderson et al. [23], Leak et al. [24], and the Eastern OldList (https://www.ldeo.columbia.edu/~adk/oldlisteast/). Modified from and reprinted by permission of Waveland Press, Inc. from Palik et al. [12],

Figure 3.2 Old-growth northern hardwood forest in New York, USA with fine-scale mosaic of tree age and size classes and deadwood abundance and decay state.

3.2.2 Historical Natural Disturbance Regime

The natural disturbance regime for northern hardwood forests is largely characterized by frequent gap- and mesoscale disturbances generated by wind, ice, insects, and disease [9]. Gap-scale events generally remove 0.5–2% of the canopy annually, with more moderate-severity disturbances (30–50% canopy removal), such as thunderstorm microbursts or ice storms, occurring every 300–400 years [10]. Given the rarity of stand-replacing events (return intervals >1000–1800 years [3]), natural landscape-scale age structures for these forests are uneven aged; a condition historically occupying 70–90% of the region [9]. Fire in these ecosystems is rare; however, several fire-adapted species, including eastern white pine (*Pinus strobus* L.) and northern red oak (*Quercus rubra* L.), are occasionally minor components and associated with physiographic settings and disturbance events that allow for their successful recruitment [11].

3.2.3 Natural Development Model

A detailed description of the natural development model for northern hardwood ecosystems is presented elsewhere [12], with the following sections providing an updated and abbreviated version to inform ecological silviculture systems for these forests (Table 3.2). The model begins with a canopy disturbance, which for these forests is at the gap or patch scale and proceeds through young, mature, and old forest stages (see Chapter 1). Given the rarity of stand-wide disturbance, the following developmental model should largely be translated and applied to the gap or patch scales, with the exception of the old forest stage, which occurs at stand-to-landscape scales.

Table 3.2 A developmental model for northern hardwood forest ecosystems.

Stage or Event	Attribute	Cause or Consequence
Disturbance and Legacy Creation	Spatially heterogeneous live and dead tree legacies, including tip-up mounds, large living trees, and advance regeneration; small (single-tree) to moderate gaps in matrix of undisturbed forest	Recurring low-severity disturbances and less frequent, mesoscale events
Preforest *Generally absent*	Shrub/tree regeneration representing diversity of mechanisms dominating gap openings	Rapid site capture after canopy disturbance through release of advance regeneration, vegetative response of surviving trees, and new seedling establishment
Young Forest *Gap or patch-level condition lasting 50–80 years*	Stratified mixture early with shorter-lived species (pin cherry, aspen, paper birch) forming upper canopy strata and slower growing mid- and shade-tolerant species (yellow birch, sugar maple, beech) occupying lower canopy positions; within patch legacies of surviving canopy trees present	Mixtures of multiple species representing range of regeneration mechanisms; differences in early height growth rate and longevity cause shift in species dominance to later successional species as stage progresses
Mature Forest *Lasting 50–100 years*	Sugar maple, beech, and yellow birch dominate; shorter-lived species decline; larger dead and decadent trees are recruited; age-class diversity develops	Windthrow and breakage of canopy dominants allow for recruitment of cohorts of shade-tolerant species; tip-up mounds serve as important foci for establishment of mid-tolerant species (yellow birch)
Old Forest *Begins after 200 years*	Dominated by sugar maple, beech, and eastern hemlock across a broad range of tree ages and canopy positions, with a lesser, mid-tolerant yellow birch component; uneven-aged structures; diverse tree sizes, including very large trees, large amounts of deadwood, including hollow snags and logs; spatial heterogeneity of structure; regeneration in patches	Frequent gap-disturbance results in diversity of live-tree size and age structures; increasing levels of heartwood decay generate complex deadwood conditions; mesoscale disturbances create spatial heterogeneity and foster tree diversity

Source: Palik et al. [12]/Reproduced with permission from Waveland Press, Inc.

3.2.3.1 Disturbance and Legacy Creation

Canopy disturbance occurs as small (single-tree fall) to moderate openings (~0.05–0.45 ha; 0.1–1.1 acres) in a matrix of undisturbed forest conditions in northern hardwood ecosystems. As a result, live and dead tree legacies are spatially heterogeneous across a given stand and include fallen trees and tip-up mound structures, as well as scattered surviving canopy trees and advance regeneration within gaps.

3.2.3.2 Preforest Stage

The preforest stage is generally not a common feature of northern hardwood forests and is quite transient. This is due to rapid site capture following canopy disturbance through release of advance regeneration, vegetative response of broken but surviving trees, and rapid establishment of new seedlings from soil seedbanks or windblown seed. There may be a short period containing a significant shrub component dominated by raspberries (*Rubus* spp.), but abundance of these species declines rapidly with rapid reestablishment of tree cover [13].

3.2.3.3 Young Forest Stage

The young forest stage includes developing regeneration from a range of tree species reflecting different regeneration mechanisms. This includes long-lived, shade-tolerant trees released from advance regeneration (sugar maple and American beech) and short-lived pioneer species originating from soil seedbanks and wind-dispersed seed, such as pin cherry (*Prunus pensylvanica* L. f.) and paper birch (*Betula papyrifera* Marsh.), respectively. As a result, the early young forest stage is generally a stratified stand with rapidly growing, short-lived species, forming upper canopy strata and slower growing mid and shade-tolerant species occupying lower canopy positions. Senescence of most short-lived tree species occurs within 50–80 years, resulting in a shift in composition toward species ultimately dominating all canopy layers later in development. Older, surviving residual trees are often scattered throughout this cohort, including mid-tolerant yellow birch (*Betula alleghaniensis* Britton), which has the lowest windthrow susceptibility of canopy trees in these forests [14].

3.2.3.4 Mature Forest Stage

The mature forest stage is characterized by a structural transition in which the death of larger canopy trees results in canopy openings that recruit new-age classes and the development of large snags and downed logs. Competitive thinning remains an important mortality agent during this period; however, localized windthrow of canopy dominants also becomes an important process. Canopy tree species becoming dominant in the later young forest stage (sugar maple and American beech) develop seedling and sapling banks released by canopy disturbance, whereas well-decayed deadwood and tip-up mounds become increasingly important for establishment of smaller-seeded species, like eastern hemlock (*Tsuga canadensis* (L.) Carrière), yellow birch, and red spruce (*Picea rubens* Sarg.).

3.2.3.5 Old Forest Stage

The old forest stage is the dominant forest condition across northern hardwood-dominated landscapes [9]. These forests are structurally and compositionally complex with shade-tolerant sugar maple, eastern hemlock, and American beech found across a broad range of tree ages [15]. Species diversity is largely maintained through large, multiple tree-fall gaps and less frequent mesoscale disturbances that generate legacies such as tip-up mounds and sufficient light conditions for yellow birch and other less tolerant species to establish and ascend to the canopy [16]. Live-tree diameter distributions reflect the aggregate influence of these disturbance and recruitment patterns and are often descending monotonic or rotated sigmoid in shape; conditions reinforced by a U-shaped mortality pattern in relation to tree size [17, 18]. Deadwood is abundant with hollow snags and logs becoming increasingly common and there is a high level of spatial heterogeneity associated with these legacies and live-tree conditions.

3.2.4 The Cycle of Disturbance and Development

Given the abovementioned rarity of stand-replacing disturbance, much of the cycle of disturbance and development for these forests is within a stand at the gap and patch scale. Patterns of structural and compositional development will largely be a function of disturbance severity, with lower severity events largely reinforcing dominance by shade-tolerant species and generating distributed deadwood and live-tree legacies and mesoscale events creating diverse and complex young forest patches embedded within undisturbed areas.

3.3 An Ecological Silvicultural System for Northern Hardwood Ecosystems

We have presented a gap-scale ecological silvicultural system for northern hardwoods elsewhere [12], with the following sections summarizing a revised version of that approach. Given much of the northern hardwood forest is either in the late young forest or mature forest stage, this ecological silvicultural system will often begin at these stages with application of silvicultural activities focused on accelerating conditions toward those associated with the old forest stage historically predominating for these forests [9].

3.3.1 Disturbance and Legacy Creation

The processes of disturbance and legacy creation occur predominantly at the gap scale in northern hardwood forests with the general goal of retaining structural legacies, but also creating stand conditions favorable for recruiting a new cohort of trees (Figure 3.3a). As initial entries often occur in structurally and compositionally homogenous stands in the young and mature forest stages, desired outcomes for initial treatments are the introduction of structural and compositional complexity using multi-aged regeneration methods.

The regeneration method employed will largely be a function of current forest conditions in terms of dominant tree species and overall stand quality. For example, in cases where the understory is strongly dominated by shade-tolerant species, such as sugar maple or American beech, regeneration methods like group selection with legacy retention or continuous cover irregular shelterwoods might be applied to create the range of gap sizes necessary to recruit less-tolerant canopy species. Similarly, these methods might be more appropriate in forests with lower timber quality stemming from past exploitive harvests, as greater volumes are removed and more opportunities are afforded to shift growing space to individuals and species with long-term economic potential relative to other methods (cf. [19]).

Regardless of the regeneration methods used, there are several key structural and compositional outcomes that are the focus of initial entries in an ecological silvicultural system (Table 3.3). First is recruitment of a new-age class of trees to begin transitioning toward the multi-aged structures characteristic of the old forest stage. Second is the advancement of the large-tree component of the stand through thinning treatments, specifically crop-tree release, to portions of the forest outside of regenerating areas (Figure 3.3b). Third is recruitment and retention of dead and live-tree legacies in harvest gaps and in matrix areas receiving thinning treatments through designation of leave trees and girdling and deliberate felling of (often) lower value overstory trees (Figure 3.4). Finally, representation of intolerant and mid-tolerant canopy species should be increased through creation of suitable microsites and light environments in areas where new age classes are being recruited, as well as through retention of these species, where possible, to maintain seed sources on site.

3.3.2 Preforest Stage

The preforest stage is of limited duration given the rapid capture of canopy gaps by advance regeneration and new seedlings. As a result, silvicultural activities during this stage are largely focused on creating the range of resource levels and microsites necessary for recruiting a diversity of tree species. Season of harvest during the disturbance and legacy creation stage will largely influence what treatments are necessary. With summer or fall harvests, logging equipment often generates

Figure 3.3 (a) Group selection harvest initiating new cohort with (b) crop-tree release applied to surrounding matrix areas to accelerate large tree structures in northern hardwood forests in Vermont currently in mature forest stage. (c) Mosaic of structural and compositional conditions 10 years following continuous cover irregular shelterwood in northern New Hampshire.

Table 3.3 An ecological silvicultural system for northern hardwood ecosystems.

Stage[a]	Years[b]	Potential actions	Outcomes
Disturbance and Legacy Creation[c]	0	Group selection, variable-density thinning, or continuous cover irregular shelterwood	Continuity of structure; initiate preforest stage at gap-scale
		Crop-tree release of dominant/codominant trees to accelerate large-tree development	
Preforest	0–5	Removal of understory American beech where thickets have developed	Release advance regeneration and create conditions for new seedling establishment; set stage for tree regeneration
		Mechanical, patch scarification to expose mineral soil seedbeds for yellow birch	

Table 3.3 (Continued)

Stage[a]	Years[b]	Potential actions	Outcomes
Young Forest (early)	5–10	Release treatments to encourage development of yellow birch and sugar maple where American beech competition is an issue	Facilitate/establish new cohort of trees at gap scale
Young Forest (later)	50–70	Crop-tree release of dominant and codominant sugar maple and yellow birch Variable-density thinning in spatially homogeneous stands[d] Creation of new single-tree and multi-tree canopy gaps at 15–20 year intervals in stands with patchy distribution of structural conditions In gaps, expose mineral soil, reduce beech, regenerate new cohorts	Accelerate tree growth; initiate new cohort (gap scale) to create spatial heterogeneity in structure and composition
Mature Forest	70–150	Crop-tree release of dominant/codominant sugar maple/yellow birch initiated earlier in stage Group selection, variable-density thinning, or continuous cover irregular shelterwood at stand-scale[d] if first entry Decadence/deadwood creation through girdling of canopy trees as part of crop-tree release and felling of cull Large group or patch openings with legacy tree retention as part of variable-density thinning or initial treatment in conversion to multi-cohort condition In gaps, expose mineral soil, reduce beech, regenerate new cohorts	Enhance spatial heterogeneity; release and accelerate growth of desired trees; enhance structural complexity; initiate new cohorts
Old Forest	+150	Decadence/deadwood creation Crop-tree release in developing cohorts Continuous cover irregular shelterwood or hybrid single-tree/group selection to create single-tree and multiple-tree gaps at 15–20 year intervals In gaps, expose mineral soil, reduce beech, regenerate new cohorts	Enhance structural complexity; restore/maintain uneven-aged structure; emulate natural dynamics

[a] A regeneration harvest initiates the developmental sequence; however, these stages are largely occurring at the gap or mesoscale in these forests. Moreover, based on the predominant condition of these forests, the first entry is typically occurring when the stand-level condition is either nearing the end of the young forest stage or in the mature forest stage.
[b] Timing does not reflect strict year(s) of activity but rather reflects how long a particular stage might last.
[c] For these forests, the scale of the disturbance/legacy creation event is generally at the gap-scale as part of group or patch selection with legacy-retention or gap components of variable-density thinning.
[d] Initiation of developmental stages is at the gap or mesoscale in a stand. Retention of live trees and deliberate felling and retention of some canopy trees for deadwood within harvest gaps is a critical component of disturbance and legacy creation stage.
Source: Palik et al. [12]/Reproduced with permission from Waveland Press, Inc.

Figure 3.4 Deadwood creation in (a) group opening to reflect pulses associated with wind events and (b) scattered felling of cull trees in matrix areas to created distributed source of downed logs associated with background mortality.

the incidental scarification required to create exposed microsites for new seedlings of light-seeded species, such as yellow birch. Mechanical site preparation to create mineral soil seedbeds may be necessary if harvests occur on snow-covered ground.

Past partial harvesting of northern hardwood forests can generate significant thickets of advance beech regeneration, particularly in areas with a history of beech bark disease. As a result, mechanical removal of beech regeneration during this stage is often necessary to establish other species. These site preparation treatments are necessary to create the competition-free microsites historically afforded by tip-up mounds and well-decayed logs.

3.3.3 Young Forest Stage (Early)

The early young forest stage requires little silvicultural intervention, as this is a period of rapid cohort development. Mechanical release of regenerating yellow birch saplings can be applied if there is limited representation of this species in the new cohort; however, the best course of action from a silvicultural and ecological perspective during this period is generally to do nothing, particularly if attention has been paid to creating suitable seedbed conditions during the previous two stages.

3.3.4 Young Forest Stage (Later)

The focus of silvicultural treatments later in the young forest stage is on accelerating growth of large trees through thinning, namely crop-tree release (cf. [4]). These treatments generally occur 50–70 years after the disturbance and legacy stage, with the intent of creating future legacy and crop trees at the cohort scale. Crop-tree release may also be applied as a first entry across pole and small sawtimber-sized, even-aged second-growth stand to create a distributed source of large canopy trees. In these cases, thinning would be applied at the stand-scale in conjunction with regeneration methods, such as group selection, variable-density thinning, or continuous cover irregular shelterwoods, to transition to a spatially heterogeneous, multi-aged condition (see Section 3.3.1).

3.3.5 Mature Forest Stage

The mature forest stage is the prevailing current condition for northern hardwood forests. Treatments during this stage are intended to further transition structure and composition toward those found in the old forest stage by recruiting new age classes, accelerating large-tree development, recruiting and retaining dead and live-tree legacies, and increasing representation of intolerant and mid-tolerant species.

Variable-density thinning (see Chapter 1) is one approach to achieve the abovementioned objectives, particularly if there are >100 canopy trees per hectare (40/acre) or 40% of current stocking with future commercial potential (cf. [7]). Ideally, this approach entails creating canopy gaps ranging from 0.04 to 0.2 ha (0.1–0.5 acre) across 10–20% of the stand area, with gap placement favoring areas with a concentration of cull or advance regeneration. However, the distribution and abundance of acceptable growing stock and advance regeneration are rarely uniform in northern hardwood forests due to past management and the influence of natural disturbances. To accommodate this variability and provide for future operability, a general rule of thumb when implementing variable-density thinning is to restrict harvest gaps to no more than 30% of the stand area and maintain residual basal areas of 14–16 $m^2\ ha^{-1}$ (61–70 $ft^2\ ac^{-1}$) between gap areas at initial entries. This general guideline maintains future economic and ecological potential, particularly related to large live trees while transforming stand structures from even-aged to two-aged conditions that include mid-tolerant species in the new cohort [7].

Regardless of regeneration method used, within-gap live-tree retention and deadwood creation are important tactics to include at this stage (Figure 3.5). Crop-tree release can be applied to the remainder of the stand to accelerate large-tree development, with those trees selected for release reflecting individuals and species with the greatest potential for long-term growth. Subsequent entries during this stage might occur on 15–20 year intervals and would concentrate on creating new canopy gaps in areas receiving thinning treatments in the previous entry. These gaps could

Figure 3.5 Legacy retention within (a) large patch opening emulating mesoscale event, and (b, c) group selection harvests, as well as scattered dead (d) and (e) dying canopy trees throughout matrix areas.

either be placed in areas away from initial gaps, to reflect the often-isolated nature of canopy gaps or used to expand previously created gaps to reflect larger coalescence of disturbed areas over time after mesoscale events [20].

3.3.6 Old Forest Stage

Old northern hardwood forests are rare in the contemporary landscape and silvicultural activities in existing forests might be limited to release treatments that reduce invasive plant species or protection measures that minimize browse on regeneration. In contrast, the long-term goal of ecological silvicultural applied to this forest is a return to old forest stage conditions. Silvicultural treatments for forests achieving this state would be applied as a mosaic of the treatments described for the previous developmental stages. The main focus is regeneration harvests representing a combination of single-tree and group selection openings that emulate frequent, gap-scale disturbances. These harvests would maintain the multi-aged structure of the old forest stage through frequent (15–20 year) initiation of the preforest and young forest stages in space and time, using gaps that included larger openings (>1 ha; 2.5 acre), along with permanent legacy retention to reflect outcomes of mesoscale events.

3.4 Climate Change Considerations

Northern hardwood silviculture has been adapting to novel stressors for almost a century given the history of non-native species (NNS) introductions across this region. This includes several insects and pathogens, like beech bark disease, hemlock woolly adelgid, and emerald ash borer, capable of functionally eliminating key canopy species (Table 3.1). These stressors can interact with other changing dynamics, including the impacts of introduced earthworms, elevated herbivory levels, and invasive plant species, to generate novel and significant barriers. Climate change impacts, including increasing growing season temperatures and droughts, and declining snow packs, further compound these challenges, requiring application of ecological silviculture through an adaptation lens [21].

In many regards, the conditions generated by silviculture following natural developmental models, including functional complexity tied to diverse structural and compositional conditions, provide important response and recovery mechanisms for sustaining the many values and services these forests provide in the face of changing environmental conditions [21]. Nevertheless, many species that are cornerstones of structure and function for these forests are vulnerable to impacts of changing climate and NNS, creating the potential for crossing challenging and undesirable future ecological thresholds. As such, beyond the resistance and resilience mechanisms provided by complexity created via the ecological silviculture system presented in this chapter, there will increasingly be a need to transition aspects of these forests to increase the range of functional responses and adaptation mechanisms represented. This may include deliberate planting or seeding of future-climate- and NNS-adapted species that are either currently present a lower levels of abundance in this region (e.g. assisted population enrichment with northern red oak and eastern white pine [22]) or expected to gain new, suitable habitat in the future as climate changes (e.g. assisted range expansion of, for example, bitternut hickory (*Carya cordiformis* (Wangenh.) K. Koch)) and sweet birch (*Betula lenta* L.) (Figure 3.6). Beyond selecting species expected to be adapted to future conditions, these regeneration efforts should focus on sustaining key functions threatened by decline of canopy species historically occurring in these ecosystems (e.g. planting hard mast-producing species to replace American beech). These adaptation plantings could occur

Figure 3.6 Planted hybrid American chestnut (left) and white pine (right) in large patch opening with legacy retention to increase future-climate-adapted component.

in larger regeneration openings created as part of the ecological silviculture system described above and would serve as unique functional nodes providing diverse adaptive responses and seed sources over time [21].

3.5 Summary

Although historical approaches to northern hardwood silviculture contained elements of the natural developmental model described above, namely the promotion and maintenance of multi-aged structures, there are key differences in the outcomes between timber-focused and ecological silviculture, including, with the latter, a higher diversity of canopy species, larger live and dead tree legacies, and a greater range of canopy disturbance severities in space and time. Key aspects of an ecological silviculture system that build from the natural developmental model include:

1) Initial treatments applied to create spatial heterogeneity in structure and composition, including gap-scale regeneration harvests to transition to multi-aged conditions and crop-tree release to accelerate large tree development.
2) Legacy retention and creation as part of all regeneration harvests, including within-gap retention of large canopy trees and felling and leaving of deadwood legacies in gaps to emulate pulses of large woody material.
3) Creation of a range of canopy gaps through regeneration harvests over time to reflect the historical size and frequency of gap- and mesoscale disturbance.
4) Regeneration harvests favoring all component tree species, including site preparation treatments or seasonal harvesting considerations to create competition-free, exposed mineral soil seedbeds for recruitment of small-seeded species.
5) Designation and management of both ecological and economic "crop-trees" for recruitment of large tree structures in a range of spatial configurations.
6) Long-term transition from initial treatments focused on increasing spatial heterogeneity to gap-scale harvests that maintain a within-stand mosaic of developmental stages and structures as development progresses.

References

1 Webster, C.R. and Lorimer, C.G. (2005). Minimum opening sizes for canopy recruitment of midtolerant tree species: a retrospective approach. *Ecological Applications* 15: 1245–1262.

2 Kenefic, L.S. and Nyland, R.D. (2007). Cavity trees, snags, and selection cutting: a northern hardwood case study. *Northern Journal of Applied Forestry* 24: 192–196.

3 Lorimer, C.G. and Frelich, L.E. (1994). Natural disturbance regimes in old-growth northern hardwoods-implications for restoration efforts. *Journal of Forestry* 92: 33–38.

4 Singer, M.T. and Lorimer, C.G. (1997). Crown release as a potential old-growth restoration approach in northern hardwoods. *Canadian Journal of Forest Research* 27: 1222–1232.

5 Keeton, W.S. (2006). Managing for late-successional/old-growth characteristics in northern hardwood-conifer forests. *Forest Ecology and Management* 235: 129–142.

6 Hagenbuch, S., Manaras, K., Shallow, J. et al. (2011). *Silviculture with Birds in Mind: Options for Integrating Timber and Songbird Habitat Management in Northern Hardwood Stands in Vermont.* Montpelier, VT: Audubon Vermont and Vermont Forests, Parks, and Recreation. 24 p.

7 Bédard, S., Guillemette, F., Raymond, P. et al. (2014). Rehabilitation of northern hardwood stands using multicohort silvicultural scenarios in Québec. *Journal of Forestry* 112: 276–286.

8 D'Amato, A.W. and Catanzaro, P.F. (2022). *Restoring Old-Growth Characteristics to New England's and New York's Forests.* University of Massachusetts, Cooperative Extension Landowner Outreach Pamphlet. 36 pp.

9 Lorimer, C.G. and White, A.S. (2003). Scale and frequency of natural disturbances in the northeastern US: implications for early successional forest habitats and regional age distributions. *Forest Ecology and Management* 185: 41–64.

10 Frelich, L.E. and Lorimer, C.G. (1991). Natural disturbance regimes in hemlock-hardwood forests of the upper Great-Lakes region. *Ecological Monographs* 61: 145–164.

11 Frelich, L.E. (2002). *Forest Dynamics and Disturbance Regimes.* New York: Cambridge University Press.

12 Palik, B.J., D'Amato, A.W., Franklin, J.F., and Johnson, K.N. (2020). *Ecological Silviculture: Foundations and Applications.* Waveland Press.

13 Donoso, P.J. and Nyland, R.D. (2006). Interference to hardwood regeneration in northeastern North America: the effects of raspberries (*Rubus* spp.) following clearcutting and shelterwood methods. *Northern Journal of Applied Forestry* 23: 288–296.

14 Canham, C.D., Papaik, M.J., and Latty, E.F. (2001). Interspecific variation in susceptibility to windthrow as a function of tree size and storm severity for northern temperate tree species. *Canadian Journal of Forest Research* 31: 1–10.

15 Leak, W.B. (1975). Age distribution in virgin red spruce and northern hardwoods. *Ecology* 56: 1451–1454.

16 Hanson, J.J. and Lorimer, C.G. (2007). Forest structure and light regimes following moderate wind storms: implications for multi-cohort management. *Ecological Applications* 17: 1325–1340.

17 Leak, W.B. (1996). Long-term structural change in uneven-aged northern hardwoods. *Forest Science* 42: 160–165.

18 Lorimer, C.G., Dahir, S.E., and Nordheim, E.V. (2001). Tree mortality rates and longevity in mature and old-growth hemlock-hardwood forests. *Journal of Ecology* 89: 960–971.

19 Strong, T.F., Erdmann, G.G., and Niese, J.N. (1995). Forty years of alternative management practices in second-growth, pole-size northern hardwoods I: tree quality development. *Canadian Journal of Forest Research* 25: 1173–1179.

20 Raymond, P., Bédard, S., Roy, V. et al. (2009). The irregular shelterwood system: review, classification, and potential application to forests affected by partial disturbances. *Journal of Forestry* 107: 405–413.

21 D'Amato, A.W. and Palik, B.J. (2021). Building on the last "new" thing: exploring the compatibility of ecological and adaptation silviculture. *Canadian Journal of Forest Research* 51: 172–180.

22 Palik, B.J., Clark, P.W., D'Amato, A.W. et al. (2022). Operationalizing forest-assisted migration in the context of climate change adaptation: examples from the eastern USA. *Ecosphere* 13: e4260.

23 Anderson, H.W., Boysen, E.P., Dey, D.C., and Rice, J.A. (2001). Natural regeneration of hardwoods. In: *Regenerating the Canadian forest: Principles and Practice for Ontario* (ed. R.G. Wagner and S.J. Colombo), 393–421. Markham, ON: Fitzhenry and Whiteside Ltd.

24 Leak, W.B., Yamasaki, M., and Holleran, R. (2014). *Silvicultural Guide for Northern Hardwoods in the Northeast*, USDA Forest Service General Technical Report NRS-132. USDA. 46 p.

4

Ecological Silviculture in Douglas-fir–Western Hemlock Ecosystems

Abraham Wheeler[1], *Jerry F. Franklin*[2], *and Stephanie J. Wessell*[3]

[1] *OR/WA State Office, USDI Bureau of Land Management, Portland, OR, USA*
[2] *School of Environmental and Forest Sciences, University of Washington, Seattle, WA, USA*
[3] *Northwest Oregon District Office, USDI Bureau of Land Management, Salem, OR, USA*

4.1 Introduction

Broadly defined, the Douglas-fir (*Pseudotsuga menziesii* (Mirb.) Franco)–western hemlock (*Tsuga heterophylla* (Raf.) Sarg.) forest region occupies the west side of the Cascade Range from southern British Columbia, Canada south through western Washington and Oregon, U.S., excepting some areas inland from the Coast Range in southwestern Oregon. The heavily forested region is noted for immense old-growth forests dominated by diverse and impressive coniferous species. The region's moist, temperate climate has cool, wet winters and relatively dry, warm summers.

Forests are a defining feature of the region important for many values, including wood production, watershed protection, wildlife habitat, and recreation. Douglas-fir–western hemlock (hereafter DF-WH) ecosystems have long been valued for timber production; harvesting initially was of virgin forests but has shifted predominantly to management of younger forests (mostly plantations) over the last 50 years.

During this same period, the importance of the DF-WH ecosystem for other values, such as wildlife habitat, recreation, and carbon sequestration, has increased dramatically, resulting in controversy over timber harvesting on public lands and forest practices on private lands. Extensive clearcutting, broad-scale herbicide applications, and impacts on streams exemplify social issues about timber-focused management. Recently, climate change and associated increases in wildfires are increasing social concerns.

Societal concerns and greatly expanded scientific knowledge of DF-WH ecosystems have combined to stimulate development and application of ecologically based management to better support the full array of forest functions.

Ecological Silvicultural Systems: Exemplary Models for Sustainable Forest Management, First Edition. Edited by Brian J. Palik and Anthony W. D'Amato.

4.2 Characteristics of Douglas-fir–Western Hemlock Ecosystems

4.2.1 Tree Species Composition

While Douglas-fir and western hemlock characterize the DF-WH ecosystem, several other coniferous and hardwood tree associates also may be present. This compositional variability relates to different environments. Western redcedar (*Thuja plicata* Donn ex D. Don), grand fir (*Abies grandis* (Douglas ex D. Don) Lindl.), bigleaf maple (*Acer macrophyllum* Pursh), red alder (*Alnus rubra* Bong.), black cottonwood (*Populus trichocarpa* Torr. & A. Gray), Oregon white oak (*Quercus garryana* Douglas ex Hook.), and Pacific dogwood (*Cornus nuttallii* Audobon ex Torr. & A Gray) are common at low elevations, along with Sitka spruce (*Picea sitchensis* (Bong.) Carrière) close to the Pacific Ocean. Cascara (*Frangula purshiana* (DC.) A. Gray) and bitter cherry (*Prunus emarginata* (Douglas ex Hook.) D. Dietr.) are common short-lived, early successional tree species. Pacific yew (*Taxus brevifolia* Nutt.) is a common understory species. Western white pine (*Pinus monticola* Douglas ex D. Don) and lodgepole pine (*Pinus contorta* Douglas ex Loudon) can occur anywhere from sea level to high mountain areas. Higher elevation associates include noble fir (*Abies procera* Rehder), Pacific silver fir (*Abies amabilis* (Douglas ex Loudon) Douglas ex Forbes), mountain hemlock (*Tsuga mertensiana* (Bong.) Carrière), and Alaska cedar (*Callitropsis nootkatensis* (D. Don) Oerst. ex D.P. Little). Ponderosa pine (*Pinus ponderosa* Lawson & C. Lawson), sugar pine (*Pinus lambertiana* Douglas), incense cedar (*Calocedrus decurrens* (Torr.) Florin), Port Orford cedar (*Chamaecyparis lawsoniana* (A. Murray bis) Parl.), California black oak (*Quercus kelloggii* Newberry), canyon live oak (*Quercus chrysolepis* Liebm.), and tanoak (*Notholithocarpus densiflorus* (Hook. & Arn.) P.S. Manos, C.H. Cannon, & S.H. Oh) are associates in southern portions of the region.

4.2.2 Natural Disturbance Regime and Developmental Model

4.2.2.1 Natural Disturbance

Infrequent wildfire represents the dominant natural disturbance regime in DF-WH ecosystems [1]. Landscape-altering wildfires can cover tens to hundreds of thousands of hectares and always include a significant high-severity component. Such fires are associated with episodes of hot, dry east winds. Historical examples include the Yacolt Burn in Washington (1902) and the Tillamook Burn in Oregon (1933), which impacted over 200,000 and 140,000 ha, respectively. Return intervals for large fires increase from south to north in the region; in northern areas historical fire-return intervals of 400+ years were characteristic [2, 3]. Ocean-generated windstorms also are important; the Columbus Day windstorm of 1962 was the largest single forest disturbance recorded in the region. Volcanic eruptions also can affect large areas, such as the May 1980 eruption of Mount St. Helens in Washington.

4.2.2.2 Preforest Stage

Following a stand-replacement disturbance, DF-WH ecosystems undergo long developmental sequences, assuming they are not interrupted by additional major disturbances [4, pp. 115–124; 5]. Immediately following the disturbance, the site is occupied by an early successional ecosystem referred to as the preforest stage.

The preforest stage is characterized by high biodiversity and structural complexity. Herbs and shrubs dominate in the open, sunny environment, providing food resources supporting high species diversity, and numerous complex food webs [6, 7]. Early successional habitat specialists are numerous, including plants that host nitrogen-fixing organisms, such as legumes, alder (*Alnus* Mill.), and ceanothus (*Ceanothus* L.) species. Insects, such as bees, moths, and butterflies, songbirds, small mammals (e.g. rodents and rabbits), and large herbivores (e.g. deer and elk), are important preforest residents and a prey base for predators. This stage may last several decades due to slow establishment and early growth of conifers [6, 8]. Standing dead trees (snags), logs and other woody structures provide a rich structural legacy in the preforest stage along with individual and patches of surviving live trees.

4.2.2.3 Young Forest Stage

Development of a predominantly closed tree canopy initiates the young forest stage in which high tree density results in intense inter-tree competition. Douglas-fir generally dominates because of its rapid growth under sunlit conditions. Some short-lived (e.g. bitter cherry and cascara) and longer-lived hardwoods, (e.g. red alder, bigleaf maple, and Pacific dogwood) may be present. Competitive-based mortality dominates in the young forest with most mortality in smaller, shaded trees [9]. The overstory tree canopy produces low light levels in the understory resulting in rapid declines in shrub and herb diversity. Structural legacies from the pre-disturbance stand continue to play important ecological roles.

4.2.2.4 Mature Forest Stage

As the forest undergoes substantial thinning, it gradually enters the mature forest stage at 80–120 years following the stand-replacing disturbance. Major structural changes occur during the approximately 100-year duration of this stage [3]. Dominant Douglas-fir trees attain most of their remaining height and canopy spread and begin developing more complex canopies by growing epicormic branches. Agent-based tree mortality becomes more common, resulting mostly from root, butt, and stem rots, Douglas-fir bark beetle (*Dendroctonus pseudotsugae* Hopkins), and wind throw. This mortality tends to be contagious rather than occurring in isolated trees. Mortality of trees in the overstory creates canopy gaps providing greater light and moisture for mid and understory plants. This allows cohorts of shade-tolerant associates, such as western hemlock, to grow into the mid and upper canopy. Mortality of larger trees also begins the important process of rebuilding stocks of standing dead and downed wood, since dead wood inherited from the previous forest has largely rotted away.

4.2.2.5 Old Forest Stage

At 200–250 years, the forest gradually takes on more and more attributes of the old forest stage, which includes complex forest structure [3]. This stage is characterized by significant spatial heterogeneity of trees (e.g. gaps and dense tree clusters) and essentially vertically continuous canopy from ground to top of dominant trees. The canopy is "bottom loaded," in contrast to the "top loaded" canopy found in the young forest; that is, foliage is predominantly in the lower half of the canopy rather than in the top half. Both agent-based and competitive-based mortality are common in the old forest stage, the latter occurring in clusters of shade-tolerant trees developed in canopy gaps. Stocks of coarse woody debris include abundant large snags and logs. Old forests typically have very high levels of stored carbon [10].

4.3 Essential Elements of an Ecological Silvicultural System for Douglas-fir–Western Hemlock Ecosystems

Ecological silviculture seeks to sustain the integrity of forest ecosystems while managing them to produce a full array of economic, environmental, and cultural values by emulating natural developmental processes [5, pp. 172–182]. In the following paragraphs, we describe our perspective on the essential elements of an ecological silvicultural system for DF-WH ecosystems.

4.3.1 Landscape Context

Silvicultural treatments are planned as part of landscape-level management units. Stands are not the primary units for planning management activities. Ecological forest management planning focuses on developing larger forest management units that incorporate a diversity of stand conditions over time. The management units are collectively multi-aged, composed of mixtures of species, and include both openings and intact forest patches. A related goal is ensuring that all parts of the landscape contribute to conservation of desired biodiversity – the concept of managed landscapes as being *shades of green* [10].

4.3.2 Special Features

Areas of special ecological value are identified during landscape-level planning and appropriately managed. Landscapes generally include such areas, which require management that conserves those special values [10]. Among the most common of these are aquatic features, including streams, rivers, lakes, ponds, springs, and wetlands. These areas receive careful attention in applications of ecological forestry. Silvicultural treatments will vary with context; some treatments are aimed at improving conditions through active management (e.g. increased available sunlight and moisture for disturbance-associated species, or reducing fuel loading), and others are directed towards protecting refugia and limiting disturbance (e.g. maintaining shade, dead wood, and favorable microclimates for disturbance-averse species).

4.3.3 Spatial Heterogeneity

Variability in stand density (e.g. clustering) is a characteristic of natural forest stands. Spatial heterogeneity is important to provide diverse conditions within the stand (e.g. areas that are both denser and more open than the average condition of the stand). Such variability sustains more biotic diversity, including soil organisms, than stands of uniform density and structure. The larger the area being treated, the greater the importance of spatial heterogeneity.

4.3.4 Structural Retention

Significant structural retention at harvest is important as both intact forest patches (aggregates or skips) and dispersed retained trees in heavily harvested portions of the management unit. Retention provides for continuity in structure, function, and biota between forest generations. Medium and large aggregates provide closed-forest habitat refugia to sustain disturbance-averse species and functions in the management unit; these aggregates often are associated with aquatic features, such as streams and springs. Harvested areas should have well-distributed retention of (1) individual and small clusters of trees and (2) snags and logs. Individual retention trees and aggregates can be converted into snags or snag patches to emulate natural processes. Finally, any individual older trees, patches of older forest, or trees providing desired habitat function present in management units should be prioritized for retention.

4.3.5 Species Diversity

Diverse species composition is a goal in managed stands, including intentional incorporation of hardwoods and shrubs. Creating stands composed of diverse species is an important way to cultivate forests adapted to climate change, including increased disturbances. Douglas-fir typically will be one of the species but not as a monoculture. Many associated tree species can provide ecological diversity and commercial value with the locale and landowner objectives determining the choices.

Deciduous hardwoods and shrub species are desirable from multiple standpoints. These include enriching biotic diversity, conserving water, providing resistance and resilience related to wildfire and other disturbances, cultural value, and aesthetics. Hardwoods and shrubs have structural forms, leaves, and wood decay patterns that contrast with conifers, and provide a diversity of flowers, foliage, fruits, seeds, and habitat. Additionally, many deciduous hardwoods utilize less water on an annual basis than coniferous evergreen trees, which can contribute to greater streamflow. Deciduous hardwoods are generally less flammable than conifers and can increase the resistance of forest stands to wildfire as well as improve their resilience. Most hardwoods sprout following disturbances, whereas conifers must regenerate from seed. Other flowering plants, such as shrubs and herbs, also help forest ecosystems respond quickly to disturbances by vegetative propagation. Rapid resilient responses help minimize erosion and leaching of soil nutrients. Accelerated reestablishment of green plants also restores the flow of high-energy carbohydrates that sustain the soil's rich biota; extended periods lacking such energy inputs can cause functional and biotic losses, especially on nutrient-limited soils [11].

4.3.6 Limited Use of Chemicals

Minimal use of herbicides and other chemicals is important in ecological management of the DF-WH ecosystems. Herbicides can accelerate tree establishment and early growth but also may shorten duration, productivity, and diversity of the preforest stage, along with increasing negative impacts on soil biodiversity and functioning. While targeted use of herbicides to control non-native plants may be warranted, forgoing heavy herbicide use allows for development of diverse early successional ecosystems and their numerous and complex food webs.

4.3.7 Regeneration Approaches

Utilization of both natural and artificial regeneration allows introduction of diversity in genetic and species composition when regenerating forest vegetation. In DF-WH ecosystems, natural regeneration almost invariably occurs on management units with significant retention, and this regeneration can significantly increase genetic variability present in the new stand [12]. Seeding or planting can complement natural regeneration, especially for species lacking on-site seed sources, or to incorporate seed sources believed to be better adapted to the predicted future environment.

4.3.8 Silvicultural System

An ecologically based multi-aged silvicultural system in DF-WH ecosystems represents a repeated cycle of treatments in perpetuity (see Table 4.1). This system is best conceptualized as multi-aged because it aims to develop three or more age classes of trees in management units over time.

4.3.8.1 Emulating Natural Disturbance

Ecologically-based regeneration harvests in DF-WH ecosystems are modeled to emulate natural disturbances and are often called variable retention harvests, although the harvesting regime involves more than just retention of structural and biotic elements of the harvested forest [13].

Table 4.1 Example of an ecologically-based multi-aged silvicultural system in Douglas-fir–western hemlock ecosystems.

Treatment type	Eligible forest stage (stand ages)	Desired outcomes	Potential treatment design elements
Regeneration harvest; variable retention harvest	Young, mature, and old forest stage[a] (40–150+ years)	Timber harvest, complex early successional ecosystems, species diversity, habitat creation, culturally valuable species, regulation of water yield, rapid growth of retained trees, provide refugia	Structural and compositional variability, significant retention of preharvest basal area, retain large old trees, protect ecologically valuable areas, well-distributed arrangement of retention trees, variable introduction of fire, snag creation
Reforestation	Preforest stage (0–15 years)	Sustained timber yield, complex early successional ecosystems, influence species diversity, introduce valuable shade-intolerant plants, improve slope stability	Strategic mix of natural and artificial reforestation, variable-density planting, reforestation of conifers, hardwoods, and other desired species
Pre-commercial thinning	Preforest and Young forest stage (15–40 years)	Influence stand composition and density, prolong early successional conditions, reduce fire hazard	Variable-density thinning, favor-desired species and hardwoods, release ecologically and culturally important species, maintain openings
Commercial thinning, variable-density thinning	Young and mature stage (40–150 years)	Sustained timber harvest, forest health, habitat development, influence stand composition and density	Skips and gaps, develop future wildlife habitat trees, treat hazardous fuels, snag creation

[a] Note that in most of the region, the proportion of managed stands in the old forest stage is relatively low, therefore it is unlikely that regeneration harvest would be implemented in those forests in the near term. However, if a system of very long intervals between regeneration harvest is adopted, then eventually those areas developed into older forests would be eligible for harvest, which is why we include them here.

Key elements of ecologically-based variable retention harvest treatments include:

1) Long intervals (80 years or more) between regeneration harvests[1];
2) Harvests that create large, predominantly open areas;
3) Substantial structural retention (live and dead trees) at harvest, including intact aggregates and dispersed retention in harvested portions of management units;
4) Stands composed of multiple tree and shrub species, including intentional inclusion of hardwoods;
5) Identification and appropriate management of areas of special ecological concern, such as riparian zones and other aquatic features.

1 Note that long intervals between regeneration harvests initially do not require a minimum regeneration harvest age for management units. This distinction is especially important in managed forests that are relatively young, where regeneration harvesting in young forests provides a short-term bridge to achieving desired forest conditions at the landscape level.

4.3.8.2 Long Recovery Periods

Long intervals between regeneration harvests are fundamental to developing many aspects of ecosystem function and are particularly critical in realizing the growth potential of Douglas-fir forests over the long term. Douglas-fir requires many decades to establish and reach high growth rates; culmination of mean annual increment[2] occurs at 80–120 years on productive sites [14, p. 113], with high growth rates continuing for many decades beyond culmination. Short intervals between regeneration harvest (less than 50 years) result in relatively lower timber yields over the long term. In multi-aged ecological silvicultural systems, long-tree-level rotations result in development of larger and older trees that provide more complex habitat, including more durable snags and logs. Finally, long rotations also result in a higher proportion of the managed forest landscape being in an older forest state, which enhances water yields, reduces impacts of rain-on-snow events, carbon storage, and wildlife habitat.

4.3.8.3 Large, Open Harvest Areas

Regeneration harvests that create large, predominantly open areas provide the opportunity for development of complex early successional ecosystems and allow for establishment and growth of shade-intolerant tree species, including Douglas-fir. These open areas should be large enough (one hectare or larger) to allow for full development of shade-intolerant plants (e.g. abundant production of fruits, seeds, flowers, nectar, and forage). Larger openings provide areas free of strong forest edge influences needed by edge-averse early successional species. *Predominantly open* means that most of the harvested area would have nearly full sunlight, albeit with individual and small clusters of dispersed green trees and well-distributed snags and downed dead trees.

4.3.8.4 Ecological Thinning

Ecologically-based thinning in DF-WH ecosystems also is modeled on natural disturbances and often is referred to as variable-density thinning. Variable-density thinning is used to direct development of existing young or mature forests, especially when existing conditions are relatively homogeneous.

Key attributes of ecologically based variable-density thinning treatments include:

1) Create, enhance, and sustain structural heterogeneity in the stand, rather than creating homogeneity;
2) Stimulate mid-story and understory development as well as growth of some dominant trees by thinning from above (i.e. removal of some dominant and codominant trees) as well as thinning from below (i.e. removal of suppressed and intermediate trees);
3) Include both skips (unharvested areas) and gaps (areas with most if not all trees removed) to increase heterogeneity and habitat diversity within the stand;
4) Retain trees that function as important wildlife habitats as well as potential crop trees; and
5) Create stands with high compositional diversity reflective of the natural ecosystem.

---

2 Mean annual increment refers to the average growth of a stand each year (board feet of sawtimber in this example); it can be calculated by dividing the total net timber volume by the age of the stand and is typically expressed on a per acre basis. Culmination of mean annual increment describes a period of stand development when average stand growth has plateaued. This may be referred to as the biological rotation age, or the harvest age at which long term sustained yield of timber would be maximized.

Variable-density thinning approaches are appropriate from the earliest pre-commercial thinning to help create, enhance, or sustain heterogeneity from the outset. For example, thinning can utilize approaches pioneered in the Individuals-Clumps-Openings (ICO) system, which can produce specific spatial outcomes comparable to conditions in natural stands [15].

Thinning from above enhances survival and development of mid-story trees and ground layer vegetation while also accelerating growth of remaining canopy trees. Representative silvicultural goals might include sustaining desired but suppressed hardwoods, stimulating growth of shade-tolerant conifers into the mid or upper canopy, and sustaining understory shrubs and herbs. A spatial offset to the south is necessary when removing overstory trees to get direct sunlight on an area of interest.

4.3.9 Ecological Silvicultural Tradeoffs

Collectively, the essential elements just described incorporate all elements of ecological silviculture – continuity, complexity/diversity, timing, and context (see Chapter 1). Representation of these ecological silviculture elements in real-world applications will vary depending upon landowner management objectives, current forest and landscape conditions in the area to be treated, and available resources. Landowners will apply each along a sliding scale depending on their unique objectives and willingness to accept tradeoffs between various ecological or cultural forest outputs and timber yield. These tradeoffs are illustrated conceptually in Figure 4.1, as a function of the percentage of site resources allocated to attainment of each landowner objective, including commercial timber production. Timber yields might be reduced accordingly, but it is important to note that the purportedly timber-focused model of short rotation even-aged management (e.g. 50 years or less) is designed to maximize return on capital, not timber production.

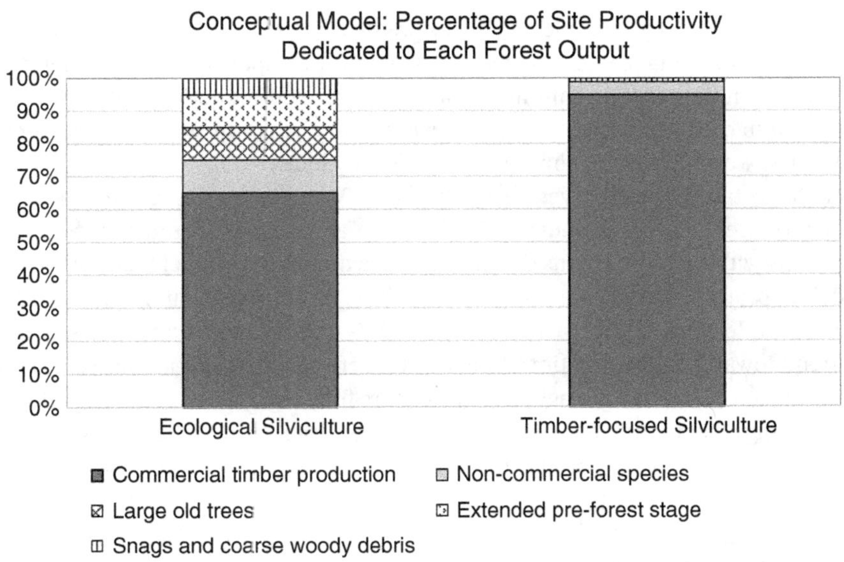

Figure 4.1 Conceptual model showing inherent tradeoffs in adopting an ecological silvicultural system in managed forests. An ecological silviculture approach is contrasted with a timber-focused approach in terms of percentage of site resources (soil, water, nutrients, and sunlight) dedicated to production of each desired forest output. The actual proportion dedicated to each output would depend on the extent to which silvicultural practices are modified and would reflect the relative weighting of each landowner objective.

4.4 Real-World Application of an Ecological Silvicultural System for the DF-WH Ecosystem

A landscape-level application of ecological silviculture occurs on 2.3 million acres of federal lands administered by the Bureau of Land Management (BLM) in western Oregon. Management goals include providing habitat for an array of native species, watershed protection, and sustained-yield timber production. BLM's 2016 resource management plan applies ecological forestry principles on all forest lands. Some areas are reserved for protection and enhancement of threatened and endangered species habitat and water quality. Other lands are part of the harvest land base, where ecologically based silvicultural systems are used to sustain timber production and achieve other ecological objectives [16].

The harvest land base is subdivided into moist and dry forests based on potential vegetation type, with uneven-aged silviculture adopted on dry forests, and multi-aged silvicultural systems adopted for moist forests, in both cases with the goal of emulating natural disturbance regimes. As moist forests, DF-WH ecosystems are managed on long rotations for sustained-yield timber production and provision of complex early successional ecosystems, diverse late-successional forests, and diverse forest structural stages over time and space.

Variable retention harvesting prescriptions retain between 5% and 30% of preharvest live tree basal area outside of riparian reserves, with both hardwoods and conifers counting towards retention targets (see Figure 4.2 for example images). All trees established prior to year 1850 and greater than 40 inches diameter at breast height are retained (except where felling is necessary for safety or operations, in which case felled trees are retained as coarse woody debris). Management direction also specifies retention of existing snags and coarse wood, and creation of snags, tailored to site-specific conditions.

Both artificial and natural reforestation are utilized. Relatively low levels of regeneration are required within five years of harvest (average 53–61 trees ha^{-1} minimum). Herbicide use is limited to non-native plants.

Variable-density thinning is implemented to produce timber, increase or maintain vegetative species diversity, and enhance development of structural complexity, among other reasons (see Figure 4.3 for example images). Thinning treatments are required to leave at least 5% of area in skips, a maximum of 10% in group selection openings, and result in stand-average density conditions between 25% and 45% of maximum relative density index.

Modeling of sustained yield harvest levels by the BLM indicates ecologically based multi-aged silviculture produces 23% less periodic timber harvest per acre when compared to even-aged silvicultural systems with rotation lengths optimized for maximum timber yield (i.e., biological rotation age). Modeling projections indicate that the percentage of managed stands containing structural legacies or multiple canopy layers will trend from 30% in 2013 to greater than 80% in 100 years. With the trend towards longer rotations, the interval between regeneration harvest is projected to be between 100 and 125 years once age classes are more fully regulated (roughly equal acres in each age class grouping). This would eventually result in about one-third of the harvest land base being in either a mature or old forest stage [17].

4.5 Climate Change Considerations

Changing environmental conditions create substantial uncertainty about future conditions in DF-WH ecosystems. Increased temperatures, including hotter and drier summers, are predicted along with increased potential for wildfires. Some species, such as western hemlock and western redcedar, may be impacted more than others.

Figure 4.2 Variable retention harvest on BLM-administered lands in western Oregon: (a) unit with variable arrangement of retention trees, BLM Roseburg District, 2014; (b) overhead view postharvest of unit with 8% live basal area retained, excluding riparian reserves. BLM Northwest Oregon District, 2020; (c) retention trees selected for their unique crown characteristics. BLM Northwest Oregon District, 2020; and (d) concentrated retention trees designed to protect advanced regeneration. BLM Northwest Oregon District, 2020. *Source:* (b-d) Kenneth J. Ruzicka.

Ecological silvicultural approaches described here help adapt the DF-WH ecosystem to ongoing changes. Reduced stand densities and increased within and between stand variability improve tree vigor and landscape-level resistance to disturbances such as drought, wildfire, and insect infestations. Purposely managing to increase structural complexity and compositional diversity to support native organisms enriches food webs, improving organisms' ability to persist despite climate change. Hardwoods and shrubs provide for rapid initial recovery following wildfires. As older trees are more resistant to wildfire and drought, an increased proportion of the managed landscape in older forest conditions contributes to increased landscape-level resistance, as well as to increased carbon storage. This, coupled with an extended preforest stage, may assist in improving water yields.

If consistently hotter and drier conditions predicted by climate models materialize, fuel management in these historically moist forests will increase in importance, as will reforestation with disturbance-resistant species. Research on management to enhance resistance and resilience is needed. Facilitated movement of genotypes or species could be necessary to maintain forest functionality.

Figure 4.3 Variable-density thinning on BLM-administered lands in western Oregon. Coos Bay District, 2022: (a) LiDAR (Light Detection and Ranging) generated overhead image; darker dots represent taller trees; (b) east side of treatment area designed for individual tree culturing; and (c) west side of treatment area thinned to 20% of maximum relative density index. *Source:* Andrew Spencer.

4.6 Summary

Ecologically based silvicultural approaches for DF-WH ecosystems are well-developed and provide a credible alternative to traditional timber-focused silviculture that has been historically dominant in the region. Great variability in the specific silvicultural treatments is possible, allowing landowners to adapt design elements to their ownership, improving outcomes related to their own unique management objectives and landscape context. This chapter is intended to inspire landowners and natural resource professionals to modify their current timber-focused silvicultural practices to achieve management outcomes that not only sustain timber harvests but also sustain the array of other values provided by these rich and highly productive ecosystems.

References

1 LANDFIRE (2016). Fire Return Interval Layer, LANDFIRE 2.0.0, U.S. Department of the Interior, Geological Survey, and U.S. Department of Agriculture. www.landfire.gov/fri.php (accessed 12 May 2022).

2 Agee, J.K. (1996). *Fire Ecology of Pacific Northwest Forests*. Washington, D.C.: Island Press.

3 Franklin, J.F., Spies, T.A., Van Pelt, R. et al. (2002). Disturbances and structural development of natural forest ecosystems with silvicultural implications, using Douglas-fir forests as an example. *Forest Ecology and Management* 155 (2002): 399–423.

4 Spies, T.A., Hessburg, P.F., Skinner, C.N. et al. (2018). Old growth, disturbance, forest succession, and management in the area of the Northwest Forest Plan. In: *Synthesis of Science to Inform Land Management Within the Northwest Forest Plan Area*, PNW-GTR-966 (ed. T.A. Spies, P.A. Stine, R. Gravenmier, et al.), 95–243. Portland: USDA Forest Service.

5 Palik, B.J., D'Amato, A.W., Franklin, J.F. et al. (2021). *Ecological Silviculture*. Long Grove, IL: Waveland Press.

6 Swanson, M.E., Franklin, J.F., Beschta, R.L. et al. (2011). The forgotten stage of forest succession: early-successional ecosystems on forest sites. *Frontiers in Ecology and the Environment* 9 (2): 117–125.

7 Swanson, M.E., Studevant, N.M., Campbell, J.L. et al. (2014). Biological associates of early-seral pre-forest in the Pacific Northwest. *Forest Ecology and Management* 324: 160–171.

8 Franklin, J.F. and Hemstrom, M.A. (1981). Aspects of succession in the coniferous forests of the Pacific Northwest. In: *Forest Succession: Concepts and Application* (ed. D.C. West, H.H. Shugart, and D.B. Botkin), 212–229. New York: Springer-Verlag.

9 Lutz, J.A. and Halpern, C.B. (2006). Tree mortality during early forest development: a long-term study of rates, causes, and consequences. *Ecological Monographs* 76 (2): 257–275.

10 Franklin, J.F., Johnson, K.N., and Johnson, D.L. (2018). *Ecological Forest Management*. Long Grove, IL: Waveland Press.

11 Dominski, A.S. (1971). Accelerated nitrate production and loss in the northern hardwood forest ecosystem underlain by podzol soils following clear cutting and addition to herbicides. Ph.D thesis. Yale University.

12 Dodson, E.K., Burton, J.I., and Puettmann, K.J. (2014). Multiscale controls on natural regeneration dynamics after partial overstory removal in Douglas-fir forests in western Oregon, USA. *Forest Science* 60 (5): 953–961.

13 Franklin, J.F. and Donato, D.C. (2020). Variable retention harvesting in the Douglas-fir region. *Ecological Processes* 9 (8): https://doi.org/10.1186/s13717-019-0205-5.

14 Franklin, J.F., Berg, D.R., Thornburgh, D.A. et al. (1997). Alternative silvicultural approaches to timber harvesting: variable retention harvest systems. In: *Creating a Forestry for the 21st Century* (ed. K.A. Kohm and J.F. Franklin), 111–139. Washington, DC: Island Press.

15 Churchill, D.J., Larson, A.J., Jeronimo, S.M.A. et al. (2016). *The ICO Approach to Quantifying and Restoring Forest Spatial Pattern: Implementation Guide. Version 3.3.* Vashon, WA: Stewardship Forestry and Science.

16 USDI BLM (2016). *Northwestern and Coastal Oregon Record of Decision and Resource Management Plan*, 1. Portland, OR: Bureau of Land Management, 23 and 59–63.

17 USDI BLM (2016). *Proposed Resource Management Plan/Final Environmental Impact Statement*, 307–345. Portland, OR: Bureau of Land Management.

5

Ecological Silviculture for Longleaf Pine Woodlands in the Southeastern U.S.

Steven B. Jack[1], Benjamin O. Knapp[2], and R. Kevin McIntyre[3]

[1] *Boggy Slough Conservation Area, T.L.L. Temple Foundation, Lufkin, TX, USA*
[2] *School of Natural Resources, University of Missouri, Columbia, MO, USA*
[3] *Jones Center at Ichauway, Newton, GA, USA*

5.1 Introduction

Longleaf pine (*Pinus palustris* Mill.) ecosystems were historically one of the most extensive forest types in North America, covering approximately 37.2 million hectares, primarily in the Coastal Plain region of the Southeastern U.S. Longleaf pine (LLP) occurs on a wide range of site types and soils, from xeric Entisols to saturated Spodosols, and a wide range of community types are currently recognized. Primary LLP forests were typically described as having a moderately stocked open woodland structure with multiple age classes of trees, and diverse herbaceous, grass-dominated understories. A regime of frequent, low-intensity fire maintained the characteristic structure of these ecosystems. A unique faunal community, with many species now federally listed as threatened or endangered, characterized LLP ecosystems [1].

Aggressive commercial harvest in the late nineteenth and early twentieth centuries greatly reduced the extent of LLP, with only about 1.3 million hectares remaining by the 1990s [2]. Most of the former LLP forests were regenerated into dense plantations of loblolly (*Pinus taeda* L.) and slash (*Pinus elliottii* Engelm. var. *elliottii*) pines. Today, intensive management of loblolly and slash pine plantations for fiber production dominate upland forests of the southern U.S. Coastal Plain.

However, interest in conservation and restoration of LLP has grown dramatically, driven in part by policies related to wildlife species that declined in tandem with the LLP ecosystem. Over the last 20 years, restoration efforts have increased LLP area to approximately 1.9 million hectares [3]. Managers and owners of both public and private forest lands often have multiple objectives that include both conservation and economic returns, and LLP fits well with an ecological silviculture approach to management.

Ecological Silvicultural Systems: Exemplary Models for Sustainable Forest Management, First Edition. Edited by Brian J. Palik and Anthony W. D'Amato.

5.2 Characteristics of Longleaf Pine Ecosystems

5.2.1 Natural History

LLP is unique among the southern yellow pines. It is the longest-lived species, it occurs naturally across a wide gradient of site types, and its prior dominance of the southeastern U.S. landscape engendered a strong association with the region's cultural history. Perhaps most relevant to ecological silviculture are two aspects of the species' natural history: the resilience to frequent fire and its unique regeneration ecology.

The ubiquity of fire across the landscape of the southeastern U.S. Coastal Plain was a major factor in the dominance of LLP in the region, with ignitions from both Native Americans and lightning [4]. Today, prescribed fire is our modern management surrogate for this natural process that was, and is, so important in LLP woodlands. Early research [5, 6] indicated that LLP woodlands are a fire subclimax community that, in the absence of fire, would naturally succeed to a mixed hardwood forest. Frequent fire in LLP woodlands controls the composition and structure at the community level and was historically a structuring process at larger scales, with the interplay of fire and soil moisture driving landscape pattern and process across a moisture gradient [7].

Regeneration dynamics of LLP are similar to other pine species where fire is common. Seed and cone production is episodic, with heavy seed crops occurring across the range on average every 8–10 years [8]. Frequent low-intensity fire controls competition and provides bare mineral soil for successful germination of seeds. Long considered shade-intolerant [9], LLP seedlings show characteristics of both tolerant and intolerant species and can persist under a range of light environments as advance reproduction [10].

LLP seedlings do have a unique characteristic, however. Seedlings that survive for one to two years after germination enter what is known as a "grass stage," putting most of their resources into root system development for another 3 to more than 10 years [11]. Grass-stage seedlings with well-developed root systems can respond to release quickly and initiate height growth when canopy trees are removed through natural disturbance or silvicultural treatments [12]. Although LLP trees have several physiological adaptations to frequent, low-intensity fire, the species is best described as resilient to fire and not "fire-proof." For instance, recently germinated seedlings (less than one year) are vulnerable to fire, as are shorter (≤2 m) saplings when terminal buds are elongating.

5.2.2 Natural Disturbance

The most common natural disturbance in LLP forests is frequent, low-intensity fire. Historical fire return intervals range from every one to three years [13] on more productive sites to as many as seven years on less productive sites [14]. Fires severe enough to kill canopy trees are rare due to low fuel loads and the typically wide spacing of mature trees in a woodland setting.

The most frequent canopy disturbances that shape structure and stocking in LLP forests are lightning and small windthrow events, with annualized overstory mortality documented as ranging from 0.05 to 1.0 trees ha^{-1} [15, 16]. Large-scale disturbances that kill most canopy trees are much less frequent and tend to be wind-related (e.g. tornadoes and hurricanes) [17, 18]. It is difficult to generalize about the unique impacts of large-scale wind events because hurricane impacts typically diminish as storms move inland and tornado effects are highly variable.

5.2.3 Composition and Structure

LLP ecosystems are characterized by open yet complex structure, with a high diversity of vascular plants in the understory layer (Figure 5.1). Due to frequent fire, the overstory is strongly dominated by LLP, although mixtures with other pine species or hardwoods such as oaks occur on some sites. Basal area can exceed $30\,m^2\,ha^{-1}$ but a range of 10 and $20\,m^2\,ha^{-1}$ is more typical. Mature trees are often widely spaced, with some percentage of canopy gaps maintained by local disturbance events [15]. Large, old trees provide important wildlife habitat, as do large snags, which may range from 4 to 15 trees ha^{-1} [19, 20]. LLP seedlings and saplings often aggregate within canopy gaps, such as those formed by grouped lightning mortality or wind, creating a multi-aged forest composed of spatially aggregated patches of even-aged cohorts [21]. Within these even-aged patches, seedlings and saplings can occur in high densities that reduce over time through local processes of competition and stand development, as well as from fire mortality (Figure 5.2). In mature woodlands, stand-level diameter distributions may follow negative exponential or rotated sigmoid patterns [19, 21], although bimodal patterns with low numbers of trees in the range of 20–40 cm diameter at breast height have also been reported [22, 23]. This bimodal distribution occurs due to the high local densities of small trees in patches interspersed within a matrix of large, mature canopy trees (Figure 5.2).

Figure 5.1 Characteristic structure and composition of *mature* and *old* longleaf pine ecosystems: (a) mature overstory is dominated by widely spaced longleaf pine, understory is dominated by herbaceous vegetation with abundant bunchgrasses; (b) patches of regenerating longleaf pine occur within canopy openings in mature stand; (c) old stage longleaf pine forest with large trees of many ages and heterogeneous spacing.

Figure 5.2 (a) Natural longleaf pine seedlings (b) frequently form dense patches of reproduction within canopy openings (c) and develop as even-aged sapling cohorts through stages of localized, heavy-within-patch competition.

The composition and structure of the understory is determined by fire frequency and intensity. Frequent fire limits recruitment of woody vegetation into the midstory, thereby increasing light availability to understory plants while also removing accumulated biomass and increasing establishment sites for regeneration on the forest floor [24]. Plant associations vary widely across the LLP range and site types, but are noted for high species richness and large numbers of endemic species. The understory is typically dominated by herbaceous species, including bunchgrasses important to fire behavior (e.g. *Andropogon* spp., *Aristida* spp., *Sorghastrum* spp., *Sporobolus* spp.), although low-stature woody species may also be common [25]. Associated hardwood tree species commonly occur in the understory but are maintained as seedlings or seedling sprouts with frequent fire; interruptions in fire-return intervals or spatially variable fire behavior may provide windows for recruitment of these species. Large dead wood on the ground does not commonly persist due to the frequent fire regime, but woody debris in the form of logs can provide refugia for some plants and animals and, along with pinecones, can create localized areas of intense heat during burning [26].

Composition and structure of LLP ecosystems, as summarized above, varies based on local site conditions. For example, regions with greater frequency of large-scale disturbances (e.g. hurricanes) may have more extensive canopy release events for even-aged cohorts of recruitment. Areas with either less frequent or more variable fire return intervals may have increased abundance of hardwood species, resulting in mixed-species stands. Edaphic conditions and site productivity affect local plant associations and rates of stand development. Low-productivity sites often support stands of lower density and may have less competitive pressure on LLP from other tree species.

5.3 Development Model

The complex, multi-cohort age structure of LLP ecosystems develops through many small-scale disturbance events over extended time periods (Figure 5.3). In the absence of stand-replacing disturbances, local canopy disturbances drive processes of regeneration and recruitment. Frequent

	Disturbance	Pre-forest	Young forest	Mature forest	Old forest
Maintenance Existing LLP-dominant canopy	Small-scale, localized canopy disturbances* (one to a few trees) LLP seedling recruitment and high competition among saplings in gaps Mosaic of age and development stages Accumulating CWD**			Even- or multi-aged LLP canopy Younger LLP cohorts Diverse grass dominated ground cover Snags & CWD	Complex multi-aged Open-canopy structure Sparse midstory Diverse ground cover Snags & CWD
Restoration LLP Plantation establishment	No canopy Woody debris Heavy competition for LLP seedlings	No canopy Less woody debris Less competition	One age LLP canopy Ground cover in varying stages of development Accumulating CWD	>1 age class LLP Natural LLP regeneration Grass dominated ground cover Snags & CWD	Multi-aged system Low-moderate canopy Sparse midstory Diverse ground cover Snags & CWD
Restoration Conversion of mature stands of other pine species to LLP	Other pine species Often closed canopy Excessive midstory Poor ground cover	Not closed canopy, gaps present Minimal midstory Variable ground cover Minimal CWD	Even- or multi-aged Mixed species canopy and regeneration Variable ground cover Accumulating CWD	Even- or multi-aged LLP dominated canopy Natural LLP regeneration Grass dominated ground cover Snags & CWD	Complex and multi-aged Lower canopy stocking Sparse midstory Diverse ground cover Snags & CWD

*Elements of early development stages are present at small scale (not whole stand)
**CWD = course woody debris

Figure 5.3 Generalized structure and composition of longleaf pine (LLP) forests through different developmental stages and from different starting conditions. Rows represent forest development from different starting conditions; columns highlight ecological characteristics found at stages of stand development following canopy disturbance (first column).

fire is a critical component for ecosystem function throughout the stages of development. Consequently, an understory dominated by herbaceous plants generally occurs and is supported across all stages of development, with potential exceptions during restoration scenarios (plantation establishment or stand conversions) after canopy closure and before thinning treatments. We include these restoration starting conditions in the stand development model (adapted from Chapter 1) because they occur commonly in the contemporary landscape (Figure 5.3) and often are the starting condition for applying an ecological silviculture approach.

5.3.1 Disturbance and Legacy Creation

In LLP ecosystems, the scale of canopy disturbance varies from infrequent stand-replacement events to canopy gaps created by the more frequent mortality/removal of one-to-many trees. LLP trees are more resistant to blowdown than associated tree species during hurricanes, resulting in legacies that include a mixture of residual live trees, standing dead or broken trees, and increased coarse woody debris [27]. Local wind disturbance may also create canopy openings that include interior residual live and standing dead trees, and mortality from lightning creates canopy gaps, often with retention of the snags. Although the scale of these disturbance events is variable, the response is similar across scales: canopy disruptions release advance reproduction cohorts of seedlings or saplings in the different-sized canopy gaps, and these young trees develop as even-aged cohorts.

5.3.2 Preforest Stage

The preforest stage may be short or not occur at all in LLP ecosystems. With a consistent frequent fire regime, characteristic structure is maintained through multiple developmental stages. Prior to overstory disturbance, LLP seedlings may develop as advance reproduction that remains small until released from overstory competition. Following the release event, continued frequent fire restricts development of competing tree species, allowing LLP seedlings and saplings to dominate the regenerating cohort.

5.3.3 Young Forest Stage

During the young forest stage, the new cohort of LLP trees, whether in small patches or more widely dispersed throughout the stand, compete for growing space. Over several decades, mortality reduces the density of the new cohort as the dominant trees gain canopy positions. These dynamics can occur following severe disturbance events, resulting in largely even-aged stands with remnant legacy trees from the prior canopy, or within small canopy gaps where patches of seedlings and saplings differentiate into crown classes through competition and variable resource availability (Figure 5.2).

5.3.4 Mature Forest Stage

During the mature forest stage, the primary cause of mortality shifts from competition within the even-aged cohorts to local, stochastic events. Seed production during mast years often results in the establishment of abundant new seedlings throughout the stand, particularly if synchronized with recent burns (Figure 5.1a, b). Canopy tree mortality events and resulting canopy gaps release established even-aged cohorts, allowing younger cohorts to move into other canopy positions. The horizontal structure of the stand becomes more heterogeneous as canopy gaps of different sizes are formed during this stage.

5.3.5 Old Forest Stage

The old forest stage is characterized as a multi-aged, structurally complex mosaic of widely spaced overstory trees interspersed with even-aged cohorts of young trees developing in patches. Spatial patterning shows strong aggregation of small-diameter trees, becoming more uniform or randomly distributed as tree size increases. Mature trees can live for hundreds of years, resulting in overstory trees that may vary widely in age from their neighbors (Figure 5.1c).

5.4 Prevailing Silvicultural Systems

There is a somewhat limited history of silvicultural management for LLP following the exploitation of the original forests [1, 7] and with the increased focus on faster-growing species to feed the growing commercial forest industry in the southern U.S. [2]. Despite the characterization of natural LLP forests as multi-aged [20, 21], even-aged silvicultural systems were traditionally favored due to LLP's grass stage seedling morphology and shade intolerance. Specifically, natural regeneration using a shelterwood approach was the predominant silvicultural system. Shelterwood protocols were well documented in research from the USDA Forest Service [9, 12] and were applied successfully by both commercial and private landowners.

More limited research was conducted on uneven-aged silvicultural systems in LLP that better mimic the disturbance regime and resultant structures found in older, natural LLP forests and as described in the previous section. Early work from the USDA Forest Service focused on application of BDq selection techniques [28], with later work emphasizing the Pro-B approach to individual tree selection [29]. Both approaches emphasize residual structural targets that provide for continued, sustainable harvests. Another selection-based uneven-aged system that was widely applied on private lands in the southeast U.S. is the Stoddard–Neel approach [22, 30]. This approach considers retention of trees that contribute ecological benefits and characteristics (and may be considered "non-commercial") in harvested stands, incorporates a form of variable retention harvest (VRH, see Chapter 1), does not prescribe a residual structure, and has been suggested as a viable ecological silviculture approach [20, 26].

With the development of reliable artificial regeneration techniques (in particular containerized seedlings) in the late twentieth century, LLP plantations have been widely established and are now the predominant silvicultural technique used for restoration of the species. Availability of seedlings and improved establishment techniques (along with substantial incentive programs) have resulted in approximately 688,000 ha of LLP plantation establishment across the South since 1995, and today these plantations represent the majority of younger age class LLP forests. Artificial regeneration has also been used for planting longleaf seedlings under other canopy species [30], a process that utilizes existing mature forest structures while gradually converting to LLP.

5.5 An Ecological Silvicultural System for Longleaf Pine

The assumption in designing an ecological silvicultural system for LLP is that the desired or maintenance condition – one that promotes plant and animal diversity, provides habitat for species of conservation concern, and is sustainable – is a mature or old forest with a multi-aged, open-canopy structure in the overstory, a sparse midstory, and a diverse ground cover. The silvicultural treatments needed to achieve these conditions are dependent on the starting structure and development stage (Figure 5.4). Forests requiring "restoration," such as the currently abundant young plantations across the species' range, will require a longer series of treatments than those woodlands already in a maintenance condition. Application of frequent prescribed fire is the one constant treatment for all development trajectories across all developmental stages. Also, whatever the starting point, once stand development reaches the "mature" and "old" stages, silvicultural practices are similar (Figure 5.4).

	Disturbance	Pre-forest	Young forest	Mature forest	Old forest
Maintenance Existing LLP-dominant Canopy	Frequent Rx fire*			Frequent Rx fire Utilize selection harvests Release LLP regeneration and overtopped LLP midstory	Maintain frequent Rx fire Selection harvests Remove undesirable tree spp. Release LLP regeneration and overtopped LLP midstory
Restoration LLP Plantation Establishment	Complete canopy removal either by harvest or major nat. disturbance	Rx fire Site prep Plant longleaf pine	Maintain Rx fire Intermediate thinning for ground cover and structural heterogeneity (row thin or VDT)**	Maintain frequent Rx fire Selection harvest to reduce canopy densities and release LLP advance reproduction	Maintain frequent Rx fire Selection harvests Remove undesirable tree spp. Release LLP regeneration and overtopped LLP midstory
Restoration Conversion of mature stands of other pine species to LLP	Reduce canopy density Create/ expand canopy gaps Reduce/ control midstory Start frequent Rx fire	Frequent Rx fire Plant LLP seedlings in gaps or throughout the stand if canopy trees not present	Frequent Rx fire Intermediate thinning to remove competition Maintain open-canopy conditions in mature cohorts	Frequent Rx fire Selection harvests for lower canopy densities Preferentially retain LLP Release LLP regeneration	Maintain frequent Rx fire Selection harvests Remove undesirable tree spp. Release LLP regeneration and overtopped LLP midstory

*Rx fire = prescribed fire

**VDT = variable density thinning (see Chapter 1)

Figure 5.4 Summary of ecological silvicultural practices for longleaf pine (LLP) forests during different developmental stages and from different starting conditions. Rows represent forest development from the different starting conditions; columns highlight silvicultural practices utilized at stages of stand development following canopy disturbance (first column). Selection harvests in older stands utilize variable retention harvests (see Chapter 1) for individual tree selection, or group selection with reserves.

5.5.1 Ecological Silviculture to Maintain Mature Forests

Ecological silvicultural approaches tend toward uneven-aged selection systems in forests with a predominantly LLP canopy that is in a maintenance condition. Of the single tree selection approaches previously mentioned, the Stoddard–Neel approach, with a focus on holistic ecosystem characteristics, is generally more aligned with principles of ecological silviculture (e.g. spatial heterogeneity, legacies) than the other approaches (BDq and Pro-B) that are more focused on timber production [20, 30]. Whichever method or approach is used, trees are selected and marked for removal based upon tree vigor and form, release of advance LLP reproduction, and distribution of fine fuels for continued burning. Attention must be paid to development of the periodic good seed crops so appropriate recruitment sites are available when seed is plentiful. Because the objective is to simulate the spatial heterogeneity of remnant old-growth stands, creating an even or regular distribution for residual trees is not a high priority. Removal of single trees or small groups of trees (i.e. VRH) emulates the natural canopy disturbances caused by lightning or small-scale wind events, and postharvest residual stocking is often in the range of 9–14 m^2 ha^{-1} to provide the open-canopy condition that promotes a diverse ground cover community and seedling growth.

Group selection approaches, sometimes with reserve trees within the group openings, have also been suggested and utilized to mimic the dominant small-scale disturbance pattern in mature stands. The goal is to keep the group openings small (0.1–0.2 ha) to allow adequate seed dispersion into the gaps and to maintain sufficient fine fuels within the openings to carry fire. One study [23] found that group selection resulted in spatial patterns more like old-growth forests than did single tree selection, at least after one harvest entry in an even-aged stand. A potential downside to the group selection approaches, however, is development of significant woody competition within the gaps, due to lower fuel loading and reduced fire intensity, which can hinder establishment or release of LLP seedlings.

5.5.2 Ecological Silviculture for Restoration

As previously discussed, many situations requiring restoration of LLP will begin with plantation establishment in areas without an existing canopy (Figure 5.5). In an ecological silviculture context, the use of plantations for restoration should be viewed as an initial step in the trajectory toward the maintenance condition shown in Figure 5.3, not as the primary management objective.

In these cases, early management is typically very similar to plantation establishment for other southern pines – site preparation treatments to reduce competition, planting seedlings, development to canopy closure, and after some number of years, a first thinning. The initial thinning operation often entails removal of every third, fourth, or fifth row with some tree removals in remaining rows to reduce the basal area by one-third to one-half. Alternatively, trees do not have to be planted in rows, and variable-density thinning can be utilized to increase heterogeneity more quickly; both alternatives, however, increase operational complexity. Subsequent thinning operations reduce canopy tree density to promote establishment of natural LLP reproduction and, eventually, the release of these seedling cohorts to grow to canopy positions and promote development of a multi-aged condition (rather than restarting the process at a set rotation age). Reducing the canopy density also allows more light to reach the ground and promotes the desired herbaceous ground cover. After development reaches the mature and old stages, ongoing management proceeds as previously described. One difference in this plantation establishment approach for LLP, as compared to other pine species, is the application of prescribed fire throughout the development

Figure 5.5 Progression of plantation development toward mature stage with younger, established cohorts of longleaf pine reproduction: (a) two-year-old longleaf seedlings planted in cutover stand; (b) 10-year-old longleaf pine plantation just prior to canopy closure; (c) longleaf pine plantation after first thinning (approximately 18-years old); and (d) 35-year-old longleaf pine plantation with abundant reproduction cohorts following two thinnings and canopy mortality from two windstorms and a hurricane.

process – LLP plantations can be burned as soon as the seedlings are big enough to tolerate fire (two to five years) [11, 21] and a regime of frequent fire is used through all stages of development.

A different, and somewhat intermediate, restoration situation is provided where the canopy is dominated by pine species other than LLP and a slow conversion that maintains the mature tree structural component is preferred. The first step is to reduce canopy density, preferentially retaining LLP if present, to establish reproduction. This reproduction can be from seed if LLP trees are present in the canopy stratum, or can be established by planting LLP seedlings in different patterns (Figure 5.6) [30]. Subsequent harvest entries will be conducted in a manner that releases the established LLP cohorts, favors retention of LLP in the canopy, and increases spatial heterogeneity of the canopy trees (e.g. VRH). Overstory trees of the undesired species can also be retained long term to eventually increase the number of snags and amount of large woody debris in the stand. At mature stages of development, the canopy should be dominated by LLP, and management will follow the same guidelines presented for mature and old stages in Figure 5.4.

Success of the gradual conversion approach is affected by site quality, species composition within the canopy, and maintenance of a frequent fire regime through their effects on the competitive status of the LLP reproduction. Competition from other plants, and in particular woody plants, is typically much higher on good sites than on poor sites, and more effort is required to control this

Figure 5.6 Longleaf pine seedlings planted under an 80-year-old slash pine canopy. Seedlings were planted in natural and harvest-created canopy gaps.

competition on the better sites. For example, some form of site preparation may be required in the areas where LLP will be planted on high-quality sites [30]. In addition, the composition of the canopy affects the seed rain from competing pine species, with loblolly pine having the greatest impact due to its prolific seed production. Frequent prescribed fire can help to alleviate both effects by controlling woody competition and the seedlings of other pine species which are not fire tolerant.

5.6 Climate Change Considerations

Expected future conditions for the southeastern region of the U.S. include increased temperatures and more variable precipitation, with periods of both heavy rainfall and prolonged drought [31]. LLP has been classified as moderately adapted to future conditions, and future range projections for LLP indicate the potential for increased abundance within its current range and limited northern range shifts [32]. Because extreme weather events, including droughts and wind events, are expected to increase, LLP, which is both drought tolerant [5, 10] and generally more resistant to damage from wind events than associated tree species [27], appears to be well-suited to predicted climatic conditions.

While direct effects of climate change on LLP may be less intense than for other species, some indirect effects may adversely affect LLP forests. Increased abundance and distribution of invasive species, including plants and pathogens, may introduce challenges to maintaining structure and composition. Given the tight coupling of the plant composition and the frequent fire regime, changes to plant community composition may also have implications for prescribed burning. Furthermore, projected future climate conditions may reduce potential prescribed burn days [33] and the ability to maintain the frequent fire regimes necessary for maintaining LLP ecosystems.

Ecological silviculture approaches, however, help to mitigate these uncertainties. The age and structural complexity created through the ecological silviculture approach increase resistance and resilience to extreme disturbance events. For example, large-scale wind events such as hurricanes predominantly impact mature trees, but maintaining the multi-aged structure with patches of younger trees that are released by such events would accelerate stand development trajectories following these disturbances.

5.7 Summary

The current extent of LLP ecosystems is greatly reduced from historical prevalence within the species' natural range, so there are many fewer mature and old stands to manage. Yet there is great interest in restoration of LLP on the landscape, which has resulted in greatly increased area of LLP plantations, and these forests will require much time and effort to move toward the desired multi-aged structure and composition of mature and old stands that provide high conservation values. Ecological silviculture approaches for stands that reach the mature and old stages of development utilize selection-based systems, whether single tree or some type of group selection, to sustain the multi-aged and spatially heterogeneous structures characteristic of the natural model for this ecosystem. In general, selection-based systems adapted from the Stoddard–Neel approach or group selection with reserves provide structural characteristics aligned more closely with natural models fundamental to ecological silviculture [26, 30]. Whichever selection system is utilized, a regime of frequent fire is critical to meet ecological and conservation goals.

References

1 Ware, S., Frost, C., and Doerr, P.D. (1993). Southern mixed hardwood forest: the former longleaf pine forest. In: *Biodiversity of the Southeastern United States: lowland Terrestrial Communities* (ed. W.H. Martin, S.G. Boyce, and A.C. Echternacht), 447–493. New York, NY: Wiley.

2 Outcalt, K. and Sheffield, R. (1996). *Longleaf Pine Forest: Trends and Current Conditions*, Resource Bulletin RB-SRS-9. Asheville, NC: U.S. Department of Agriculture, Forest Service, Southeastern Forest Experiment Station. 28 p.

3 USDA Forest Service, Forest Inventory and Analysis Program (2022). *Forest Inventory EVALIDator Web-Application Version 1.8.0.01*. St. Paul, MN: U.S. Department of Agriculture, Forest Service, Northern Research Station.

4 Bartram, W. (1792). *The Travels of William Bartram* (ed. F. Harper). New Haven, CT: Yale University Press.

5 Wahlenberg, W.G. (1946). *Longleaf Pine: Its Use, Ecology, Regeneration, Protection, Growth, and Management*. Washington, DC: Charles Lathrop Pack Forestry Foundation.

6 Chapman, H.H. (1932). Is the longleaf type a climax? *Ecology* 13 (4): 328–334.

7 Landers, J.L., Van Lear, D.H., and Boyer, W.D. (1995). The longleaf pine forests of the southeast: requiem or renaissance? *Journal of Forestry* 93 (11): 39–44.

8 Maki, T. (1952). Local longleaf seed years. *Journal of Forestry* 50 (4): 321–322.

9 Croker, T.C. and Boyer, W.D. (1975). *Regenerating Longleaf Pine Naturally*, Research Paper SO-105, 21. New Orleans, LA: U.S. Department of Agriculture, Forest Service, Southern Forest Experiment Station.

10 Jack, S.B. and Pecot, S.D. (2018). Regeneration dynamics, competition, and seedling response. In: *Ecological Restoration and Management of Longleaf Pine Forests* (ed. L.K. Kirkman and S.B. Jack), 71–88. Boca Raton, FL: CRC Press.

11 Boyer, W.D. (1979). *Natural Regeneration of Longleaf Pine*, Technical Publication SA-TP3, 6–8. U.S. Department of Agriculture, Forest Service, Southeastern Area, State and Private Forestry.

12 Boyer, W.D. (1993). Regenerating longleaf pine with natural seeding. In: *Proceedings of the Tall Timbers Fire Ecology Conference, No 18, The Longleaf Pine Ecosystem: Ecology, Restoration and Management* (ed. S.M. Hermann), 299–309. Tallahassee, FL: Tall Timbers Research Station.

13 Stambaugh, M.C., Guyette, R.P., and Marschall, J.M. (2011). Longleaf pine (*Pinus palustris* Mill.) fire scars reveal new details of a frequent fire regime. *Journal of Vegetation Science* 22 (6): 1094–1104.

14 Platt, W.J. and Rathbun, S.L. (1993). Dynamics of an old-growth longleaf pine population. In: *Proceedings of the Tall Timbers fire ecology conference, No 18, The Longleaf Pine Ecosystem: Ecology, Restoration and Management* (ed. S.M. Hermann), 275–297. Tallahassee, FL: Tall Timbers Research Station.

15 Palik, B.J. and Pederson, N. (1996). Overstory mortality and canopy disturbances in longleaf pine ecosystems. *Canadian Journal of Forest Research* 26 (11): 2035–2047.

16 Boyer, W.D. (1979). Mortality among seed trees in longleaf pine shelterwood stands. *Southern Journal of Applied Forestry* 3 (4): 165–167.

17 Fortuin, C.C., Montes, C.R., Vogt, J.T., and Gandhi, K.J.K. (2022). Predicting Risks of Tornado and Severe Thunderstorm Damage to Southeastern U.S. Forests. Landscape Ecology 2022.

18 Keim, B.D., Muller, R.A., and Stone, G.W. (2007). Spatiotemporal patterns and return periods of tropical storm and hurricane strikes from Texas to Maine. *Journal of Climate* 20 (14): 3498–3509.

19 Varner, J.M. and Kush, J.S. (2004). Remnant old-growth longleaf pine (*Pinus palustris* Mill.) savannas and forests of the southeastern USA: status and threats. *Natural Areas Journal* 24 (2): 141–149.

20 Mitchell, R.J., Hiers, J.K., O'Brien, J.J. et al. (2006). Silviculture that sustains: the nexus between silviculture, frequent prescribed fire, and conservation of biodiversity in longleaf pine forests of the southeastern United States. *Canadian Journal of Forest Research* 36 (11): 2724–2736.

21 Platt, W.J., Evans, G.W., and Rathbun, S.L. (1988). The population dynamics of a long-lived conifer (*Pinus palustris*). *The American Naturalist* 131 (4): 491–525.

22 Moser, W.K., Jackson, S.M., Podrazsky, V., and Larsen, D.R. (2002). Examination of stand structure on quail plantations in the Red Hills region of Georgia and Florida managed by the Stoddard–Neel system: an example for forest managers. *Forestry: An International Journal of Forest Research* 75 (4): 443–449.

23 Cannon, J.B., Bigelow, S.W., Hiers, J.K., and Jack, S.B. (2022). Effects of silvicultural selection treatments on spatial pattern and dynamics in a *Pinus palustris* Mill. woodland. *Forest Ecology and Management* 505: 119888.

24 Hiers, J.K., O'Brien, J.J., Will, R.E., and Mitchell, R.J. (2007). Forest floor depth mediates understory vigor in xeric *Pinus palustris* ecosystems. *Ecological Applications* 17 (3): 806–814.

25 Peet, R.K. (2007). Ecological classification of longleaf pine woodlands. In: *The Longleaf Pine Ecosystem: Ecology, Silviculture, and Restoration* (ed. S. Jose, E.J. Jokela, and D. Miller), 51–93. New York: Springer.

26 Mitchell, R.J., Hiers, J.K., O'Brien, J., and Starr, G. (2009). Ecological forestry in the Southeast: understanding the ecology of fuels. *Journal of Forestry* 107 (8): 391–397.

27 Rutledge, B.T., Cannon, J.B., McIntyre, R.K. et al. (2021). Tree, stand, and landscape factors contributing to hurricane damage in a coastal plain forest: post-hurricane assessment in a longleaf pine landscape. *Forest Ecology and Management* 481: 118724.

28 Farrar, R.M. Jr. and Boyer, W.D. (1991). Managing longleaf pine under the selection system – promises and problems. In: *General Technical Report SE-70; Proceedings of the Sixth Biennial Southern Silvicultural Research Conference; Memphis, TN* (ed. S.S. Coleman and D.G. Neary), 357–368. Asheville, NC: U.S. Department of Agriculture, Forest Service, Southeastern Forest Experiment Station.

29 Brockway, D.G., Loewenstein, E.F., and Outcalt, K.W. (2014). Proportional basal area method for implementing selection silviculture systems in longleaf pine forests. *Canadian Journal of Forest Research* 44 (8): 977–985.

30 Jack, S.B. and McIntyre, R.K. (2018). Restoring and managing the overstory: an ecological forestry approach. In: *Ecological Restoration and Management of Longleaf Pine Forests* (ed. L.K. Kirkman and S.B. Jack), 175–205. Boca Raton, FL: CRC Press.

31 Mitchell, R.J., Liu, Y., O'Brien, J.J. et al. (2014). Future climate and fire interactions in the southeastern region of the United States. *Forest Ecology and Management* 327: 316–326.

32 Peters MP, Prasad AM, Matthews SN, Iverson LR. Climate Change Tree Atlas, Version 4: U.S. Department of Agriculture, Forest Service, Northern Research Station and Northern Institute of Applied Climate Science; Delaware, OH; 2020. https://www.nrs.fs.fed.us/atlas (Accessed 1 April 2022).

33 Kupfer, J.A., Terando, A.J., Gao, P. et al. (2020). Climate change projected to reduce prescribed burning opportunities in the south-eastern United States. *International Journal of Wildland Fire* 29 (9): 764–778.

6

Ecological Silviculture for Southeastern US Pine-Oak Forests

Justin L. Hart[1], J. Davis Goode[1], and Daniel C. Dey[2]

[1] *Department of Geography and the Environment, University of Alabama, Tuscaloosa, AL, USA*
[2] *Northern Research Station, USDA Forest Service, Columbia, MO, USA*

6.1 Introduction

In the southeastern United States, pine (*Pinus*) and oak (*Quercus*) species may co-occur across a wide range of site types and exist in a diversity of assemblages. The geographic focus of this chapter is the vernacular region of the United States Upland South (Figure 6.1). In this region, oak species diversity is high, and these species occur in almost every forest type. Although oaks can share dominance with all pine species native to this region, our focus is on mixtures of oak species that most frequently occur in association with shortleaf pine (*Pinus echinata* Mill), loblolly pine (*Pinus taeda* L.), and Virginia pine (*Pinus virginiana* Mill.), where under management these taxa may co-occur in the overstory. The oaks most frequently found in association with these pines include northern red oak (*Quercus rubra* L.), southern red oak (*Quercus falcata* Michx.), scarlet oak (*Quercus coccinea* Muenchh.), black oak (*Quercus velutina* Lam.), white oak (*Quercus alba* L.), chestnut oak (*Quercus montana* Willd.), and post oak (*Quercus stellata* Wangenh.).

Stands that contain mixtures of pines and oaks in this region can be classed as pine-oak or oak-pine dominant and the species of each genus may vary in their basal area contribution. Recently, the term mixedwoods has been applied to these assemblages. Mixedwood forests are those that contain a mixture of hardwood and softwood species, with neither exceeding 75–80% of stocking based on basal area or canopy cover [1]. The stand types discussed in this chapter fit this definition of mixedwoods. We make no distinctions between pine-oak and oak-pine ecosystems. In such stands, species from other genera may share canopy dominance with these taxa, including hickories (*Carya* spp.), tulip-poplar (*Liriodendron tulipifera* L.), sweetgum (*Liquidambar styraciflua* L.), and red maple (*Acer rubrum* L.) among others.

6.2 Characteristics of Pine-Oak Ecosystems

Pine-oak ecosystems are variable in species richness, canopy cover, tree density, basal area, and other factors. This variability is attributed in part to the diversity of site conditions that support pine-oak mixtures, land use, and management history and the range of species from these genera

Ecological Silvicultural Systems: Exemplary Models for Sustainable Forest Management,
First Edition. Edited by Brian J. Palik and Anthony W. D'Amato.

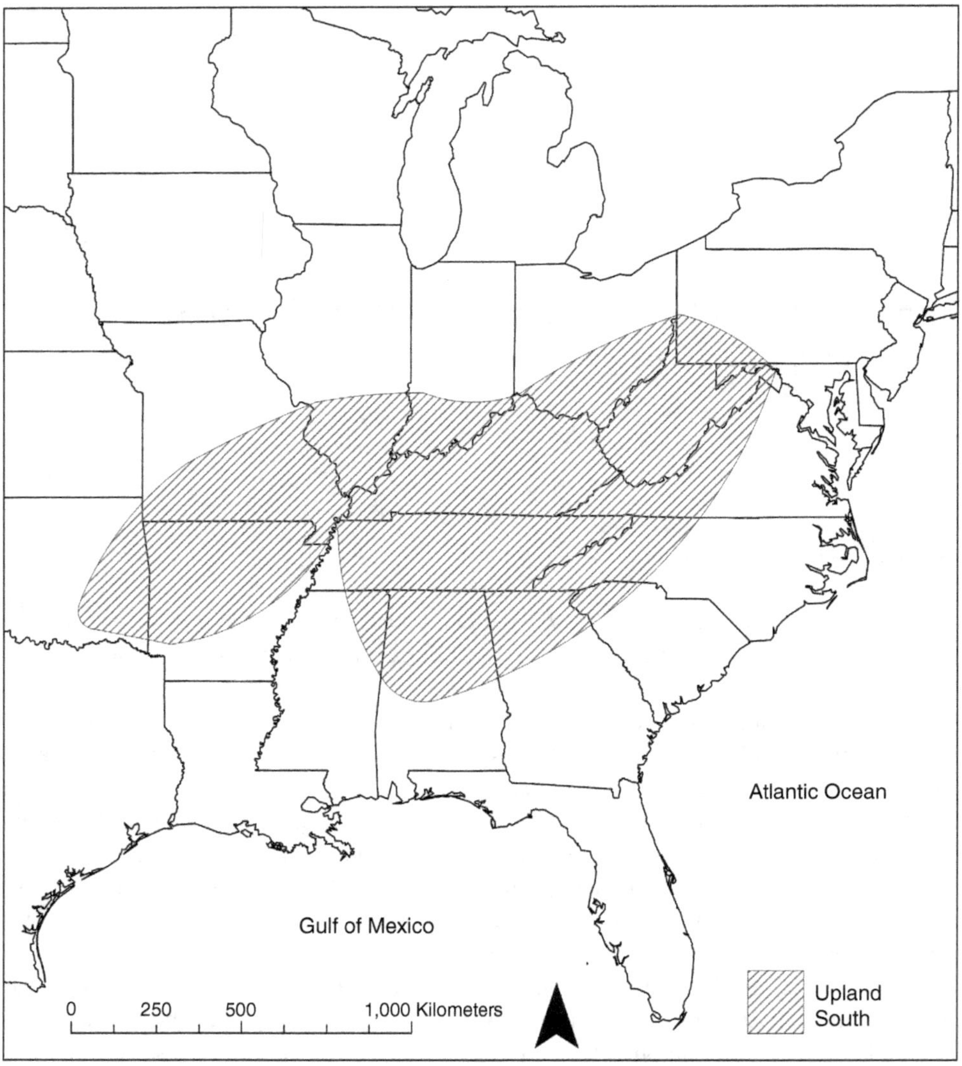

Figure 6.1 The vernacular region known as the Upland South of the eastern United States defined by landforms, history, and culture.

that may be dominant. Pine-oak mixtures can occur as savannas, woodlands, or forests depending on the management goals and disturbance regime (Figure 6.2). As previously stated, species diversity in this region is high and the differences in shade tolerance, growth rate, and life span among the oaks and pines are important drivers of species regeneration and competition dynamics (Table 6.1).

6.2.1 Silvics of Common Species

In general, the pines are more shade-intolerant, faster-growing, and disturbance-dependent species, relative to the co-occurring oaks. Loblolly pine, shortleaf pine, and Virginia pine all require bare or nearly bare mineral soil for seed germination and seedling establishment. Germination is

Figure 6.2 Pine-oak mixedwoods in the southeastern United States may occur as savannas (a), woodlands (b), or forests (c) depending on the management and disturbance regime.

Table 6.1 Shade tolerance, moisture affinity class, and approximate maximum lifespan for selected pine and oak species that commonly co-occur in pine-oak mixedwoods of the southeastern United States.

Species	Shade tolerance	Moisture affinity class	Maximum life span (years)
Black oak	Intermediate	Subxeric	250
Chestnut oak	Intermediate+	Subxeric	425
Loblolly pine	Intolerant	Submesic	250
Northern red oak	Intermediate	Submesic	325
Post oak	Intolerant	Xeric	400
Scarlet oak	Intermediate	Xeric	215
Shortleaf pine	Intolerant	Subxeric	325
Southern red oak	Intermediate–	Subxeric	145
Virginia pine	Intolerant	Xeric	130
White oak	Intermediate+	Submesic	470

Intermediate + indicates closer to the tolerant class and Intermediate – indicates closer to the intolerant class.

epigeal and occurs in the spring. Shortleaf pine, which is increasingly prioritized within pine-oak ecosystems by managers, is well-adapted to frequent fire because it can resprout from a basal crook that contains dormant buds. The unique ability of shortleaf pine to resprout after shoot mortality is competitively advantageous compared to other pine species in frequent-fire systems. Pines are noted for their rapid bark growth from early ages which confers protection of the cambium from fire, an advantage they gain earlier in life than hardwoods. Growth rates of pine species are inversely related to shade. Generally, pines growing in the open can outgrow oaks given equal initial tree sizes, but well-developed oak advance reproduction as large seedling sprouts can rapidly suppress young pine regeneration [2]. Loblolly pine tends to dominate on more mesic sites and Virginia pine on more xeric sites. The pine species discussed here are considered drought tolerant and their drought tolerance may be enhanced by management practices, such as thinning. Virginia pines have been documented to live for over 100 years, loblolly pines for over 200 years, and shortleaf pines for over 300 years.

In mixed pine-oak stands, oak species are more shade tolerant than co-occurring pines. Chestnut oak and white oak are generally more shade tolerant than other oak species considered in this chapter. Southern and northern red oak and black oak are intermediate in shade tolerance, and scarlet oak and post oak are shade intolerant. Sources of oak reproduction include seed, seedling sprouts, and stump sprouts. In general, oaks have slow shoot growth as new seedlings as they preferentially allocate growth to roots. Consequently, they have low regeneration potential compared to their competitors in open environments.

Since most oaks decrease in their ability to produce stump sprouts as they grow larger in diameter and age, the main competitive source of oak regeneration is large advance reproduction, i.e. seedling sprouts that accumulate over multiple acorn crops and grow with sufficient light in the understory [3]. Many oak species can develop as large advance reproduction under 30–50% full sunlight in forest understories, and the survival and growth of the more intolerant species improve with increasing levels of light.

Oaks have fire-adapted traits that permit them to persist in frequent fire regimes such as root-centric carbon allocation and dormant, adventitious buds clustered around the root collar that is often buried in soil [4, 5]. Older, larger-diameter oaks have relatively thick bark that protects their cambium from fire. Many oak species are resistant to drought and have flammable litter characteristics that promote fire during dry periods. Species in the red oak group normally live about 80–150 years. Those in the white oak group are much longer lived with life spans that can extend to over 400 years. Oak species from xeric environments such as post oak, white oak, and chestnut oak are more drought resistant and persistent in the face of recurring droughts than are red oak species [6], and white oak species are in general more adapted to droughts than red oaks.

6.2.2 Disturbance Ecology

Pine-oak stands are considered mid-successional, and these mixtures cannot typically be maintained without a combination of canopy disturbance and surface fire. Wind-related canopy disturbance is common in mixed pine-oak ecosystems. Oak decline, caused by complex interactions of environmental stresses and pests, can result in substantial mortality of mature oaks, especially those in the red oak group. Ice storms occur in mixed pine-oak ecosystems and in some locations occur at relatively short intervals (30–40 years). Native insects produce canopy disturbances by causing mortality in mature pines with southern pine beetle (*Dendroctonus frontalis* Zimmermann) being the most common and widespread. Fire may result in canopy tree mortality, but is primarily a disturbance of the subcanopy strata.

Canopy disturbances in mixed pine-oak ecosystems range from highly localized events that remove trees individually to catastrophic events that remove most overstory vegetation. Intermediate-severity disturbances are those that are larger in extent than canopy gaps, but are not stand replacing. The return interval of these disturbances in this region ranges from 20 to 50 years [7, 8]. Disturbance impacts on species composition, stand structure, and ecosystem function vary by disturbance agent, spatial extent, frequency, and intensity. Some disturbance agents are species or genus-specific and those that are not, may disproportionately damage or kill some taxa over others.

Fire is an important disturbance agent in mixed pine-oak ecosystems. Low-intensity fire reduces leaf litter, exposes bare mineral soil, and prevents the development of a duff layer. Fire also causes shoot mortality of small diameter (<10 cm diameter) woody stems. Resprouting is an important aspect of regeneration and competition in pine-oak ecosystems. Although they can resprout, frequent fire inhibits recruitment into larger size classes. Thus, a sufficiently long fire-free period may be needed to recruit oak-pine into the overstory [9]. Fire behavior is in part a function of fuel composition, structure, loading, and flammability, and these characteristics may be highly variable in pine-oak stands. Flammability of litter varies by species and overstory composition. The pines discussed here have highly flammable litter, and so too do oaks common in xeric environments such as white oak, turkey oak, southern red oak, and post oak [10]. Frequent fires can increase herbaceous diversity and cover especially when combined with overstory mortality. Historical fire frequency in this region averaged 4–15 years before European colonization, with spatial and temporal variation in fire severity, season, size, and frequency [11]. Almost universally, fire frequency increased throughout the eastern United States with European settlement with mean fire intervals ranging from one to five years until the fire suppression period of the early 1900s [12].

6.3 Development Model

Development of pine-oak stands is strongly influenced by the pre-disturbance stand condition. Especially important in the pre-disturbance stand is the presence or absence of oak advance reproduction. We note that many contemporary pine-oak ecosystems did not develop following natural disturbance and instead established on abandoned agricultural lands or in abandoned pine plantations. The natural development model presented here begins with canopy disturbance and legacy creation and proceeds through young, mature, and old forest stages. We note that regenerating disturbances in pine-oak stands are more typically intermediate-severity events (i.e. intra-stand scale) and are not stand replacing. Thus, a single pine-oak stand may contain neighborhoods in different developmental stages.

6.3.1 Disturbance and Legacy Creation

Development after canopy disturbance in pine-oak stands is strongly influenced by ecological memory as hardwood rootstocks typically persist in the understory as advance reproduction. Thus, the presence of hardwood rootstocks or other advance reproduction will strongly influence stand development. In pine-oak ecosystems, natural stand-scale catastrophic disturbances are not common. However, natural intermediate-severity disturbances typically occur within stands once every 20–50 years. These disturbances result in legacies of deadwood structures and surviving trees. Sizes, shapes, and configurations of natural disturbance-created openings vary widely depending upon the disturbance agent and disturbance severity. For example, wind disturbances typically remove the largest stems and in pine-oak systems, those are often pine species. These disturbances often create patches of dead trees and residual trees within disturbed neighborhoods may be aggregated or dispersed.

6.3.2 Preforest Stage

The preforest stage in pine-oak ecosystems is relatively short with no true herb-dominated period, given the ability of many species to sprout or establish advance reproduction. The duration of this period may be influenced by the fire regime. Open canopy conditions in these systems often support high species richness and cover of ground flora.

6.3.3 Young Forest Stage

Although pine species typically grow faster than co-occurring hardwoods in open environments, sprouting from well-developed hardwood advance reproduction may negate the height growth advantage of pines that regenerate from seed. When grown from seed, young oaks that typically allocate resources preferentially to root growth can be overtopped by taxa with rapid rates of shoot growth. However, oak shoots originating from well-developed root systems grow faster than those originating from seed when released by disturbance. In addition, mature pine-oak stands may have advance oak reproduction in the understory owing to the greater shade tolerance of oaks. Shortleaf pine, with its ability to resprout, may develop into large advance reproduction under certain low-density stand conditions, however, its reproduction cannot survive in the low light environment of fully-stocked stands.

The density and size of oak stems in the regeneration layer may be promoted by fire in these systems if understory light levels exceed 20% of full sunlight, levels too low for pine reproduction development. If oak advance reproduction is present at the time of canopy disturbance, those stems have a height growth advantage over pine stems that will originate from seed. In these scenarios, naturally regenerated pine should still occur on the site, but will typically be less abundant than oaks. Site conditions are important as pine regeneration is usually more competitive on lower-quality sites. The pine component may exhibit spatial clumping coincident with neighborhood-scale environmental conditions. Regardless, fire and canopy disturbance are required to maintain pine regeneration and recruitment in what may otherwise develop into an oak-dominated system.

6.3.4 Mature Forest Stage

As stands mature, oaks and other hardwoods establish in the pine understory as pines in the canopy senesce. Once the oaks and associated hardwoods establish, the two-aged stands develop as stratified mixtures at taxon-specific growth rates. In this scenario, fire may inhibit the recruitment of the more shade-tolerant oaks and associated hardwoods and relegate them to seedling and sapling size classes. Fire alone will not maintain pine dominance in these systems except on the most xeric sites, but fire frequency is important for pine recruitment into the overstory. Canopy disturbance events are also important to allow sufficient light to reach the regeneration layer. Thus, the combination of fire and canopy disturbance is crucial for maintenance of pine-oak stands. Clearly, intra-stand spatial patterns are important, as fire behavior will vary through a stand with canopy disturbance events occurring stochastically. For example, even in a stand subjected to frequent prescribed fire, neighborhoods of oaks may avoid shoot mortality from high temperature and recruit to size classes sufficiently large (>10 cm dbh) so that they can survive low-intensity fire. In fact, a relatively short fire-free period, at stand or intra-stand scales, may be sufficient time for oaks to reach a size threshold for shoots to survive typical temperatures of prescribed fires. Where fire frequency is greater than three years and overstory density is low enough to permit >50% of full sunlight in the understory, shortleaf pine can develop into large advance reproduction, much like oak; and they will have increased competitive capacity after an overstory disturbance that releases regeneration. Another pathway to pine regeneration is in the presence of recurring fires after overstory removal. Pines develop thick bark more rapidly than hardwoods and can achieve resistance to topkill at an earlier age. Hence, if fires occur after pine reproduction reaches fire-tolerant size, it will discriminate against hardwood competitors [13, 14].

Figure 6.3 In natural, old pine-oak mixedwoods of the southeastern United States, pines often occur as clusters (i.e. pine-dominated neighborhoods) in what are otherwise oak-dominated stands.

6.3.5 Old Forest Stage

In older, natural mixed pine-oak systems, the pine component is often spatially clustered within the stand (Figure 6.3). The clustering of pine into distinct neighborhoods may result from stochastic events or biophysical site conditions [15]. For example, large canopy openings on xeric sites or those subjected to frequent, low-intensity fire may provide regeneration sites for pines. In the absence of stand-wide catastrophic disturbance, pine-oak stands may develop uneven-aged structures. Although the hardwood canopy component may represent a single age class, the pine component may represent multiple cohorts that established in accord with intra-stand scale canopy disturbances. Of course, these stands can be composed of multiple oak cohorts as well, as old pine-dominated neighborhoods succeed in oak dominance. Older stands are characterized by large living stems as well as large snags and downed logs. Pit and mound microtopography is common from windthrow and spatial patterns of species composition within stands (i.e. pine neighborhoods, oak neighborhoods) are often apparent.

6.4 Ecological Silvicultural Systems for Pine-Oak Ecosystems

The silvicultural approach used to create and/or maintain pine-oak stands depends principally upon the current prevailing condition (Table 6.2). Managers that wish to create pine-oak stands typically begin with pine stands, including pine plantations, that lack an oak component or oak stands that lack a pine component. Clearly, different silvicultural systems are required based on initial stand composition and structure. In both instances, an ecological silvicultural system based on natural development would likely begin in the young to mature forest stages. As stated above, the canopy disturbances that lead to regeneration and legacy creation in pine-oak systems often occur at intra-stand scales (Table 6.3). Thus, preforest and young forest stages may occur in neighborhoods within mature and old stands.

Table 6.2 An ecological silvicultural system for mixed pine-oak ecosystems during early stages of development based on initial stand conditions.

	Initial condition					
	Pine dominated		Oak dominated		Mixed pine-oak	
Stage	**Potential actions**	**Outcome**	**Potential actions**	**Outcome**	**Potential actions**	**Outcome**
Disturbance and legacy creation (0 y)	Variable-density thinning	Release oak advance reproduction	Patch clearcut with reserves	Create canopy openings for pine regeneration	Patch clearcut with reserves; Patch seedtree with reserves; Patch shelterwood with reserves	Create canopy openings for pine and oak regeneration
	Girdle and fell trees	Creation of dead downwood and snags	Girdle and fell trees	Creation of dead downwood and snags	Girdle and fell trees	Creation of dead downwood and snags
	Prescribed fire	Expose bare mineral soil; reduce competition from fire-sensitive species	Prescribed fire	Expose bare mineral soil; reduce competition from fire-sensitive species	Prescribed fire	Expose bare mineral soil; reduce competition from fire-sensitive species
Preforest (1+y)	Chemical or mechanical competition control if needed	Reduce competition from shade-tolerant/fire-sensitive species	Outplant pine in canopy openings	Establish a pine cohort	Cluster plant pine seedlings if needed	Supplement natural pine regeneration in openings
					Chemical or mechanical competition control if needed	Reduce competition from shade-tolerant/fire-sensitive species
Young forest (1–80 y)	Prescribed fire	Expose bare mineral soil; reduce competition from fire-sensitive species	Prescribed fire	Expose bare mineral soil; reduce competition from fire-sensitive species	Prescribed fire	Expose bare mineral soil; reduce competition from fire-sensitive species
			Crop tree release pine as needed	Recruit pine individuals to the canopy		

Table 6.3 An ecological silvicultural system for mixed pine-oak ecosystems during mature and old forest stages.

Stage	Potential actions	Outcome
Mature forest (80–150 y)	Patch clearcut with reserves; patch seedtree with reserves; patch shelterwood with reserves	Regeneration of desired species in patches
	Cluster plant pine as needed in canopy openings	Supplement natural pine regeneration if required
	Prescribed fire	Reduce competition from fire-sensitive species; expose bare mineral soil
Old forest (150+ y)	Patch clearcut with reserves; patch seedtree with reserves; patch shelterwood with reserves	Regeneration of desired species in patches
	Cluster plant pine as needed in canopy openings	Supplement natural pine regeneration if required
	Prescribed fire	Reduce competition from fire-sensitive species; expose bare mineral soil

6.4.1 Disturbance and Legacy Creation

An ecological silvicultural system to convert pine stands, including pine plantations, into mixed pine-oak forests would take advantage of the developing hardwood understory. Oaks are often present in pine stands along with a diversity of other hardwoods. Red maple, a shade-tolerant species, is common and often dominant in pine understories and is a major competitor of oak. If oak advance reproduction is present, a localized, partial removal cutting may be used to thin the pine canopy and recruit oaks to larger size classes and eventually result in a two-aged stand. Thinning alone also releases other hardwoods in the understory including highly competitive red maple. If not controlled by prescribed fire or herbicide, red maple may dominate the future overstory.

Variable-density thinning (see Chapter 1) and group selection with reserves may be modeled after natural canopy disturbances. Intermediate-severity disturbances provide an appropriate template as these types of events may remove trees individually throughout a stand or in variably sized and shaped patches [8, 16]. The return interval of intermediate-severity disturbances is typically 20–50 years and provides a template for repeated stand entries. The long-term focus should be on maintaining a strong pine component. A shelterwood with reserves approach would work to recruit oak but may diminish the pine component so that long-term maintenance of pine in the stand could be a challenge. Larger pine trees would be retained in such a system as they would represent a key source of mature tree structure and future seed sources.

In oak stands that lack a strong pine component, a silvicultural system modeled after natural disturbance and development patterns may include patch clearcut or group selection with reserves. In existing pine-oak stands, an ecological silvicultural system may include seedtrees with reserves and shelterwood harvesting with different configurations of reserve tree spatial patterns. Combinations of regeneration methods may be used in large stands to account for initial stand conditions, variations in abiotic conditions, and other factors influencing regeneration such as invasive species, deer herbivory, and interfering ground flora. The sizes and shapes of the patches should be variable to account for the range of openings created by natural canopy disturbance events and to create the resource environment needed for regeneration of desired species (Figure 6.4). For example, wind disturbance often creates relatively small patches of less than 5 ha,

Figure 6.4 The size and shape of natural, intermediate-severity canopy disturbances are highly variable in pine-oak mixedwoods and these disturbances provide a template for silvicultural systems. In this image, natural canopy disturbance openings greater than $100\,m^2$ are highlighted.

however, these storms may also result in a low frequency of large openings of over 100 ha [17]. The shapes of natural canopy disturbances are also variable and range from circular openings to linear patterns [8]. The openings should be focused on microsites best suited for pine dominance, such as ridgelines and upper slope positions, south aspects, or other sites with convex curvature and soils where pine is inherently more competitive. In contrast, microsites such as fens, seeps, and springs that contain sensitive species or abiotic features may be avoided.

6.4.2 Preforest Stage

In pure hardwood stands, the pine component must be artificially regenerated in the harvest-created openings. Regardless, if regeneration is natural or artificial, site preparation is important and would normally include prescribed fire to reduce deep litter and competing vegetation. Herbicide may be required to reduce competition from hardwood stems. Site preparation treatments should be tailored based on site conditions and the specific pine species of interest. In harvest-created openings, outplanting of containerized shortleaf pine seedlings has been shown to achieve greater growth and survival compared to bare root seedlings, especially on nutrient-poor and dry sites [18]. In harvest openings, shortleaf pine seedlings may be planted at high densities (e.g. $<1 \times 1$ m; [19]). Newly planted pine seedlings are vulnerable to fire until they are established for a couple of years in adequate light environments. Thus, fire may be withheld until shortleaf pine seedlings reach a diameter of 2–10 cm [13] and height of 4–5 m [20] to reduce risk of topkill. If prescribed fire must be used to control competing vegetation, a three to five-year burn cycle maybe used to develop vigorous shortleaf pine sprout reproduction. Dey and Hartman [14] found that new shortleaf pine seedlings and advance reproduction of small basal diameter had a lower probability of survival after fire compared to hardwoods of similar size. Chemical treatment of hardwood competition may be necessary to increase height and diameter growth of planted shortleaf pine and could be used in conjunction with prescribed fire to decrease fire-induced mortality of shortleaf pine by reducing competing vegetation and potential fuel.

6.4.3 Young Forest Stage

Low-intensity surface fire is an important component of the disturbance regime in pine-oak stands and thus, a prescribed fire program is important in the young forest stage. Prescribed fire may be withheld until tree reproduction reaches size thresholds so that fire-related damage will be insignificant as discussed above. Tending of young regeneration and crop tree release at canopy closure may be needed to sustain pine in dominant crown positions to ensure desirable pine stocking levels.

6.4.4 Mature Forest Stage

In mature pine stands, prescribed fire can be used effectively either alone or in combination with overstory harvesting and other silvicultural practices to control the composition and development of hardwood regeneration in pine stand understories and promote oak dominance. Low-intensity dormant season fires can reduce midstory density by topkilling small hardwood stems, but typically increase the density of regeneration-sized stems due to sprouting of fire-injured midstory trees [21]. However, repeated fires over a decade can substantially reduce stem density, especially in the understory of fully-stocked, unthinned overstories [22]. Schweitzer et al. [23] reported that the hardwood midstory was largely eliminated by repeated burning and hardwood regeneration was greater with a heavy pine thinning, leaving $11\,m^2\,ha^{-1}$ of basal area, after three prescribed fires over nine years compared to less intense treatments. They found that thinning and repeated burning did favor development of larger oak reproduction, but the initially dominant red maple advance reproduction was still equally or more dense and larger than oak regeneration. Red maple is a prolific sprouter following the death of the shoot and its tolerance to shade allows it to recover between fires under moderate overstory densities.

In pure pine stands devoid of natural oak regeneration, oaks must be artificially regenerated by underplanting. Guidelines for success in oak plantings have been published (e.g. [24–26]). However, oak advance reproduction typically exists in mature pine stands in this region.

A common approach to establish a pine component in a mature oak stand is to prepare the site with prescribed fire to reduce woody competition, which may require multiple burns. The stand would then be thinned to increase light in the regeneration layer sufficient for pine survival and growth, and then pine seedlings would be underplanted. Further reductions in overstory density may be done around pockets of pine reproduction. Hardwood competition around individual pines or patches of pine reproduction can be controlled by mechanical or chemical thinning to release pine and promote its recruitment into the overstory. This approach however is not based on natural development patterns (i.e. hardwoods normally do not naturally stratify above pines) and typically treats stands uniformly without regard for the spatial patterns that are often apparent in old pine-oak stands.

In mature oak-dominated stands that have a minor pine component, silvicultural treatments may be used to increase pine dominance. For example, group seedtree harvests may be used to regenerate pine around existing sexually mature pine stems. In these situations, prescribed fire is important to expose bare mineral soil and reduce competition. Pine seed crops are episodic and operations may be timed to coincide with good seed production events. Even when mature pine trees exist on the site, supplemental planting of spots may be used to increase pine seedling density in openings. In general, we recommend focusing on the pine-dominated neighborhoods within stands instead of treating entire stands uniformly. Pine-dominated neighborhoods can be expanded with future entries. Subsequent treatments may create openings tangentially or concentrically to pine neighborhoods to expand neighborhood size.

6.4.5 Old Forest Stage

To account for inherent spatial patterns of species composition and structure typical in old pine-oak stands, we suggest that silvicultural systems use an approach that acknowledges spatial aspects of tree composition and that creates conditions that meet the silvical needs of the desired species without unduly promoting competition. Regeneration harvests will be essential in old pine-oak stands to ensure pine is regenerated and maintained in the system. Thus, disturbance and legacy creation at appropriate spatiotemporal scales is crucial to perpetuate old pine-oak stand conditions. We suggest harvest entries focus on intra-stand spatial patterns. For example, individual openings may range from 0.5 to 1 ha and be clustered at distances up to 200 m in groups of 2–5 openings [16]. We suggest these openings should contain reserve trees and these entries should occur at 20–50 year intervals based on the frequency of natural canopy disturbances. During entries, ecologically important features, such as large snags and large and emergent trees, should be retained. The long-term focus in old pine-oak stands must be on maintaining the pine component. The maintenance of mature pine trees for seed production and a prescribed fire program to enhance pine regeneration potential are critical. Fire effects are important to increase pine regeneration potential by exposing bare mineral soil and reducing hardwood competition.

6.5 Climate Change Considerations

In the southern United States, temperature is expected to increase, and droughts are projected to become more frequent and severe. Atmospheric disturbances that may result in catastrophic forest damage are expected to become more common. Pine-oak stands, because of their relatively diverse species assemblages and complex structures, should be more resistant and resilient to these perturbations and stressors, as some species may be better adapted to future climate conditions. Forest communities with high species diversity represent a wide range of life history strategies and functional traits. As plant species diversity increases, so does the probability that resident species have resistance to or will recover quickly from a perturbation. Promoting pine-oak mixedwoods increases stand diversity and landscape diversity in forested regions that are dominated by mature fully-stocked hardwood forests or monoculture pine plantations. Tree resistance to drought may be enhanced by reducing stand density such as in fire-maintained savannas and woodlands [27] and reducing stand density by thinning to retain drought-tolerant pines and oaks may increase resistance and recovery to drought [28]. Mixedwoods of pine and oak produce highly flammable litter and promote the return of a frequent fire regime. Fire-maintained mixedwoods should be well adapted to fire disturbances, which are expected to increase in the future.

6.6 Summary

Key points to consider based on natural disturbance and development patterns in pine-oak ecosystems include:

1) Intra-stand spatial variability of the species assemblage should be acknowledged. In old, natural pine-oak ecosystems, neighborhood-scale patterns of species composition and stand structure are often apparent.

2) Silvicultural systems that do not treat stands uniformly acknowledge intra-stand variability and would be consistent with natural patterns of disturbance in pine-oak ecosystems.
3) A silvicultural system modeled after natural disturbance and development patterns in pine-oak stands may include patch clearcut, seedtree, and shelterwood harvests with reserves, as well as variable-density thinning. These systems allow for selections based on spatial patterns of species assemblages and microsite variability within stands. These systems also result in multi-aged stands and relatively complex structures at the stand scale.
4) Prescribed fire is critical to create and maintain pine-oak ecosystems in promoting desired regeneration, managing competing vegetation, and modifying vegetation structure.
5) Prescribed fire application should be adjusted according to stand developmental processes and management goals. Prescribed fire should not be strictly implemented at a regular interval but, should be applied deliberately as needed in accordance with stand developmental processes and the silvics of desired species to achieve desired fire effects.
6) Regardless of the initial stand condition, the long-term focus of pine-oak ecosystem management should be on maintaining a pine component. Pine regeneration and recruitment is the bottleneck in mixedwood management.
7) Pine-oak ecosystems contain taxa with a wide range of life history and functional traits and are structurally complex, which confer enhanced ecosystem adaptation potential.

References

1 Kenefic, L.S., Kabrick, J.M., Knapp, B.O. et al. (2021). Mixedwood silviculture in North America: the science and art of managing for complex, multi-species temperate forests. *Canadian Journal of Forest Research* 51: 921–934.

2 Kabrick, J.M., Knapp, B.O., Dey, D.C., and Larsen, D.R. (2015). Effect of initial seedling size, understory competition, and overstory density on the survival and growth of *Pinus echinata* seedlings underplanted in hardwood forests for restoration. *New Forests* 46: 897–918.

3 Johnson, P.S., Shifley, S.R., Rogers, R. et al. (2019). *The Ecology and Silviculture of Oaks*. Boston, MA: CABI.

4 Dey, D.C. (2014). Oak regeneration ecology and dynamics. In: *Proceedings, Wildland Fire in the Appalachians: Discussions Among Managers and Scientists*, General Technical Report SRS-199 (ed. T.A. Waldrop), 3–11. Asheville, NC: USDA Forest Service, Southern Research Station.

5 Varner, J.M., Kane, J.M., Hiers, J.K. et al. (2016). Suites of fire-adapted traits of oaks in the southeastern USA: multiple strategies for persistence. *Fire Ecology* 12: 48–64.

6 Novick, K., Jo, I., D'Orangeville, L. et al. (2022). The drought response of eastern US oaks in the context of their declining abundance. *BioScience* 72: 333–346.

7 Hart, J.L. and Cox, L.E. (2017). Incorporating intermediate-severity disturbances in oak stand development. *Forests* 8: 284.

8 Hart, J.L. and Kleinman, J.S. (2018). What are intermediate-severity forest disturbances and why are they important? *Forests* 9: 579.

9 Arthur, M.A., Alexander, H.D., Dey, D.C. et al. (2012). Refining the oak-fire hypothesis for management of oak-dominated forests of the eastern United States. *Journal of Forestry* 110 (5): 257–266.

10 Varner, J.M., Kane, J.M., Kreye, J.K., and Shearman, T.M. (2021). Litter flammability of 50 southeastern North American tree species: evidence for mesophication gradients cross multiple ecosystems. *Frontiers in Forests and Global Change* 4: 727042.

11 Guyette, R.P., Stambaugh, M.C., Dey, D.C., and Muzika, R.M. (2012). Predicting fire frequency with chemistry and climate. *Ecosystems* 15: 322–335.

12 Stambaugh, M.C., Marschall, J.M., and Abadir, E.R. (2020). Revealing historical fire regimes of the Cumberland Plateau, USA, through remnant fire-scarred shortleaf pines (*Pinus echinata* Mill.). *Fire Ecology* 16: 24.

13 Walker, L.C. and Wiant, H.V. (1966). *Silviculture of Shortleaf Pine*, Bull. 9. Nacogdoches, TX: Stephen F. Austin State College School of Forestry. 60 p.

14 Dey, D.C. and Hartman, G. (2005). Returning fire to Ozark Highland forest ecosystems: effects on advance regeneration. *Forest Ecology and Management* 217: 37–53.

15 Goode, J.D., Hart, J.L., Dey, D.C. et al. (2021). Spatial patterns of canopy disturbance and shortleaf pine in a mixedwood forest. *Forest Science* 67: 433–445.

16 Goode, J.D., Narayanan, A., Phillips, D.L. et al. (2022). Intermediate-severity disturbance impacts in a mixedwood forest: a multi-scale analysis. *Forest Ecology and Management* 526: 120582.

17 Cannon, J.B., Hepinstall-Cymerman, J., Godfrey, C.M., and Peterson, C.J. (2016). Landscape-scale characteristics of forest tornado damage in mountainous terrain. *Landscape Ecology* 31: 2097–2114.

18 Schnake, D.K., Roberts, S.D., Willis, J.L. et al. (2021). Overstory retention and stock type impact survival and growth of underplanted shortleaf pine beneath a hardwood canopy. *Forest Science* 67: 219–230.

19 Clabo, D. and Clatterbuck, W.K. (2020). Restoration of shortleaf pine (*Pinus echinata*)-hardwood mixtures in low quality mixed upland hardwood stands using cluster planting and natural regeneration. *Forests* 11: 457.

20 Baker, J.B. 1992. Natural regeneration of shortleafpine. In: Brissette J.C.; Barnett, J.P. comps. Proceedings of the Shortleaf pine regeneration workshop; 1991 October 29-31; Little Rock, AR. Gen. Tech. Rep. S0-90. New Orleans, LA: U.S. Department of Agriculture, Forest Service, Southern Forest Experiment Station: 102–112.

21 Rose, P.H., Dey, D.C., Phillips, R.J., and Waldrop, T.A. (2013). A meta-analysis of the fire oak literature: does prescribed burning promote oak reproduction in eastern North America? *Forest Science* 59 (3): 322–334.

22 Fan, Z., Ma, Z., Dey, D.C., and Roberts, S.D. (2012). Response of advance reproduction of oaks and associated species to repeated prescribed fires in upland oak-hickory forests, Missouri. *Forest Ecology and Management* 266: 160–169.

23 Schweitzer, C.J., Dey, D.C., and Wang, Y. (2016). Hardwood-pine mixedwoods stand dynamics following thinning and prescribed burning. *Fire Ecology* 12: 85–104.

24 Dey, D.C., Jacobs, D., McNabb, K. et al. (2008). Artificial regeneration of major oak (*Quercus*) species in the eastern United States – a review of the literature. *Forest Science* 54 (1): 77–106.

25 Dey, D.C., Gardiner, E.S., Schweitzer, C.J. et al. (2012). Underplanting to sustain future stocking of oak (*Quercus*) in temperate deciduous forests. *New Forests* 43 (5-6): 955–978.

26 Clark, S.L. and Schlarbaum, S.E. (2019). Artificial regeneration in the Southern Appalachians. In: *Proceedings of the Oak Symposium: Sustaining Oak Forests in the 21st Century through Science-based Management*, e-Gen Tech. Rep. SRS-237 (ed. S.L. Clark and E.J. Schweitzer), 95–100. Asheville, NC: U.S. Department of Agriculture Forest Service, Southern Research Station.

27 Knapp, E.E., Bernal, A.A., Kane, J.M. et al. (2021). Variable thinning and prescribed fire influence tree mortality and growth during and after a severe drought. *Forest Ecology and Management* 479: 118595.

28 D'Amato, A.W., Bradford, J.B., Fraver, S., and Palik, B.J. (2013). Effects of thinning on drought vulnerability and climate response in north temperate forest ecosystems. *Ecological Applications* 23 (8): 1735–1742.

7

Ecological Silviculture for Lowland Wet Conifer Forest Lake States

Marcella A. Windmuller-Campione[1], Laura F. Reuling[1], Robert A. Slesak[2], and Randy K. Kolka[3]

[1] *Department of Forest Resources, University of Minnesota Twin Cities, MN, USA*
[2] *USDA Forest Service, Pacific Northwest Research Station, Olympia, WA, USA*
[3] *USDA Forest Service, Northern Research Station, Grand Rapids, MN, USA*

7.1 Overview

The North American boreal forest spans over 600 million hectares and includes a diversity of ecosystems and forest communities [1]. The boreal forest reaches its southwestern range limit within the Great Lakes Region (GLR) in the United States (Michigan, Minnesota, and Wisconsin, USA) and Canada (Ontario). Within the GLR, one of the dominant southern boreal forest communities is the lowland conifer ecosystem, which has been variously referred to as lowland conifer ecosystem, swamp, bog, peatland, marsh, waabashkiki (Ojibwe), and mashkiig (Ojibwe) [2, 3]. From a European-American perspective, some of these names imply a negative relationship to these forests – they were viewed as places filled with biting insects, with limited potential for agriculture or homesteading and, as such, were seen to have little value [4]. However, lowland wet conifer forest ecosystems have numerous social, ecological, economic, and cultural benefits [1].

Within the GLR, lowland conifer ecosystems cover 35.5 million hectares and range from 1 to 2-ha pocketed features within upland ecosystems to expansive wetland complexes covering thousands of hectares [5, 6]. The lowland conifer forest ecosystems include open vegetation communities (non-forested systems), low productivity forests (defined as a site index less than 7 m at 50 years), and closed canopy forest communities [3]. The closed canopy forest communities are dominated by three overstory species, black spruce (*Picea mariana* (Mill.) B.S.P.), eastern larch (*Larix laricina* (Du Roi) K. Koch), and northern white cedar (*Thuja occidentalis* L.). While eastern larch and northern white cedar have slightly more southerly distributions compared to black spruce, these species have their greatest dominance within the lowland conifer system because of their ability to tolerate the harsh, high moisture conditions associated with lowland conifer forests. These species may co-occur in mixed lowland conifer ecosystems or occur as more single-species communities based on disturbance history and site characteristics [6].

The extensive forest communities in the GLR have a generational history of management [7, 8]. Ojibwe tribes have been the stewards of these forests for generations and continue to steward these

Ecological Silvicultural Systems: Exemplary Models for Sustainable Forest Management,
First Edition. Edited by Brian J. Palik and Anthony W. D'Amato.

important ecosystems. Black spruce has numerous traditional uses across its range including use of the roots to sew baskets and snowshoes. Within Ojibwe culture, both black spruce and eastern larch are used medicinally [9] and these ecosystems also contain many culturally important food sources, including blueberries and cranberries (*Vaccinium* spp.). These ecosystems provide habitat for numerous wildlife species such as moose (*Alces alces*), white-tail deer (*Odocoileus virginianus* Zimmermann), beaver (*Castor canadensis* Kuhl), grey wolves (*Canis lupus* Linnaeus), and the boreal chickadee (*Poecile hudsonicus* Forster) [10].

Eastern larch and black spruce are also managed for conventional forest products, but markets vary regionally and through time (e.g. black spruce is utilized heavily in Minnesota but lowland conifer species are currently less utilized for conventional markets in Wisconsin and Michigan; [11]). Northern white cedar is mostly unmanaged in this region because of concerns related to an inability to regenerate this species due to white-tail deer herbivory and other factors [12]. Forest harvesting is often limited to the winter under frozen ground conditions [13]; harvesting during nonfrozen conditions is difficult for equipment and can cause extensive damage to soils and hydrology. The combination of limited economic markets and seasonal access issues has resulted in less consistent and less overall silvicultural research in lowland conifer ecosystems than in more economically valuable upland forests in the GLR, such as red pine (*Pinus resinosa* Ait.) or aspen (*Populus* spp.) ecosystems (but see [14]).

This chapter will focus on the black spruce and eastern larch forest types since these forests are currently more likely to be actively managed using silvicultural systems. The vast majority of the chapter will be focused on closed canopy ecosystems, but low-productivity black spruce stands will also be discussed to provide context for the broader lowland conifer ecosystem. Additionally, since we are still learning foundational information about the growth, development, and stand dynamics of these ecosystems, the structure of our chapter will look a little different than others in this book. For some sections, we may have limited information (e.g. recovery period after a regeneration harvest) and include aspects that require additional investigation. In other sections, we will share preliminary findings; finally, in some cases, we will share potential directions or hypotheses which will require future study.

7.2 Glacial History

A major defining characteristic of lowland conifer forests in the GLR is the glacial history and its effects on associated soils [6]. The Wisconsinan glaciation, which occurred about 11,600 YBP, encompassed the entire boreal forest region and much of the GLR. Generally, it was the last lobe that left the surface materials and landforms that now influence the GLR landscape [15]. Given the variety of transport mechanisms associated with glacial activity, surface sediments in these forests are variable and include coarse outwash sands, lacustrine clays, and glacial till and moraine features that have variable textures. As the glaciers receded from the GLR, black spruce and eastern larch, which have pollen records of being present in modern-day Florida and Louisiana, respectively, migrated north and west [16].

Post glaciation, lowland conifer communities are generally derived from two processes, terrestrialization, and paludification [5]. Terrestrialization occurs in small kettle lakes that formed from ice blocks that calved off glaciers, were subsequently buried, then melted following the retreat of glaciers. These shallow lakes were filled in with vegetation over time as plant productivity was slightly greater than decomposition, mainly because of the waterlogged and cool conditions. During paludification, expanses of relatively dense mineral soils (e.g. lake-derived clays), inhibit the movement of water, leading to the accumulation of organic matter and the formation of peat, which further restricts the movement of water, promoting waterlogged conditions. Bogs are peatlands that usually form through the terrestrialization process and are acid in nature, low in nutrients, and disconnected from the regional water table (WT). Fens are peatlands that are typically formed by the paludification process and are generally higher in pH, more nutrient rich, and connected to regional WTs. In both types of peatlands, the depth of accumulated peat can be upwards of 10 m [17].

7.3 Plant Community Composition

The plant composition of lowland conifer forest ecosystems varies based on WT and soil acidity [6] (Table 7.1). Forested acid peatlands (i.e. bogs) are dominated by black spruce with eastern larch as a smaller component. In more nutrient-rich, higher pH conditions, the importance of eastern larch increases relative to black spruce, and the occurrence of other tree species including black ash (*Fraxinus nigra* Marsh), northern white cedar, American elm (*Ulmus americana* L.), red maple (*Acer rubrum* L.), and yellow birch (*Betula alleghaniensis* Britton) increases.

Table 7.1 Characteristic plant species of lowland conifer systems in the western Lake States.

				Importance by forest type	
Scientific name	**Common name (English)**	**Common name (Ojibwe)**	**Growth form**	**Forested rich peatland**	**Forested acid peatland**
Larix laricina	Eastern larch	Mashkiigwaatig	Tree	●●●	●●●
Picea mariana	Black spruce	Zesegaandag	Tree	●●	●●●
Alnus incana	Speckled alder	Wadoop	Shrub	●●	–
Betula pumila	Bog birch		Shrub	●●●	●●●
Rhamnus alnifolia	Dwarf alder		Shrub	●●	●●●
Andromeda glaucophylla	Bog rosemary		Ericaceous shrub	●	●●
Chamaedaphne calyculata	Leatherleaf		Ericaceous shrub	●	●●●
Kalmia polifolia	Bog laurel		Ericaceous shrub	●	●●
Rhododendron groenlandicum	Bog Labrador tea		Ericaceous shrub	●●	●●●
Vaccinium angustifolium/ myrtilloides	Lowbush and velvetleaf blueberry	Miinan	Ericaceous shrub	●	●●
Vaccinium oxycoccos	Small cranberry	Mashkiigimin	Ericaceous shrub	●●	●●●
Caltha palustris	Marsh marigold		Forb	●●	–

(*Continued*)

Table 7.1 (Continued)

Scientific name	Common name (English)	Common name (Ojibwe)	Growth form	Importance by forest type: Forested rich peatland	Importance by forest type: Forested acid peatland
Campanula apariondes	Marsh bellflower		Forb	●●	–
Epilobium spp.	Willow herbs		Forb	●	–
Rumex orbiculatus	Greater water dock		Forb	●●	–
Sarracenia purpurea	Purple pitcher plant		Forb	–	●
Carex disperma	Soft-leaved sedge		Graminoid	●●	●
Carex leptalea	Bristle-stalked sedge		Graminoid	●●	–
Sphagnum spp.	Sphagnum moss	Mashkiigwakamig	Moss	●●●	●●●

Importance is based on frequency and cover.
●●●, species with high importance; ●●, species with moderate importance; ●, somewhat important species; –, species generally not associated with the forest type.
Source: Ojibwe name sourced from the Ojibwe People's Dictionary https://ojibwe.lib.umn.edu/.

7.4 Historical Natural Disturbance Regime

7.4.1 Stand-Replacing Disturbance

Across eastern larch and black spruce forest communities, stand dynamics and disturbance processes are complex and include historically large stand-replacing fires, mixed severity fires, and small-scale gap dynamics due to insects, parasitic plants (dwarf mistletoe, *Arceuthobium pusillum* R. W. Smith), and wind throw [18, 19]. Historical fire return intervals for high-severity fires are extremely variable and range from as little as 120 years to over 1000 years depending on the specific forest community and the location on the landscape [20]. More frequent, low-severity fires were common in lowland conifer ecosystems that were in close proximity to more fire-prone or fire-dependent upland ecosystems [21]. Stand-replacing fires are hypothesized to have coincided with drought years, i.e. reduced moisture levels (Box 7.1) and previous fine-scale disturbance (windthrow or insects) that increased the amount of downed dead wood on the forest floor [22]. These fires likely burned at a landscape scale resulting in a mosaic of fire intensity including unburned patches.

Due to the relatively thin bark of eastern larch and black spruce, surviving fire-scarred individuals are rare, resulting in fewer dendrochronological studies of fire history compared to pine species in the region (but see [21]). Charcoal analysis from soils has been used to infer fire history, providing fire history for longer time periods (i.e. thousands of years) but at lower precision than dendrochronology [23]. Additional details on disturbance dynamics have been explored in Minnesota using the Public Land Survey (PLS) records between 1846 and 1908 to identify potential fire return intervals, which include using some fire-scarred trees [24]; this technique provides very broad estimates of fire return but has limitations including potential sampling bias. Given predicted increases in growing season drought in the region due to climate change and the connection between fire and drought in these systems, the fire dynamics of these systems may look very different in the future [25].

Box 7.1 Current and Future Impacts of Fire

As mentioned above, historically, stand-replacing fires likely occurred in lowland conifer forests in the GLR following extended periods of drought. Climate change is expected to increase the frequency and intensity of these droughts, and thus fire is also predicted to increase in frequency [25]. Recent and ongoing research on two fires in northern Minnesota may provide insight into this question. A study of the North Minnie fire in 2012 in northwest Minnesota found after two growing seasons that more nutrient-poor bogs dominated by black spruce were likely to recover quickly from low and moderate severity fire and regenerate to black spruce, while more nutrient-rich swamps dominated by eastern larch were not regenerating to larch and may shift away from forested conditions to open wet meadows [43]. However, fire severity and thus fire effects were varied and patchy across this landscape [43]. Observations of the 2021 Greenwood Fire after one growing season also support findings of heterogeneity in fire severity and impacts within lowland conifer forests (Figure 7.1). In areas where sphagnum was not completely consumed, recovery of understory vegetation was apparent even early in the first growing season after the fire (Figure 7.1). In areas with higher burn severity where more sphagnum and peat were consumed, recovery maybe slower or the trajectory of the forest maybe changed (Figure 7.1).

Figure 7.1 Early vegetation response of peatland forests to Greenwood fire in northeast Minnesota. The Greenwood fire burned in August and September of 2021, photos were taken in June 2022. In areas with lower burn severity, the ground-layer vegetation recovered rapidly (a, b), but areas of high burn severity showed minimal recovery of ground-layer vegetation early in the first growing season (c–e).

7.4.2 Gap Dynamics

Gaps dynamics are an understudied but critical component of stand dynamics within lowland conifer ecosystems. Due to high WTs and their shallowed-rooted nature, both eastern larch and black spruce are vulnerable to windthrow, which can create various-sized gaps [4]. Additionally, black spruce is susceptible to eastern spruce dwarf mistletoe (EDM), a parasitic plant that, unlike other mistletoes, kills 75% of infected host trees in 17 years creating a fine-scale pattern of disturbance [26, 27]. EDM seeds can be dispersed short distances (a few meters) by the plant itself or longer distances by wind or wildlife such as birds. It is likely that birds move the sticky seed around since EDM brooms can provide excellent nesting habitat. Once established, mistletoe spreads outward, creating mortality pockets [26, 27]. However, EDM presence and severity are difficult to predict [28]. In forests managed for maximum timber value, EDM is seen as a forest health threat, but at low to moderate levels, EDM creates opportunities for increased species and structural diversity within peatland forests [27, 29]. For example, eastern larch, balsam poplar, eastern white pine, and other species were found within EDM mortality centers, reflecting the potential for increased diversity when EDM-initiated gaps are formed [27, 29]. Alternatively, high levels of EDM can cause widespread overstory mortality and the ecosystem may shift toward more open forests or shrub-dominated communities such as willow or alder for a period of time [30]; the duration of this stage before eventual canopy closure is not well understood.

Another disturbance agent in gap dynamics within lowland conifer ecosystems is eastern larch beetle (*Dendrodoctonus simplex*, ELB) [31]. ELB is a native insect that historically fed on the phloem of recently dead or dying eastern larch, resulting in very limited mortality of living larch. Although ELB has always been a component of the disturbance dynamics of these systems, it did not historically result in large outbreaks like other *Dendrodoctonus* species (e.g. mountain pine beetle [*Dendroctonus ponderosae* Hopkins] or spruce beetle [*Dendroctonus rufipennis* Kirby]). However, since 2000, the GLR region has been experiencing unprecedented ELB population growth and subsequent widespread mortality of eastern larch [31]. As part of the ongoing outbreak in northern Minnesota, the impact of ELB has created a mosaic of severity across the landscape, including large areas that have widespread and extensive, stand-replacing mortality. Early in the outbreak, the beetle targeted larger-diameter, more productive stands [32], but has moved into smaller-diameter and higher-density stands as the outbreak continues [33]. Across the areas impacted by ELB, some eastern larch trees survive, but the reason(s) why are unclear. Unlike other *Dendrodoctonus* species [34], research on ELB has been limited, with most work focused on recent changes to the beetle's ecology and dynamics due to climate change [31]. There are currently no published management guidelines for ELB but recent work has highlighted similarities between ELB and other *Dendrodoctonus* species in terms of responses to host structure at stand and landscape scales [34].

7.5 Silvics of Black Spruce and Eastern Larch

Eastern larch and black spruce can both be considered early successional or seral species, able to regenerate under full light conditions such as those that may exist after a stand-replacing fire. Initial growth of both species can be rapid; however, given the same conditions eastern larch generally has faster initial growth than black spruce [2]. Eastern larch is conventionally classified as an extremely shade-intolerant species with limited ability to survive under any shade. However, some studies have observed eastern larch's ability to establish under some level of shade, including after insect damage from larch sawfly (*Pristiphora erichsonii* Hartig) [35]. Recent work in northern Minnesota forests impacted by ELB has documented a wide range of ages for eastern larch saplings; some of these trees were small in diameter (2.5–5 cm) but upwards of 80 years old and were likely growing in shaded understory prior to release by ELB [33].

While black spruce is noted as an early successional species, it can also commonly establish as advance regeneration in the understory and in gaps [27]. Release of black spruce advance regeneration has been observed and can occur in individuals that are greater than 50 years old [36].

The ability of black spruce to self-replace under a range of disturbance severities creates many opportunities for multiple directions or pathways for stand development and stand dynamics. Currently, there is limited information on the dynamics of these systems in the GLR and past work has often relied on diameter as a proxy for age [24], which may overlook the ability of these species to survive extended periods of shade. For example, recent work indicates that 15 cm DBH black spruce and tamarack can be anywhere between 30 and 150+ years old (Figure 7.2). Growth of these tree species may be related to several factors including microsite conditions [37] (see Box 7.2), and a lack of a consistent relationship between diameter and age has commonly been documented in forests dominated by shade-tolerant species. Given the potential for multiple developmental pathways in these systems, lowland conifer ecosystems in the GLR would benefit from more stand dynamics research that included extensive tree cores and cross sections to better understand stand development processes, which may include multiple pathways within one single stand or management unit (see [38–40] for examples of dendrochronology in black spruce in Canada).

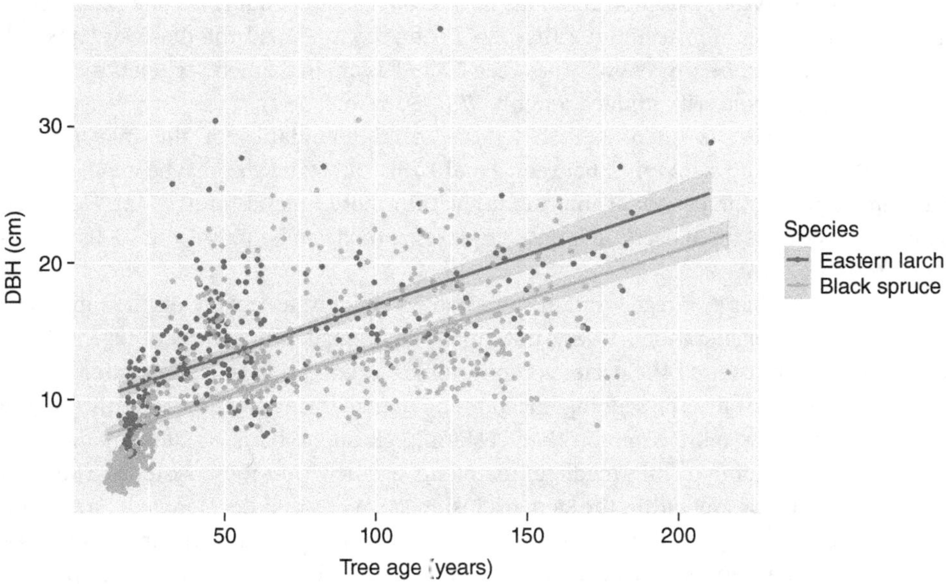

Figure 7.2 Tree age (based on tree core data) vs diameter at breast height (cm) for eastern larch and black spruce at 36 lowland conifer sites in northern Minnesota. Regression lines are based on linear regression with shading indicating 95% confidence intervals.

Box 7.2 Connection Between Hydrology and Silvicultural Treatments

Hydrology and its influence on the magnitude and duration of waterlogged or flooded conditions is a critical aspect affecting the ecosystem functions (e.g. plant productivity and carbon storage) of lowland conifer forests in the GLR. Hydrological processes in lowland conifer forests are strongly influenced by the degree of connectively to regional water tables (WT) and the timing and amount of precipitation. Fens typically have a relatively stable WT throughout the year at a depth that is dependent on the degree of connectivity to regional groundwater. In contrast, WT in most other lowland conifer forests are more strongly influenced by precipitation. In drought years, WT drawdown occurs more quickly and to a greater depth than more typical years with average annual precipitation [48]. Forest disturbance can also alter WT dynamics, largely because loss of the forest canopy alters the accumulation and melting of snow and reduces water loss via transpiration. However, one study observed limited connection between the water table elevation and black spruce growth [47].

7.6 Current/Conventional Silvicultural Approaches

As mentioned above, both eastern larch and black spruce can act as pioneer species and regenerate well under high light conditions. Regeneration occurs relatively rapidly after a large-scale disturbance and conventional management has focused on production; The Manager's Handbook for Black Spruce in the North Central States emphasizes that silvicultural systems should *produce a high sustained yield of pulpwood and other forest values as efficiently as possible* [4]. This focus on maximizing productive potential is still captured in the current conventional management for lowland conifer species within the GLR. Commonly, lowland conifer stands are treated with the clearcut regeneration harvest system on rotations of 60–100 years depending on site index. Strip clearcutting is another common technique that has been recommended for these systems [4]. Eastern larch and black spruce are generally between 12 and 30 cm in diameter when harvested; a productive black spruce stand will produce roughly 270–450 m^3 ha^{-1} [41].

The regeneration harvest is often defined by the stand boundaries of the management organization. Common stand sizes range between 4 and 16 ha (but stand size maybe much greater than 40 ha) depending on the degree of intermixing within more upland forests [42]. However, lowland conifer ecosystems often occur in large complexes, which can be 100s if not 1000s of hectares in size.

After the regeneration harvest, regeneration can establish quickly, especially on the more productive sites. In clearcut regeneration harvests in Minnesota, average stem density of regeneration exceeded 1000 stems per hectare [4]. Species composition of regeneration is dependent on available seed source. Supplemental aerial seeding is a common management practice; recently seeding of black spruce has become more common than eastern larch due to both market demands and seed availability. Little tending or intermediate treatments occur within these systems. The final harvest occurs as the stands are within the stem exclusion stage of stand development; structural diversity and species diversity are limited [27]. While not commonly used, prescribed fire was a more common site preparation tool used during the early and mid-twentieth century to provide more favorable seedbed conditions for the semi-serotinous black spruce cones [4].

The conventional management of lowland conifer ecosystems is based on a relatively narrow window of stand development and a narrow interpretation of the disturbance dynamics of these forest ecosystems (i.e. frequent stand-replacing events). The gradient of the silvics of eastern larch and black spruce and recognition of the influence of gap dynamics highlight the potential for these ecosystems to have a broader range of structural diversity and a greater potential for the use of multiple silvicultural strategies in managing lowland conifer ecosystems. Additionally, these forest ecosystems are experiencing rapid changes from the long-running ELB outbreak, fueled in part by the impacts of a warming climate on both beetle life cycles and below-ground processes. Increasing the potential silvicultural strategies within lowland conifer ecosystems can likely increase resistance and resilience to immediate forest health threats (i.e. ELB) and broader, longer-term climate-induced impacts [31].

7.7 Natural Development Model for Lowland Conifer Ecosystems

7.7.1 Disturbance and Legacy Creation

Large, stand-replacing disturbances that may include fire, windthrow, and currently wide-spread ELB mortality result in large areas with standing dead or dying trees intermixed with live tree legacies of similar diameter. Legacy trees are likely scattered, more likely to be in groups or clumps, and may be influenced by microsite characteristics.

7.7.2 Preforest Stage

Depending on the location of the landscape, the disturbance, and site history, length of the preforest stage may be highly variable [6]. In eastern larch stands that suffer heavy mortality from ELB, regeneration may be slower and the preforest stage may last more than a decade [33]. The species composition and the timing of regeneration have also been observed to be variable after fire within these systems ([43] – Box 7.1).

The variability in the preforest stage within this ecosystem is likely heavily influenced by microsite variability. Regeneration can be delayed even longer if a disturbance causes a stand to shift to open shrub or sedge, grass meadows [5]. Conventionally this is considered a "loss" of forest land, however, it is more likely representative of a large landscape pattern where forests move from closed canopy to open canopy across long periods and based on disturbances [6].

7.7.3 Young Forest Stage

The young forest stage within lowland conifer forests is rarely visited by forest managers due to the limited use of intermediate treatments. There is huge potential to increase our understanding of the range of variability between stand initiation and stem exclusion.

Under a natural development model, stand density would likely be highly variable. In areas where there was a good match between seed availability and microsite quality, stand density is likely high with limited structural diversity (e.g. moving from stand initiation to stem exclusion). However, in other portions of the stand, stand density would be lower and tree species density would be more evenly mixed with shrub species. Additionally, diseases like EDM could start the process of gap formation, creating heterogenous conditions and allowing for more continuous recruitment [30].

7.7.4 Mature Forest Stage and Old Forest Stage

We are grouping the mature and old forest stages together for lowland conifer ecosystems due to our current understanding of structural and composition variables paired with disturbance dynamics. Future research may support the need to decouple this grouping.

While there is an increase in tree size from the Young Forest to the Mature and Old Forest Stage, the biggest differences relate to the overstory structural characteristics due to the small to moderate scale disturbances dominating these stages [27, 29]. Disturbances like EDM create expanding mortality centers, which result in increased standing dead trees across the diameter distribution [27]. ELB mortality can result in both expansive areas of mortality or more pocketed features, again increasing the number of standing dead trees [33]. Windthrow can result in a shift from standing dead to down dead wood, which can be highly variable. Within the disturbed landscape, regeneration can be continuous, creating a flat-diameter distribution. Diameter is likely a poor surrogate for age since the majority of trees fall within a narrow range of diameters (10-30 cm dbh) [36].

7.8 Ecological Silviculture System

We will broadly discuss a potential model for an ecologically based silvicultural system within the lowland conifer ecosystem. With increased investment in applied research within these systems, there will likely be large gains in knowledge within the next decade or two that will allow us to expand and provide more detailed guidance (see [39] for shelterwood growth response study in black spruce in Canada). Thus, we provide a broadly based model here but acknowledge critical pieces are still being explored within the GRL. The overarching system were are proposing is an expanding gap irregular shelterwood with reserves (Table 7.2).

Table 7.2 Lowland conifer developed across different stages, years, actions, and scale.

Stage	Year	Description	Actions	Scale
Disturbance/ Legacy creation	0	Scattered residual trees that are tied to microsite	Including residual trees after a regeneration harvest	Can begin at the stand scale
Preforest	1–10+	Mix of trees, shrubs, and forbs. A stand may shift toward more open dominated by shrubs depending on disturbance intensity, site quality, and weather	Patience	Can begin at the stand scale
Young forest	3–50	Variable-density small pole trees intermixed with more open areas more dominated by shrubs and forbs	Gaps could be created to decrease time to attain mature forest	Can begin at the stand scale
Mature and old forest stage	50+ years	Trees have increased in size and gap dynamics are occurring; gap size will vary based on disturbances and will follow microsite features	Irregular shelterwood with reserves; reserves composing 40–70% and based on microsite features and location of stand within the landscape	Can begin at the stand scale
Old forest stage (cont.)	120+ years	Similar to above with regrowth providing opportunity for another entry to shift structural complexity	Moving from the stand to the landscape, the next entry should consider how residual patches will shift and gap expansion	Moving toward the landscape

7.8.1 Disturbance and Legacy Creation

Current conventional management in lowland conifer ecosystems focuses on even-aged silvicultural systems with limited legacy creation, which likely encompasses only a portion of the natural disturbance dynamics of these systems [4]. We recommend increasing the amount and distribution of residual trees. This can be a good first step if the current management is more "conventionally" focused. Residuals will provide structural complexity and seed source. Additionally, the difference in microsite, especially if residuals are clumped, can provide opportunities for habitat for more closed canopy species.

7.8.2 Preforest Stage and Young Forest Stages

We are grouping these two sections together. The vegetative response and length of time in these two stages can be highly variable based on factors related to the disturbance, weather patterns, and longer-term climatic trends. For stands with goals other than maximizing stand volume, we recommend little active management within these stages. Site occupancy by tree species may happen quite rapidly (within five years) or may be more prolonged.

If there is interest in moving a young forest stage to the mature or old forest stage of development, pre-commercial gap creation could be implemented but would be unlikely given economic and logistical constraints (e.g. winter harvest in a changing climate).

7.8.3 Mature Forest Stage and Old Forest Stages

As forests transition from the young forest stage to the mature and old forest stage, we recommend the use of an expanding gap irregular shelterwood system with reserves and managing at the landscape scale. This could another be a point of entry if starting with a mature or old forest stage and implementing ecological silviculture within this system.

The irregular shelterwood may more closely emulate the disturbance dynamics of fire, windthrow, and disease in lowland conifer systems. By varying the density, shape, and spacing of reserve trees, reserve trees may serve multiple purposes with scattered individual trees serving as future down dead wood and more clumped patches providing seed source and acting as refugia for plants and animals. Groups of reserves tree can vary in size between 0.2 and 4 ha; the placement of reserve trees should follow microsite characteristics. Leaving widely scattered individual trees or small groups (5–10 trees) within more open areas of the of the shelterwood can provide opportunity for snag creation and blow down for down dead wood. Reserves should compose between 40% and 70% of the initial stand area. The initial intensity of the shelterwood will depend on microsite features, the location of the stand within the landscape, and current species composition (e.g. more reserve trees may be left if there are higher components of rare features like northern white cedar groves or inclusions of paper birch).

As noted above, the exact prescription will be based on the individual site and the location on the landscape. The overarching goal would be to move from individual stand prescriptions to a broader landscape approach to create a complex mosaic of the different stages or natural development occurring across a lowland conifer complex. It would also still maintain a flow of economic services from these systems, but with more focus on microsite features and how disturbance may have influenced the landscape and less on stand boundaries.

While an irregular shelterwood system is more complex than a clearcut system, the implementation of irregular shelterwood cuts would be feasible given advances in technology, namely handheld mapping platforms (e.g. Arc or Avenza) that can be shared with logging professionals. The primary time investment would be related to extra surveying of stands in the growing season to document the nuances in site conditions, slight changes in elevation, and note areas where reserves maybe placed; essentially the art of silviculture. The GRL does have recent examples of moving toward *more complicated* silvicultural systems. These systems were initially met with hesitancy but are now widely implemented and have resulted in increased ecological resilience and provided greater economic returns to loggers through increased quality of wood products [42].

7.9 Climate Changes Impact on Lowland Conifer Ecosystems

Lowland conifer ecosystems within the GLR have emerged as critical ecosystems, which are predicted to be drastically impacted by a changing climate [44]. Researchers and natural resource managers are actively considering climate change. There have been huge investments to understand the full ecosystem response to increasing temperatures and CO_2 in the Spruce and Peatlands Response Under Changing Environments (SPRUCE) Experiment in northern Minnesota (Figure 7.3). Some critical initial results indicate that warming causes lowland conifer peatlands to shift from carbon sinks to carbon sources, with rates of carbon loss ranging from 5 to nearly 20 times that of historical rates of accumulation [45]. The plant communities in this experiment have also experienced significant changes with warming, with longer growing seasons and more frequent false springs allowing shrubs like blueberries to overtake black spruce and eastern larch [46].

Warming winter temperatures are also predicted to shift the timing and length of frozen ground conditions, influencing both overall ecosystem dynamics and the logistics of management in these systems (Box 7.3). While these peatland communities in the GLR are vulnerable to climate change, the inherit complexity of the system and the duality of the two dominant overstory species (both shade intolerant and tolerant) can provide opportunities to increase ecological resistance and resilience.

7.10 Summary

1) Lowland conifer ecosystems within the GRL occur along the southern edge of the boreal forest within Michigan, Minnesota, and Wisconsin in the United States and Ontario in Canada.
2) These systems are more understudied compared to other forests in the region but are actively managed for conventional and nonconventional forest products.
3) Conventional management generally focuses on even-aged regeneration harvest using clearcuts with or without reserves on relatively long rotations (60–100 years) due to slower growth rates in peat soils and focused on stand development up to stem exclusion.
4) Disturbance dynamics are complex within lowland conifer ecosystems and include stand-replacing disturbance through mixed severity and stand-replacing fires. However, small to moderate disturbances that increase structural and species complexity are also common.
5) The pre and young forest stages can last variable amounts of time (less than one year to decades) depending on the disturbance (timing, frequency, intensity) and the site characteristics.
6) The mature and old forest stage include relatively small diameter trees (10–30 cm dbh) and currently are grouped together. However, as additional research occurs on stand development in these systems, these stages maybe pulled apart.

Figure 7.3 (a) Overhead of SPRUCE (Spruce and Peatlands Response Under Changing Environments) Experiment and (b) inside an enclosure. *Source:* Oak Ridge National Laboratory, U.S. Dept. of Energy.

Box 7.3 Lowland Conifer Management Under Shifting Winter Conditions

Given that lowland conifer forests in the GLR are at the southern edge of their range, they are likely to experience large impacts of climate change on ecosystem functions. Forest harvesting in these systems is generally constrained to winter when soils are frozen and have sufficient strength to support the weight of harvest machinery and limit the amount of soil disturbance that occurs. Climate models generally indicate that snowfall will decrease and air temperatures increase in the GLR – both of which have large influence on the rate of formation and total depth to which frost develops. Recent research has shown that in peatlands the absence of snow causes frost to develop five times faster relative to when snow is present [13].

What does this mean for the timing of forest management in the future? As long as increased winter air temperatures remain below freezing, it is likely that frost will develop faster and to a greater depth given predicted decreases in snow accumulation. This will likely increase the times when suitable frost is present to conduct harvesting but may also alter other ecosystem functions that are controlled by soil temperature. For example, Friesen et al. [13] found frost as late as mid-July when snow was removed from lowland conifer forests. Depressed soil temperatures into the growing season may cause indirect effects on ecosystem functions such as site productivity. Of course, all of this depends on that rate of warming – if winter temperatures increase above freezing for prolonged periods, then frost is likely to decrease regardless how much snow cover occurs.

7) Shifting from a clearcut regeneration harvest, we propose the use of an expanding gap irregular shelterwood with reserve system that would incorporate the mixed severity disturbance dynamics of these systems and provide increased opportunities for structural and species complexity.
8) Natural resource managers could incorporate ecological silviculture principles by increasing legacy trees and shifting to natural regeneration after a conventional clearcut regeneration harvest or implementing irregular shelterwoods in a mature or old forest stand with the goal of shifting management from the stand to landscape.

References

1 Parish, F., Sirin, A.A., Charman, D. et al. *Assessment on Peatlands, Biodiversity and Climate Change*. Kuala Lumpur and Wageningen: Global Environment Centre and Wetlands International.

2 Johnston WF. *Larix laricina* (Du Roi) K. Koch. Tamarack. Silvics of North America. 1990;141–51. In R.M. Burns & B.H. Honkala (ed). *Silvics of North America: Volume 1, Conifers*. U.S. Department of Agriculture, Forest Service. Agricultural Handbook 654. https://www.srs.fs.usda.gov/pubs/misc/ag_654/volume_1/larix/laricina.htm

3 Frelich, L.E. (2002). *Forest Dynamics and Disturbance Regimes: Studies from Temperate Evergreen-Deciduous Forests*. Cambridge University Press.

4 Zasada, J.C., Gordon, A.G., Slaughter, C.W., and Duchesne, L.C. (1997). Ecological Considerations for the Sustainable Management of the North American Boreal Forests. IIASA Interim Report.

5 Wright, H.E., Coffin, B., and Aaseng, N.E. (1992). *The Patterned Peatlands of Minnesota*. University of Minnesota Press.

6 Heinselman, M.L. (1963). Forest sites, bog processes, and peatland types in the Glacial Lake Agassiz Region, Minnesota. *Ecological Monographs* 33 (4): 327–374.

7 Bussey, J., Davenport, M.A., Emery, M.R., and Carroll, C. (2016). "A lot of it comes from the heart": the nature and integration of ecological knowledge in tribal and nontribal forest management. *Journal of Forestry* 114 (2): 97–107.

8 Waller, D.M. and Reo, N.J. (2018). First stewards: ecological outcomes of forest and wildlife stewardship by indigenous peoples of Wisconsin, USA. *Ecology and Society* 23 (1): https://doi.org/10.5751/ES-09865-230145.

9 Geniusz, M.S. (2015). *Plants Have So Much to Give Us, All We Have to Do Is Ask: Anishinaabe Botanical Teachings*. University of Minnesota Press.

10 Kozich, A., Colbert, K., Jondreau, G.J. et al. (2022). Regeneration of culturally-significant conifer tree species in the L'Anse Indian Reservation: impacts from herbivory by Deer (Waawaashkeshi). *Tribal College and University Research Journal* 6: 32.

11 Vasievich, J.M., Potter-Witter, K., and Leefers, L.A. (1997). Timber Supply and Market Trends in the Lake States. United States Department of Agriculture Forest Service General Technical Report NC. pp. 290–310.

12 Palik, B.J., Haworth, B.K., David, A.J., and Kolka, R.K. (2015). Survival and growth of northern white-cedar and balsam fir seedlings in riparian management zones in northern Minnesota, USA. *Forest Ecology and Management* 337: 20–27.

13 Friesen, H.C., Slesak, R.A., Karwan, D.L., and Kolka, R.K. (2021). Effects of snow and climate on soil temperature and frost development in forested peatlands in Minnesota, USA. *Geoderma* 394: 115015.

14 Johnston, W.F. (1977, 1977). *Manager's Handbook for Black Spruce in the North Central States*, General Technical Report. NC-34. USDA Forest Service, North Central Forest Experiment Station.

15 Schaetzl, R.J. and Isard, S.A. (1996). Regional-scale relationships between climate and strength of podzolization in the Great Lakes region, North America. *Catena* 28 (1–2): 47–69.

16 Watts, W.A., Hansen, B.C., and Grimm, E.C. (1992). Camel Lake: A 40 000-yr record of vegetational and forest history from northwest Florida. *Ecology* 73 (3): 1056–1066.

17 Bridgham, S.D., Pastor, J., Janssens, J.A. et al. (1996). Multiple limiting gradients in peatlands: a call for a new paradigm. *Wetlands* 16 (1): 45–65.

18 Turetsky, M.R. and St Louis, V.L. *Disturbance in Boreal Peatlands. InBoreal Peatland Ecosystems 2006*, 359–379. Berlin, Heidelberg: Springer.

19 Thom, D. and Seidl, R. (2016). Natural disturbance impacts on ecosystem services and biodiversity in temperate and boreal forests. *Biological Reviews* 91 (3): 760–781.

20 Payette, S. (1992). Fire as a controlling process in the North American boreal forest. In: *A Systems Analysis of the Global Boreal Forest* (ed. H.H. Shugart, R. Leemans, and G.B. Bonan), 144–169. Cambridge University Press.

21 Sutheimer, C.M., Meunier, J., Hotchkiss, S.C. et al. (2021). Historical fire regimes of North American hemiboreal peatlands. *Forest Ecology and Management* 498: 119561.

22 Verry, E.S. and Janssens, J. (2011). Geology, vegetation, and hydrology of the S2 bog at the MEF: 12,000 years in northern Minnesota. In: *Peatland Biogeochemistry and Watershed Hydrology at the Marcell Experimental Forest* (ed. R. Kolka, S. Sebestyen, E.S. Verry, and K. Brooks), 93–134. CRC Press.

23 Larsen, C.P. and MacDonald, G.M. (1998). Fire and vegetation dynamics in a jack pine and black spruce forest reconstructed using fossil pollen and charcoal. *Journal of Ecology* 86 (5): 815–828.

24 MN DNR (2021). NPC silviculture strategies for forest stand prescriptions (accessed 13 June 2023). https://www.dnr.state.mn.us/forestry/ecs_silv/npc/index.html

25 Gao, P., Terando, A.J., Kupfer, J.A. et al. (2021). Robust projections of future fire probability for the conterminous United States. *Science of the Total Environment* 789: 147872.

26 Baker, F.A. and French, D.W. (1980). Spread of *Arceuthobium pusillum* and rates of infection and mortality in black spruce stands [*Picea glauca, Picea mariana*]. *Plant Diseases (USA)* 64 (12): 1074–1076.

27 Skay, R., Windmuller-Campione, M.A., Russell, M.B., and Reuling, L.F. (2021). Influence of eastern spruce dwarf mistletoe on stand structure and composition in northern Minnesota. *Forest Ecology and Management* 481: 118712. https://doi.org/10.1016/j.foreco.2020.118712.

28 Gray, E.R., Russell, M.B., and Windmuller-Campione, M.A. (2021). The difficulty of predicting eastern spruce dwarf mistletoe in lowland black spruce: model benchmarking in Northern Minnesota, USA. *Forests* 12 (7): 843. https://doi.org/10.3390/f12070843.

29 Gray, E.R., Russell, M.B., Babcock, C., and Windmuller-Campione, M.A. (2022). Structural characteristics of black spruce (*Picea mariana*) infested with eastern spruce dwarf mistletoe (*Arceuthobium pusillum*), Minnesota, USA. *Forest Ecology and Management* 523: 120495.

30 Gray, E. (2022). *Effects of Eastern Spruce Dwarf Mistletoe (Arceuthobium pusillum Peck) Infestation in Lowland Black Spruce (Picea mariana (Mill.) BSP) at Multiple Scales*. University of Minnesota.

31 Bark Beetle Management, Ecology, and Climate ChangeMcKee, F.R., Windmuller-Campione, M.A., Althoff, E.R. et al. (2022). Eastern larch beetle, a changing climate, and impacts to northern tamarack forests. In: (ed. K.J.K. Gandhi and R.W. Hofstetter), 261–300. Academic Press.

32 Crocker, S.J., Liknes, G.C., McKee, F.R. et al. (2016). Stand-level factors associated with resurging mortality from eastern larch beetle (*Dendroctonus simplex* LeConte). *Forest Ecology and Management* 375: 27–34. https://doi.org/10.1016/j.foreco.2016.05.016.

33 Shaunette, A. (2022). *Evaluating Regeneration and Stand Dynamics in Tamarack Stands Impacted by Eastern Larch Beetle in Northern Minnesota, USA*. University of Minnesota.

34 Windmuller-Campione, M.A., DeRose, J., and Long, J.N. (2021). Landscape-scale drivers of resistance and resilience to bark beetles: a conceptual susceptibility model. *Forests* 12 (6): 798.

35 Girardin, M.P., Tardif, J., and Bergeron, Y. (2002). Dynamics of eastern larch stands and its relationships with larch sawfly outbreaks in the northern Clay Belt of Quebec. *Canadian Journal of Forest Research* 32 (2): 206–216.

36 Anderson, B.D., Russell, M.B., Windmuller-Campione, M.A. et al. (2018). Development and evaluation of black spruce (*Picea Mariana* (Miller) BSP) diameter increment models across silvicultural treatments in Northern Minnesota, USA. *Forests* 9 (12): 747. https://doi.org/10.3390/f9120747.

37 Macdonald, S.E. and Yin, F. (1999). Factors influencing size inequality in peatland black spruce and tamarack: evidence from post-drainage release growth. *Journal of Ecology* 87 (3): 404–412.

38 Hember, R.A., Kurz, W.A., and Girardin, M.P. (2019). Tree ring reconstructions of stemwood biomass indicate increases in the growth rate of black spruce trees across boreal forests of Canada. *Journal of Geophysical Research: Biogeosciences* 124 (8): 2460–2480.

39 Montoro Girona, M., Morin, H., Lussier, J.M., and Walsh, D. (2016). Radial growth response of black spruce stands ten years after experimental shelterwoods and seed-tree cuttings in boreal forest. *Forests* 7 (10): 240.

40 Tremblay, M.J., Rossi, S., and Morin, H. (2011). Growth dynamics of black spruce in stands located between the 51st and 52nd parallels in the boreal forest of Quebec, Canada. *Canadian Journal of Forest Research* 41 (9): 1769–1778.

41 Vanderschaaf, C. (2012). Estimating Yields of Lowland Black Spruce for Minnesota DNR Subsection Forest Resource Management Plans (SFRMPS). https://doi.org/10.13140/RG.2.2.10623.89764.

42 Windmuller-Campione, M.A., Russell, M.B., Sagor, E. et al. (2020). The decline of the clearcut: 26 years of change in silvicultural practices and implications in Minnesota. *Journal of Forestry* 118 (3): 244–259.

43 Rowe, E.R., D'Amato, A.W., Palik, B.J., and Almendinger, J.C. (2017). Early response of ground layer plant communities to wildfire and harvesting disturbance in forested peatland ecosystems in northern Minnesota, USA. *Forest Ecology and Management* 398: 140–152.

44 Handler, S., Marcinkowski, K., Janowiak, M., and Swanston, C. (2017). *Climate Change Field Guide for Northern Minnesota Forests: Site-Level Considerations and Adaptation*, USDA Northern Forests

Climate Hub Technical Report #2. St. Paul, MN. 88p: University of Minnesota College of Food, Agricultural, and Natural Resource Sciences.

45 Hanson, P.J., Griffiths, N.A., Iversen, C.M. et al. (2020). Rapid net carbon loss from a whole-ecosystem warmed peatland. *AGU Advances* 1 (3): e2020AV000163. https://doi.org/10.1029/2020AV000163.

46 McPartland, M.Y., Montgomery, R.A., Hanson, P.J. et al. (2020). Vascular plant species response to warming and elevated carbon dioxide in a boreal peatland. *Environmental Research Letters* 15 (12): 124066.

47 Dymond, S.F., D'Amato, A.W., Kolka, R.K. et al. (2019). Climatic controls on peatland black spruce growth in relation to water table variation and precipitation. *Ecohydrology* 12 (7): e2137.

48 Stelling, J. M., Slesak, R. A., Windmuller-Campione, M. A., & Grinde, A. (2023). Effects of stand age, tree species, and climate on water table fluctuations and estimated evapotranspiration in managed peatland forests. *Journal of Environmental Management*, 339, 117783. https://doi.org/10.1016/j.jenvman.2023.117783.

8

Ecological Silviculture for Southern Appalachian Hardwood Forests

Jodi A. Forrester[1], *Tara L. Keyser*[2], *and David K. Schnake*[1]

[1] *Department of Forestry and Environmental Resources, North Carolina State University, Raleigh, NC, USA*
[2] *USDA Forest Service, Southern Research Station, Asheville, NC, USA*

8.1 The Southern Appalachian Mixed-Oak Forests

The forests of the southern Appalachians, defined here as the Central Blue Ridge Mountains, Southern Blue Ridge Mountains, and the Metasedimentary Mountains, stretch from northeast Georgia to southwestern Virginia, USA (Figure 8.1). The complex topography and humid temperate climate create growing conditions for a rich variety of forested plant communities. Forest types occur in ecological zones classified by elevation, aspect, and geology that create patterns of temperature, moisture, and productivity [1].

Forestland is the dominant cover type within the region, comprising 77% or ~2.9 million ha, and is held by a mix of ownerships including private (55%), federal (40%), and state and local (5%) entities. Within the region, 13 forest-type groups are identified, with 75% of forestland classified as oak/hickory (*Quercus*/*Carya*) (Figure 8.2a, ~2.2 million ha) and the next most abundant forest type (9%), being oak/pine (*Quercus/Pinus*). Most forests were harvested in the early twentieth century, with less than 1% of oak/hickory forestland >140 years (Figure 8.2b). Federal lands have been managed for multiple uses and private landowners increasingly value managing their properties for ecosystem services beyond timber production.

Forest management in the southern Appalachians is centered on regenerating a suite of oak species, including white oak (*Quercus alba* L.), chestnut oak (*Quercus montana* Willd.), northern red oak (*Quercus rubra* L.), scarlet oak (*Quercus coccinea* Muenchh.), and black oak (*Quercus velutina* Lam.), with the relative species dominance dependent on-site quality. Coupled with restoration efforts to regenerate and conserve oak species are objectives associated with the sustainable production of high-value hardwood sawtimber and other non-timber-related goods and services, such as understory plant diversity, wildlife populations, and wildlife habitat diversity.

The diversity of hardwood species is important for timber products within the local and regional economy and leads to complex timber values, markets, and therefore, management. Management systems for commercial hardwoods vary across ownerships, with high-grading and commercial clearcutting commonly applied to privately owned forestland and a broader suite of true silvicultural systems, including various even-aged, two-aged, and uneven-aged

Ecological Silvicultural Systems: Exemplary Models for Sustainable Forest Management, First Edition. Edited by Brian J. Palik and Anthony W. D'Amato.

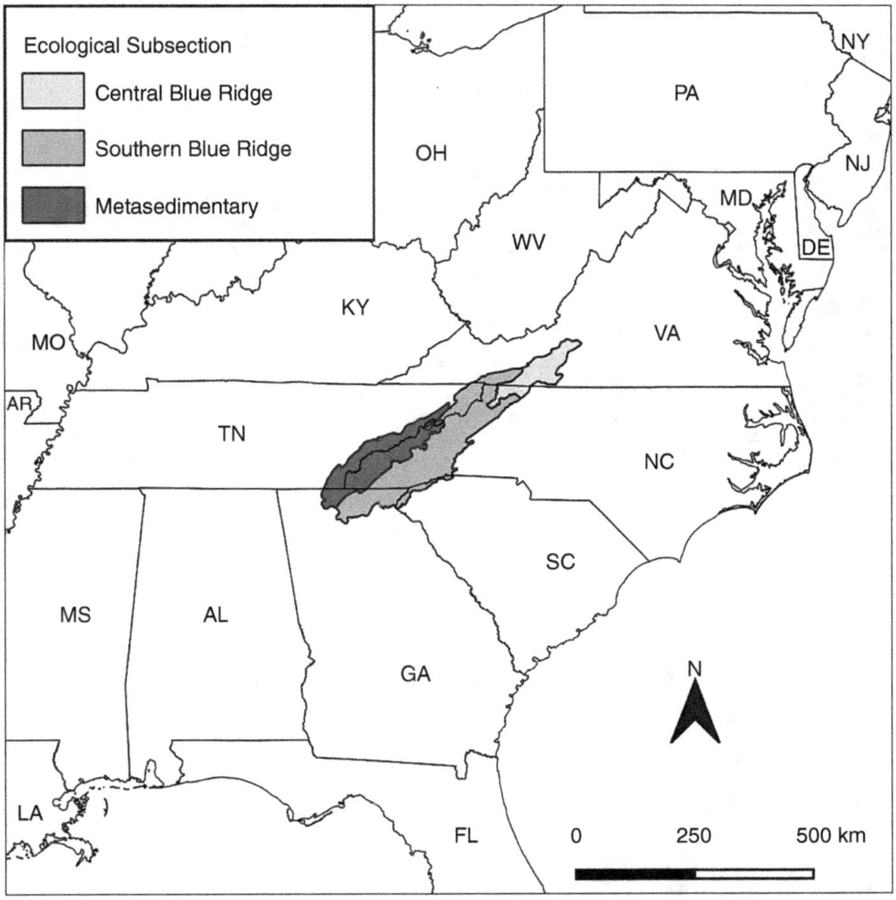

Figure 8.1 Location of Ecological Subsections forming the southern Appalachians.

regeneration methods on public lands. Commercial thinning and improvement cuts are uncommon in upland hardwoods due to inconsistent demand for pulpwood and high operational costs throughout the region.

8.2 Contemporary Forests of the Southern Appalachians

Current forests of the region are a legacy of land use practices from as far back as the 1800s. Europeans colonized the area using subsistence agriculture that involved frequent fire to clear forestland and stimulate forage production. Surrounding areas were harvested for fuel and construction materials, and over time land management practices increased in scale and frequency. Clearing practices intensified as industrialization expanded. By the early 1900s, over 50% of the forestland in the southern Appalachians was owned by private companies, and forests were cleared or severely high-graded for timber. Cutover and degraded forestland were eventually abandoned, along with traditional forest farming practices, and a portion of this land was incorporated into the National Forest System.

The introduction of chestnut blight (*Cryphonectria parasitica* (Murrill) Barr) to North America in 1904 functionally eliminated American chestnut (*Castanea dentata* (Marsh.) Borkh.) from this forest by the 1930s [2]. Before its decline, American chestnut comprised 36% of the basal area and

(a)

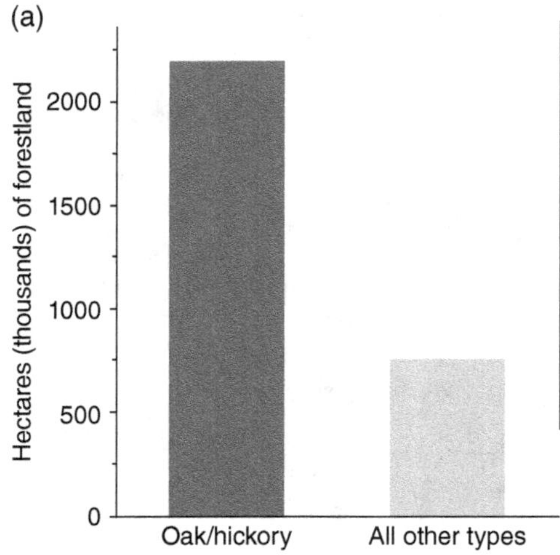

Forest type	Hectares
Other eastern softwoods	1,489
Aspen/birch	2,490
Exotic hardwoods	2,574
Oak/gum/cypress	3,737
Nonstocked	6,897
Spruce/fir	7,977
Elm/ash/cottonwood	14,728
Other hardwoods	89,847
Maple/beech/birch	91,344
White/red/jack pine	117,183
Loblolly/shortleaf pine	134,543
Oak/pine	271,921
All other types	*744,731*

(b)

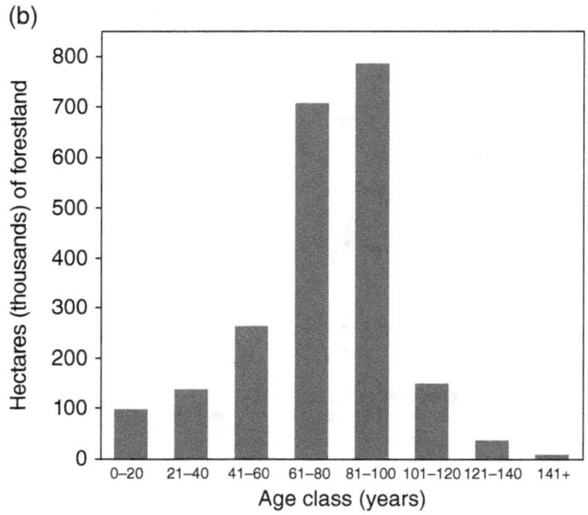

Figure 8.2 Distribution of forestland in the southern Appalachians by forest type group (a) and age class distribution of the most abundant forest type, oak-hickory (b).

22% of the density of the second-growth forests in the Blue Ridge Mountains [3]. The loss of American chestnut from the canopy released codominant and subordinate oak and hickory, which persisted through the period of frequent anthropogenic disturbance, thus allowing oak and hickory to recruit into dominant canopy positions where they remain in the contemporary landscape.

Anthropogenic fire was common throughout the eighteenth and nineteenth centuries [4]. Active fire suppression began in the mid-twentieth century which, coupled with a wetter climate and lack of other anthropogenic disturbances, allowed a suite of shade-tolerant and mesophytic hardwood species, including red maple (*Acer rubrum* L.), American beech (*Fagus grandifolia* Ehrh.), and blackgum (*Nyssa sylvatica* Marsh.), to dominate the midstory layer. This strata reduces light levels, limits air movement, and increases humidity, thus resulting in cooler and moister conditions in the understory than likely found in ecosystems historically influenced by frequent fire [5].

This history has led to contemporary landscapes dominated by forests where oak and hickory are canopy dominants but are not recruiting following disturbance events and mortality. This pattern is illustrated in tree establishment and growth release periods from a 200-year-old mixed oak stand in Pisgah National Forest [6] (Figure 8.3). Peaks in tree establishment during a transition in

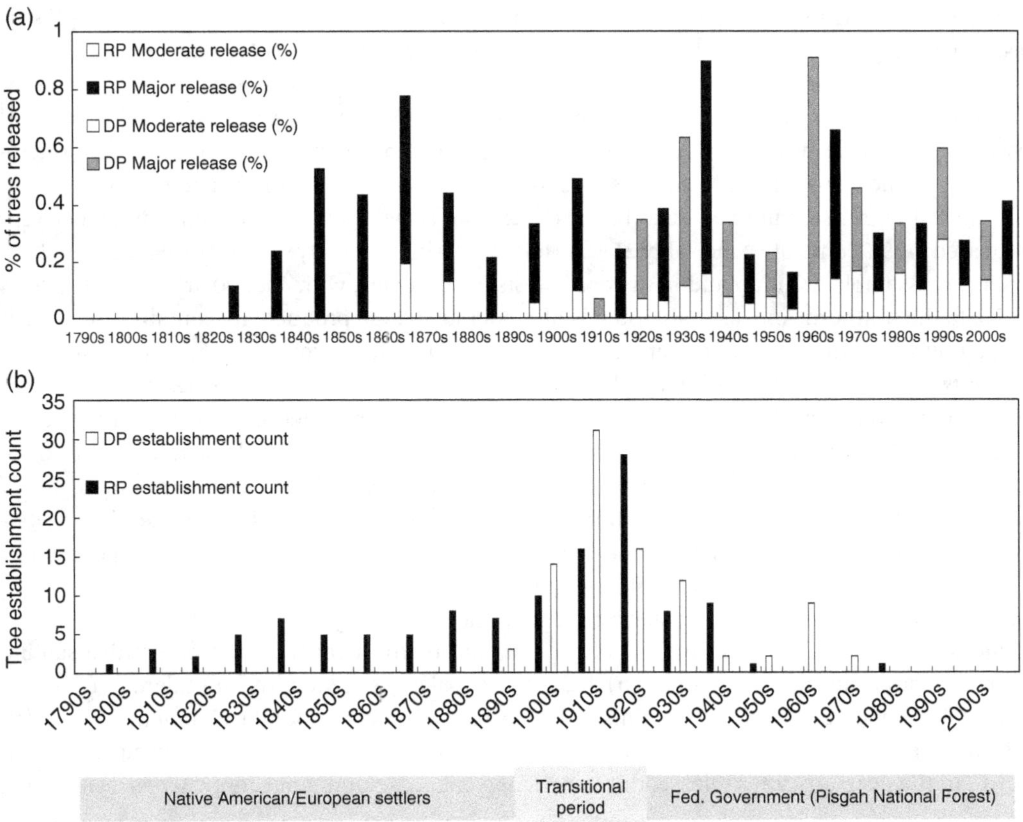

Figure 8.3 Growth release events (a) and establishment (b) of dominant species groups. Frequent burning and harvesting of this area before federal protection favored oaks and hickory (ring-porous species, RP). In contrast, after federal protection in 1914, the forest developed with less intensive management activities that reduced opportunities for oak and hickory but created conditions conducive to the establishment of more mesic species, including yellow-poplar and red maple (diffuse-porous species, DP). *Source:* Reproduced from Grover et al. [6].

ownership (1910–1930) highlight the effects of shifting forest land-use practices. These patterns are consistent across the region, resulting in a landscape of even-aged forests lacking species such as oak and hickory in the lower canopy strata compared to the mosaic of developmental stages, particularly the complex, multi-aged stands that once dominated the region [7].

8.3 Structure, Composition, and Development of the Southern Appalachian Mixed-Oak Ecosystem

8.3.1 Structure and Composition

The natural disturbance regime of mixed-oak forests is characterized by frequent, small-scale disturbances punctuated by intermediate disturbances of moderate extent and frequency. Much of the knowledge on natural dynamics of mixed-oak forests comes from the work available from cove forests, which are unique systems occurring in sheltered topographic positions on concave slopes that promote moist conditions. Canopy gap sizes vary from 25 to >1000 m^2, with most gaps <200 m^2 in size [8, 9]. Intermediate intensity disturbances create gaps ranging from single-tree-sized up to

10 ha [10], affecting forests at stand, watershed, and landscape levels. Stand reconstructions [8] suggest canopy gaps in old-growth cove forests form at an average rate of 1% per year on an area basis and are primarily caused by single-tree mortality, with an average rotation of 100 years. Age- or competition-related mortality is similar to mortality due to exogenous (e.g. drought, wind, insects/disease) disturbance events [11]. A recent study tracking tree populations 21 years after wind microbursts in second-growth mixed-oak forests corroborated that these patterns extend beyond cove forests. Mortality from small-scale disturbance (*background*) averaged 0.8% of stems dying per year compared to 39% cumulative mortality 21 years following a high-severity wind disturbance event [12].

Species assemblages in upland hardwood forests are complex, with over 90 arborescent species native to the southern Appalachians. Stands of moderate to high productivity contain a mixture of oak, along with numerous other genera and species also with intermediate shade tolerance (e.g. *Fraxinus americana* L., *Tilia heterophylla* Vent., and *Magnolia* spp.). These species have persistent reproductive strategies that rely on the presence of advance reproduction prior to and persisting through the disturbance. More mesophytic species with varying levels of shade tolerance also co-occur in these mixtures, especially as site productivity increases. The reproductive strategies of many such species, for example, rapid initial height-growth of yellow-poplar (*Liriodendron tulipifera* L.) from seed that can persist in litter for four to seven years, differ greatly from the persistence strategy common among the shade-midtolerant species.

Successful regeneration for persistent-strategy species, especially oak, is a multistep process that includes seed production, seedling establishment, and recruitment into the competitive sapling layer (i.e. large advance reproduction) [13]. Since oaks are generally mid-tolerant of shade (Figure 8.4), moderate overstory or mid-story disturbance is required for development of competitive advance reproduction (i.e. oak saplings >1.4 m tall) [14]. This oak understory is more likely to recruit to the overstory after release, thus sustaining oak as a dominant canopy tree species. Lack

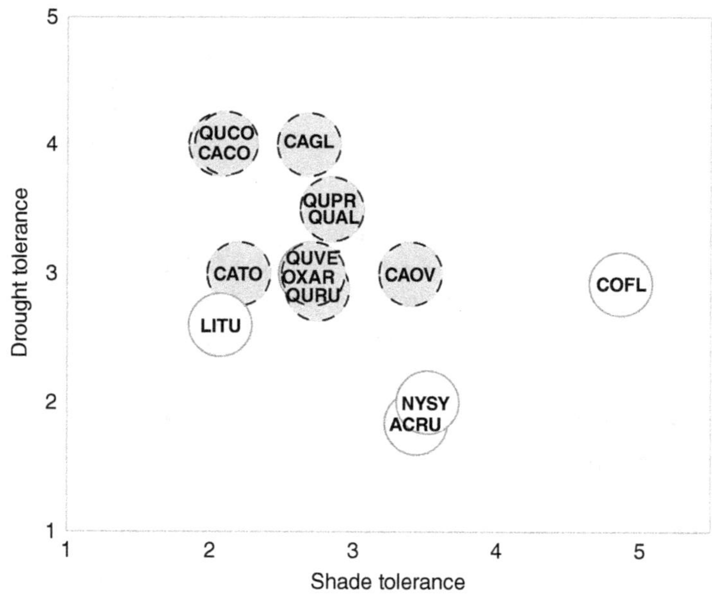

Figure 8.4 Sorting of common species as a function of their tolerance for shade and drought using rankings from Niinemets and Valladares [38]. *Quercus* and *Carya* species appear in green (upper left), while mesophytic species are hollow-fill (except OXAR, *Oxydendrum arboretum* (L.) DC.). ACRU, *Acer rubrum*; CACO, *Carya cordiformis*; CAGL, *C. glabra*; CAOV, *C. ovata*; CATO, *C. tomentosa*; COFL, *Cornus florida*; LITU, *Liriodendron tulipifera*; NYSY, *Nyssa sylvatica*; QUAL, *Quercus alba*; QUCO, *Q. coccinea*; QUGL, *Q. glabra*; QUPR, *Q. prinus*; QURU, *Q. rubra*; QUVE, *Q. velutina*. *Source:* Adapted from Niinemets and Valladares [38].

of disturbance prevents oaks from developing competitive advance reproduction and allows more shade-tolerant species to replace oak and eventually dominate the overstory canopy on moderate to good quality sites (e.g. [15]). If the canopy is opened without competitive advance oak regeneration, shade-intolerant competitors, such as seed-origin yellow-poplar, sweet birch (*Betula lenta* L.), and black cherry (*Prunus serotina* Ehrh.), quickly outcompete small, less competitive oak and hickory seedlings [14, 16] resulting in oak regeneration failure.

The interior of canopy gaps in contemporary forests is often dominated by yellow-poplar [17] and sweet birch [18], while advance reproduction of a variety of species near gap edges substantially increases in density, survivorship, and growth [19]. In oak forests of the Cumberland Plateau, oak seedlings exhibited greater growth extending 20 m from the edge of large created gaps [20], and light availability was elevated up to 12 m outside gaps [21]. Edge environments and their influence differ based on the origin type, shape, position on the landscape, or orientation.

8.3.2 Natural Development Model

A natural development model for oak-hickory forests (Table 8.1) describes the succession of the forest from disturbance through old forest stages. Following a large disturbance, multiple life-forms compete for resources and quickly fill in the understory and midstory layers of the

Table 8.1 Attributes of developmental stages in southern Appalachian oak-hickory forest ecosystems.

Stage or event	Attribute	Cause or Consequence
Disturbance and legacy creation	Spatially heterogeneous live and dead tree legacies, seedling and sapling layer representing diversity of regeneration mechanisms; small to moderate openings in matrix of second-growth even-aged forest	Results from recurring low-severity disturbances and less frequent, mesoscale events
Preforest	Advance regeneration, woody sprouts, vines, and herbs form a dense midstory and groundlayer within two years following disturbance	Rapid site capture through sprouting, seedling establishment from soil seedbanks or windblown seed, and release of advance regeneration when present
Young forest	Oak success depends on its initial crown position or persistence relative to other species	Crown closure is complete by 20 years after disturbance with well-defined crown classes in place. Self-thinning may cause approximately 80% mortality on average sites. Competitive sorting occurs with the composition of the stand at this stage being indicative of the future canopy
Mature forest	Diverse forest structure and age-class; large dead trees accumulate	Windthrow and breakage of canopy dominants allow for recruitment of new cohorts. Midtolerants recruit in large gaps following frequent burning, otherwise, intolerants or tolerants outcompete
Old forest	Dominated by mid-tolerant oak-hickory species in the overstory, with a lesser, but significant shade-tolerant and intolerant components	Long-term, frequent gap disturbance and fire result in diversity of live-tree size and age structures; complex deadwood conditions; multiple treefall gaps and mesoscale wind events create heterogeneity and foster tree diversity

developing forest. Fire at even low intensity during this early stage will give a competitive advantage to oaks, promoting seedling development by reducing fire-intolerant competitors. A moderate-severity disturbance occurring within decades after the large disturbance opens the canopy again and recruits the advance regeneration of oak and hickory that have accumulated. In the absence of fire, crown closure occurs within 20 years of the disturbance and the oak seedling bank fails to develop.

It is important to note that introduced pathogens and insects have significantly influenced stand dynamics and the developmental model. Mortality events in oak forests caused by pests and pathogens are often triggered by an inciting factor, most often drought in this region, causing stress and declines in growth. Periodic droughts have occurred historically, but enhanced hydroclimate variability has led to increased plant water stress and widespread tree mortality in recent decades [22] driven by decreases in precipitation and increases in temperature throughout the region [23].

8.4 Regenerating Upland Oak Forests in the Southern Appalachians

Regeneration harvests are the most common silvicultural activity conducted on public and private lands. Clearcutting was used extensively during the mid- to late-twentieth century to create open conditions that were thought to favor oak over less economically valuable shade-tolerant and intolerant species. Decades of research indicate that clearcutting on all but the most xeric sites fails to regenerate oak [24]. More recently, regeneration harvests on public lands have predominantly implemented a two-aged silvicultural system to meet multiple-use objectives. In a typical two-aged system, often referred to as an extended irregular shelterwood, overstory removal cuts are heavy (2–6 m^2 ha^{-1} of residual basal area), and dispersed or aggregated reserve trees are often maintained throughout the rotation of the developing cohort to provide wildlife habitat, structural diversity, and eventually deadwood in the regenerating stand.

Development of large advance oak reproduction prior to substantial reductions in overstory density is critical to successful regeneration and recruitment of oak. In mixed-oak forests, a midstory removal treatment conducted approximately 10 years prior to a regeneration harvest (i.e. oak shelterwood) [14] is a critical step in any silvicultural prescription. The midstory removal treatment reduces stand basal area from below by 20–30% using herbicides to inhibit vigorous sprout competition [25]. The reduction in density provides an ephemeral increase in light in the understory [26], which can provide sufficient resources for mid-tolerant oak and hickory seedlings to develop beneath the intact overstory [27]. These treatments simulate burning that can effectively promote the fire-tolerant oaks over mesic fire-intolerant competitors.

Group selection, an uneven-aged silvicultural method, is implemented on public lands, though it is often not the most economical option for hardwood management (Figure 8.5). Although used successfully in other hardwood forest types where it resembles the natural disturbance patterns and subsequent regeneration process [28], when group selection is implemented in mixed-oak stands, yellow-poplar outcompetes oak and hickory even in small openings (>0.07 ha). Single-tree selection, which reduces diversity and favors shade-tolerant species, is not practiced. Exploitive diameter-limit harvesting or "high-grading" is common on private lands.

Prescribed fire has been promoted for regenerating oaks, but only recently have more silvicultural prescriptions been developed and tested. Understory burning in mature stands does promote oak

Figure 8.5 Multi-aged oak-hickory forest, Bent Creek Experimental Forest, North Carolina, USA (forest matrix, 12-year-old gap, and recent harvest area/gap in view).

regeneration, but the oak fails to reach competitive sizes without repeated burns which is often difficult to achieve. In Piedmont oak-dominated hardwood stands, a prescription that combines shelterwood harvest and repeated prescribed fire has successfully promoted oak regeneration to competitive size in the understory. The shelterwood-burn method [29] has incorporated the understanding that fires of sufficient intensity are required to significantly set back competitors such as yellow-poplar, but additional vegetation control is required to ensure oak is a significant portion of the stand.

8.5 An Ecologically Based Silvicultural System for Mixed-Oak Ecosystems

Differences in regeneration strategies among the most frequent and abundant species demand careful design and implementation of silvicultural treatments that favor the regeneration and recruitment of mid-tolerants, without simultaneously making growing conditions favorable for competitors. The template for such a design may come from our growing understanding of the multiple canopy disturbances – across space and time and of variable intensity – that were necessary for the successful regeneration and recruitment of the oak and hickory in contemporary canopies.

Based on the natural developmental model of mixed-oak forests and regeneration dynamics of the important species, we propose that an expanding-gap irregular shelterwood may best create conditions similar to the natural developmental sequence in mixed-oak forests that are needed to successfully regenerate and recruit mid-tolerant species (Table 8.2). This silvicultural system is largely implemented in the mature forest stage, so the developmental sequence is at the gap scale with gap expansion creating a mosaic of developmental stages at the stand level. These actions would result in conditions characteristic of the old forest stage of the natural development model. However, actions implemented in the preforest and young forest stages will work to increase the likelihood of successful regeneration in later stages.

8.5.1 Disturbance and Legacy Creation

An irregular group shelterwood regeneration method [30, 31], seeks to create multi-cohort stands over space and time. The initial entry is similar to a group selection harvest, but subsequent entries expand the gap created with the initial entry instead of creating new independent gaps. This pattern reflects the mosaic of gap-scale conditions across a landscape affected by intermediate disturbance.

Table 8.2 An ecological silvicultural system for southern Appalachian mixed-oak forests.

Stage	Years	Potential actions	Outcome
Disturbance/ Legacy creation	0	Initial harvest for an irregular shelterwood expanding-gap system (occurs in Mature or Old Forest); create snags and leave deadwood	Continuity of structure; initiate preforest stage
Preforest	1–10	Site preparation to remove non-merchantable stems following harvest Midstory removal treatment	Midstory removal treatments are important for manipulating understory conditions to favor development of mid-tolerant species
Young forest	10–40	Competition control; crown-touch release treatments	Increase competitiveness of established mid-tolerants
	40–70	Pre-commercial thinning; continued competition control; midstory removal	
Mature forest	70–120	Initial harvest followed by expansion entries every decade up to four times; additional entries to regenerate new areas; site preparation and competition control between entries	Increase growth and canopy heterogeneity; add microsite variability; establish new cohorts; release oak and hickory regeneration with subsequent entries; create 5 age classes
Old forest	100+	Oak and hickory should represent ≥30% of stand composition; expand natural gaps or initiate new harvest sequence	Enrich species, maintain structurally complex and diverse conditions

An initial harvest entry creates gaps similar to those that result naturally from both small-scale endogenous and exogenous events and intermediate-scale disturbance events. Gaps characterized following Hurricane Opal had a median size of 0.4 ha and ranged from the size of a single canopy tree to approximately 4 ha [32]. An experimental demonstration of the expanding-gap system, selected entry sizes of 0.1 and 0.4 ha based on the natural developmental model, operability, and economics, with the initial entry removing 25% of the area.

Each entry releases growing space for trees, some of which will be captured by shade-intolerant competitors, but along edge conditions, the mid-tolerants will establish and remain competitive. These entries across the landscape create patches of species diversity and variation in forest structure similar to gap dynamics underlying the natural development model for the ecosystem. Gaps can be located adjacent to existing advance regeneration to create advantageous edge effect. Retention can be incorporated within the initial entries or expansions, depending on intentions.

8.5.2 Preforest Stage

Silvicultural activities during a preforest stage would focus on-site preparation and control of competitive saplings and invasive shrubs and vines that fill in quickly. A midstory removal treatment, as described earlier, can create the open understory conditions characteristic of the landscape before fire suppression policies and land abandonment in the early to mid-1900s. Competition control may be accomplished using prescribed fire, chemical treatments (injections or targeted sprays), or mechanical methods.

8.5.3 Young Forest Stage

Silvicultural actions during the young forest stage are similar to the preforest stage. Efforts would focus on developing oak and hickory seedlings into competitive saplings. Early in this stage, competition control around the developing regeneration is required to ensure continued presence. Crown-touch release treatments can be used to increase growing space around desirable trees.

Later, a pre-commercial thinning could provide a modest early economic return, but more to free resources for desirable species. Prescribed fire is avoided at this stage when species are most susceptible to damage. Mechanical or chemical treatments are used to control invasives.

8.5.4 Mature Forest Stage

The gap expansion entries are similar to patterns of border-tree mortality documented following gap-forming disturbances [33] and are designed to take advantage of the large advance reproduction that develops naturally within the periphery of created gaps. In addition, the expanding-gap system maintains the economic advantages of even-aged methods by reusing harvesting infrastructure established during the initial entry and provides landowners the opportunity for periodic income.

The expansions create openings of irregular patterns, modeled after natural gaps that are often more elliptical in shape than circular. Working towards an irregular shape promotes freedom to take advantage of any advance oak development in the periphery after gap creation. Wider expansions can be marked around gaps where favorable advance reproduction has developed but, smaller expansions or even delaying expansion for another cycle along gaps where advance reproduction has failed to develop. Both site preparation and competition control may be needed before or after each additional entry. Gap-level live tree retention could be incorporated as this would be expected in a downburst or hurricane patch.

The cutting cycle and the amount removed during each entry are flexible. While the baseline gap formation rate is minimally 1% per year from small disturbances, it can extend to nearly 40% of the total land area following less-frequent intermediate intensity disturbances. Using the baseline natural gap formation rate alone as a guide, entries might be designed to remove 10% of the stand every 10 years for a 100-year rotation. However, entries of this size may not produce enough volume to be economically or operationally feasible, so the entry schedule and area treated can be customized accordingly.

Each expansion entry will result in the production of an age class; if for example, four entries removing 25% of the stand each are planned, four age classes will result from each regeneration harvest, and the fifth age class will be the result of retention trees throughout the stand (e.g. inoperable or riparian areas). No harvesting is planned following the fourth entry for approximately 60 years. If timber production is among the management objectives, commercial harvest may occur at 60+ years. Initial entries would start the expanding gap cycle again, creating legacy conditions to promote regeneration of shade mid-tolerant and intolerant species.

8.5.5 Old Forest Stage

This old forest stage could be reached if stands are managed for alternative ecosystem services instead of timber production. Management for these stands may focus instead on wildlife habitat, biodiversity, carbon storage, watershed protection, resistance, and resilience to disturbances. Silviculture will focus on adding structure and maintaining or restoring species that had decreased in abundance. Activities might include prescribed fire or mechanical control of competing species in gaps where regeneration is focused. If harvesting is desired in the old forest stage, trees are marked so as to release advance regeneration of desirable species occurring near natural or prior harvested areas.

8.6 Climate Change Considerations

Oak-hickory forests already require active restoration strategies due to the altered disturbance regime under which they are established. Mixed oak-hickory forests are desirable as they have more consistent and greater adaptability and species diversity relative to mixed mesophytic forests (Figure 8.6). The compounding effects of a warmer and drier climate and range expansions of forest pests and pathogens may favor oaks over the long term [34], though various oak species will respond differently. In general, species in the white oak group are relatively more shade tolerant, longer-lived, slower growing, and more drought tolerant than species in the red oak group, resulting in higher adaptability scores for the white versus red oak group [35, 36]. Experimental evidence also indicates that two important white oak species (chestnut and white oak), when in dominant and codominant canopy positions, are not sensitive to stand density, stem size, or site productivity [37] indicating they will likely be less affected by future climate change. Management activities focusing on the regeneration, establishment, and retention of white oak versus red oak species may be a viable strategy for sustaining the growth and productivity of oak species with climate change. Ecologically based management that includes prescribed fire and thinning would reduce wildfire hazards and create more desirable oak forests.

8.7 Summary

Alternative silvicultural approaches will be necessary to ensure mixed-oak forests that provide a wide range of ecosystem services, remain an important component of southern Appalachian hardwood forests. By increasing the structural diversity of the current second-growth mixed hardwood forests, multi-cohort silvicultural systems based on the natural development model for these forests will increase floral and faunal diversity while enhancing disturbance and drought resilience. The design of the silvicultural system we present is based on the natural developmental model of the forest type, creating multiple entries of different shapes and sizes, and placed to expand off a previous entry to release the desirable mid-tolerant species that often successfully naturally regenerate in gap edge conditions. The variable-aged adjoining entries create a mix of microhabitats promoting species diversity and developing into a structurally diverse forest system. Gap-based silviculture offers an ecologically justifiable, operationally feasible, and publicly acceptable approach to upland hardwood regeneration. Such attributes are desirable to public land managers and the growing number of private landowners managing their forests to balance conservation, recreation, aesthetics, and production objectives.

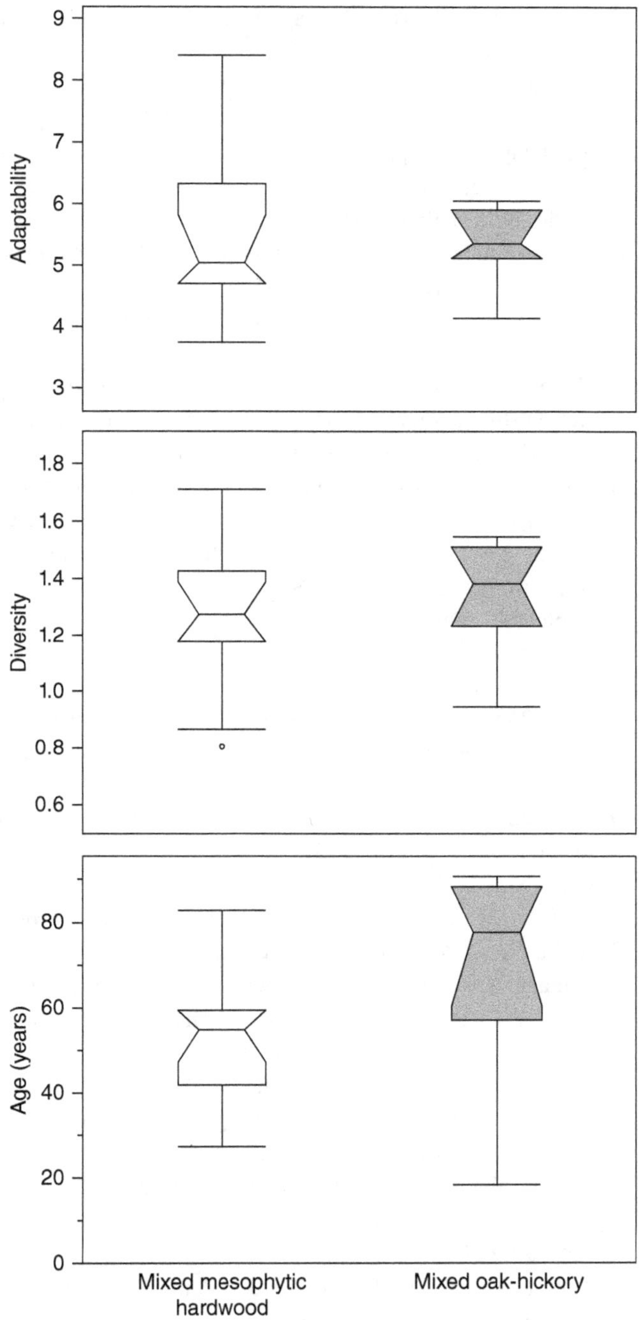

Figure 8.6 Estimates of forest stand age, adaptability score, and Shannon-Weiner diversity calculated from FIA data from 825 stands in the southern Appalachians of North Carolina. FIA data were collected between 2011 and 2021 and represent plots with diverse ownership, last land-use history, and site conditions. Stand and stock data were used to calculate basal area-weighted stand-level adaptability, similar to the methods described in Kabrick et al. [39]. Adaptability scores range from 0 to 10, with a low score indicating lower capacity to tolerate or respond to predicted changes. *Source:* Adapted from Kabrick et al. [39].

References

1 Simon, S.A., Collins, T.K., Kauffman, G.L. et al. (2005). *Ecological Zones in the Southern Appalachians: First Approximation*, Res. Pap. SRS-41. Asheville, NC: USDA, Forest Service, Southern Research Station. 41 p.

2 van deGevel, S.L., Hart, J.L., Spond, M.D. et al. (2012). American chestnut (*Castanea dentata*) to northern red oak (*Quercus rubra*): forest dynamics of an old-growth forest in the Blue Ridge Mountains, USA. *Botany* 90: 1263–1276.

3 Elliott, K.J. and Swank, W.T. (2008). Long-term changes in forest composition and diversity following early logging (1919–1923) and the decline of American chestnut (*Castanea dentata*). *Plant Ecology* 197: 155–172.

4 Flatley, W.T., Lafon, C.W., Grissino-Mayer, H.D., and LaForest, L.B. (2013). Fire history, related to climate and land use in three southern Appalachian landscapes in the eastern United States. *Ecological Applications* 23: 1250–1266.

5 Nowacki, G.J. and Abrams, M.D. (2008). The demise of fire and "Mesophication" of forests in the Eastern United States. *BioScience* 58: 123–138.

6 Grover, Z.G., Forrester, J.A., Keyser, T.L. et al. (2023). Growth response, climate sensitivity and carbon storage vary with wood porosity in a southern Appalachian mixed hardwood forest. *Agricultural and Forest Meteorology* 332: 109358.

7 Whitney, G.G. (1996). *From Coastal Wilderness to Fruited Plain: A History of Environmental Change in Temperate North America 1500 the Present.* Cambridge University Press.

8 Runkle, J.R. (1982). Patterns of disturbance in some old-growth mesic forests of the eastern North America. *Ecology* 63: 1533–1546.

9 Rentch, J.S., Fajvan, M.A., and Hicks, R.R. Jr. (2003). Spatial and temporal disturbance characteristics of oak-dominated old-growth stands in the Central Hardwood Region. *Forest Science* 49: 778–789.

10 Mou, P. and Warrillow, M.P. (2000). Ice storm damage to a mixed hardwood forests and its impacts on forest regeneration in the Ridge and Valley region of southwestern Virginia. *Journal of the Torrey Botanical Society* 127: 66–82.

11 Lorimer, C.G. (1980). Age structure and disturbance history of a southern Appalachian virgin forest. *Ecology* 61: 1169–1184.

12 Greenberg, C.H. (2021). Long-term recovery dynamics following hurricane-related wind disturbance in a southern Appalachian forest. *Forest Ecology and Management* 502: 119704.

13 Arthur, M.A., Alexander, H.D., Dey, D.C. et al. (2012). Refining the oak- fire hypothesis for management of oak-dominated forests of the eastern United States. *Journal of Forestry* 110: 257–266.

14 Loftis, D.L. (1990a). A shelterwood method for regenerating red oak in the southern Appalachian mountains. *Forest Science* 36: 917–929.

15 Lorimer, C.G., Chapman, J.W., and Lambert, W.D. (1994). Tall understory vegetation as a factor in the poor development of oak seedlings beneath mature stands. *Journal of Ecology* 82: 227–237.

16 Shure, D., Phillips, D.L., and Bostick, P.E. (2006). Gap size and succession in cutover southern Appalachian forests: an 18 year study of vegetation dynamics. *Plant Ecology* 185: 299–318.

17 Weigel, D.R. and Parker, G.R. (1997). Tree regeneration response to the group selection method in southern Indiana. *Northern Journal of Applied Forestry* 14: 90–94.

18 Thomas-Van Gundy, M., Rentch, J., Adams, M.B., and Carson, W. (2014). Reversing legacy effects in the understory of an oak-dominated forest. *Canadian Journal of Forest Research* 44: 350–364.

19 Chen, J., Franklin, J.F., and Spies, T.A. (1992). Vegetation responses to edge environments in old- growth Douglas-fir forests. *Ecological Applications* 2: 387–396.

20 Lhotka, J.M. and Stringer, J.W. (2013). Forest edge effects on *Quercus* reproduction within naturally regenerated mixed broadleaf stands. *Canadian Journal of Forest Research* 43: 911–918.

21 Patterson, C.P., Hackworth, Z.J., Lhotka, J.M., and Stringer, J.W. (2022). Light and regeneration patterns following silvicultural gap establishment in *Quercus* dominated stands of northern Cumberland Plateau, USA. *Forest Ecology and Management* 505: 119871.

22 Brzostek, E.R., Dragoni, D., Schmid, H.P. et al. (2014). Chronic water stress reduces tree growth and the carbon sink of deciduous hardwood forests. *Global Change Biology* 20: 2531–2539.

23 Ficklin, D.L., Maxwell, J.T., Letsinger, S.L., and Gholizadeh, H. (2015). A climatic deconstruction of recent drought trends in the United States. *Environmental Research Letters* 10: 044009.

24 Swaim, J.T., Dey, D.C., Saunders, M.R. et al. (2018). Overstory species response to clearcut harvest across environmental gradients in hardwood forests. *Forest Ecology and Management* 4258: 66–80.

25 Keyser, T.L. and Zarnoch, S.J. (2014). Stump sprout dynamics in response to reductions in stand density for nine upland hardwood species in the southern Appalachian Mountains. *Forest Ecology and Management* 319: 29–35.

26 Schweitzer, C.J. and Dey, D.C. (2011). Forest structure, composition, and tree diversity response to a gradient of regeneration harvests in the mid-Cumberland Plateau escarpment region, USA. *Forest Ecology and Management* 262: 1729–1741.

27 Loftis, D.L. (1990). Predicting post-harvest performance of advance red oak reproduction in the southern Appalachians. *Forest Science* 36: 908–916.

28 Runkle, J.R. (1991). Gap dynamics of old-growth eastern forests: management implications. *Natural Areas Journal* 11: 19–25.

29 Brose, P.H., van Lear, D.H., and Keyser, P.D. (1999). A shelterwood-burn technique for regenerating productive upland oak sites in the Piedmont Region. *Southern Journal of Applied Forestry* 16: 158–163.

30 Puettmann, K.J., D'Amato, A.W., Kohne, U., and Bauhaus, J. (2009). Individual-tree growth dynamics of mature *Abies alba* during repeated irregular group shelterwood (Femelschlag) cuttings. *Canadian Journal of Forest Research* 39: 2437–2449.

31 Raymond, P., Bedard, S., Roy, V. et al. (2009). The irregular shelterwood system: review, classification and potential application to forest affected by partial disturbances. *Journal of Forestry* 107: 405–413.

32 McNab, W.H., Greenberg, C.H., and Berg, E.D. (2004). Landscape distribution and characteristics of large hurricane-related canopy gaps in a southern Appalachian watershed. *Forest Ecology and Management* 196: 435–447.

33 Runkle, J.R. (1998). Changes in southern Appalachian canopy tree gaps sampled thrice. *Ecology* 79: 1768–1780.

34 Vose, J.M. and Elliott, K.J. (2016). Oak, fire, and global change in the eastern USA: what might the future hold? *Fire Ecology* 12: 160–179.

35 Matthews, S.N., Iverson, L.R., Prasad, A.M. et al. (2011). Modifying climate change habitat models using tree species-specific assessments of model uncertainty and life history-factors. *Forest Ecology and Management* 262: 1460–1472.

36 Prasad, A.M., Gardiner, J.D., Iverson, L.R. et al. (2021). A climate change atlas for 134 forest tree species of the eastern United States. https://www.fs.fed.us/nrs/atlas/tree/ (accessed 30 March 2022).

37 Keyser, T.L. and Brown, P.M. (2014). Climate-growth relationships for yellow-poplar across structural and site quality gradients in the southern Appalachian Mountains. *Forest Ecology and Management* 329: 158–165.

38 Niinemets, U. and Valladares, F. (2006). Tolerance to shade, drought, and waterlogging of temperate northern hemisphere trees and shrubs. *Ecological Monographs* 76: 521–547.

39 Kabrick, J.M., Clark, K.L., D'Amato, A.W. et al. (2017). Managing hardwood-softwood mixtures for future forests in eastern North America: assessing suitability to projected climate change. *Journal of Forestry* 115: 190–201.

9

Ecological Silviculture for Yellow Birch–Conifer Mixedwoods in Eastern Canada

Patricia Raymond and Daniel Dumais

Direction de la recherche forestière, Ministère des Ressources naturelles et des Forêts du Québec, Canada

9.1 Introduction

Ecosystems composed of yellow birch (*Betula alleghaniensis* Britt.) and conifers are abundant in eastern Canada. This rich and diversified mixedwood forest zone marks the transition between the northern hardwood forest dominated by sugar maple (*Acer saccharum* Marshall), in the south, and the boreal coniferous forest dominated by balsam fir (*Abies balsamea* (L.) Mill.) and black spruce (*Picea mariana* (Mill.) B.S.P.), in the north. In the province of Quebec, the temperate mixedwood forest corresponds to the balsam fir–yellow birch bioclimatic domain [1], located mostly between latitudes 47° and 48°N (Figure 9.1).

For two centuries, temperate mixedwood forests have supplied the forest industry of the region with various high-quality products. With beautiful landscapes, they also are frequently sought after for recreational activities. Moreover, the proximity of yellow birch–conifer forests to urban centers gives them inestimable value in providing human wellness benefits and clean air and water. Therefore, sustainable management of these ecosystems requires balancing socioeconomic and ecological needs to ensure both wood production and diverse ecosystem services.

The need to conserve biodiversity also justifies the management objective of maintaining the mixed-species composition and structure of these forests. However, it remains a forest management challenge, given the diverse ecological characteristics of the many tree and shrub species in these ecosystems [2]. Though recent research on mixedwood forest silviculture has led to better forestry practices, several knowledge gaps remain. This chapter aims to provide a framework for understanding these complex ecosystems and implementing an ecological approach based on natural dynamics.

9.2 Characteristics of Yellow Birch–Conifer Mixedwoods

9.2.1 Species Composition

In these ecosystems, yellow birch and balsam fir (Figure 9.2) dominate and grow along with companion species that have a range of ecological characteristics and whose abundance varies depending on geophysical setting and disturbance regime (Table 9.1a). These include red spruce (*Picea*

Ecological Silvicultural Systems: Exemplary Models for Sustainable Forest Management, First Edition. Edited by Brian J. Palik and Anthony W. D'Amato.

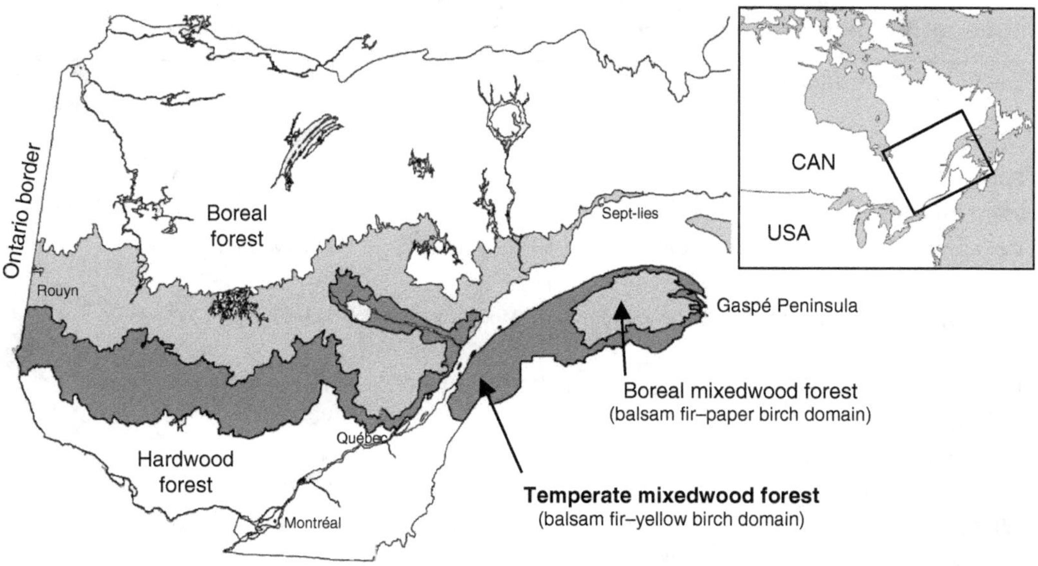

Figure 9.1 Distribution of the temperate mixedwood forest in the province of Quebec (eastern Canada).

Figure 9.2 Main tree species associated with temperate mixedwood forests in eastern Canada. (a) Mature yellow birch. (b) Mid-aged balsam fir. (c) Mature red spruce. *Source:* Daniel Dumais.

rubens Sarg.), red maple (*Acer rubrum* L.), paper birch (*Betula papyrifera* Marshall), sugar maple, white spruce (*Picea glauca* (Moench) Voss), American beech (*Fagus grandifolia* Ehrh.), eastern white cedar (*Thuja occidentalis* L.), eastern white pine (*Pinus strobus* L.), and eastern hemlock (*Tsuga canadensis* (L.) Carr.).

Temperate mixedwoods are characterized by an abundant layer of understory competing vegetation composed of noncommercial tree species and shrubs (Table 9.1b), all with a high risk of proliferation following disturbance. Competing species such as mountain maple (*Acer spicatum* Lam.), hobblebush (*Viburnum lantanoides* Michaux), striped maple (*Acer pensylvanicum* L.), and beaked hazelnut (*Corylus cornuta* Marshall) are generally well established in the understory [2].

Table 9.1 Ecological characteristics of (a) trees and (b) woody shrubs in yellow birch–conifer mixedwoods.

(a) Tree species	Shade tolerance	Favorable seedbeds	Early relative growth rate	Sprouting ability	Maximum lifespan (years)
Yellow birch (*Betula alleghaniensis*)	Intermediate	Mineral/organic mix, rotten wood	Moderate to fast	Moderate (stumps)	300
Balsam fir (*Abies balsamea*)	Very high	Moist mineral, rotten wood	Moderate	Poor (layering)	200
Red spruce (*Picea rubens*)	Very high	Rotten wood	Very slow	Poor (layering)	400
Red maple (*Acer rubrum*)	Intermediate	Varied, litter	Moderate to fast	Good (stumps)	200
Paper birch (*Betula papyrifera*)	Low	Moist mineral	Fast	Moderate (stumps)	200
Sugar maple (*Acer saccharum*)	Very high	Varied, litter	Slow	Moderate (stumps)	350
White spruce (*Picea glauca*)	High	Moist mineral, rotten wood	Moderate	None	300
American beech (*Fagus grandifolia*)	Very high	Varied	Slow (seedlings) to fast (suckers)	Good (suckering)	400
Eastern white cedar (*Thuja occidentalis*)	High	Rotten wood	Slow	Good (layering)	900
Eastern white pine (*Pinus strobus*)	Intermediate	Moist mineral	Slow to moderate	None	450
Eastern hemlock (*Tsuga canadensis*)	Very high	Rotten wood	Very slow	Very poor (layering)	800

(b) Woody shrubs	Shade tolerance	Main propagation mode	Early relative growth rate	Post-cut abundance[a]	Maximum height and particularities
Mountain maple (*Acer spicatum*)	Intermediate to high	Sprouting	Slow (seedlings) to fast (sprouts)	+	8 m. High nutrient needs; layering.
Hobblebush (*Viburnum lantanoides*)	High	Suckering	Slow (seedlings) to moderate (suckers)	+ (partial cut) – (clear-cut)	2 m. Good sprouting ability; layering.
Striped maple (*Acer pensylvanicum*)	High	Seedling (preestablished)	Moderate	+ (partial cut) – (clear-cut)	15 m. Good sprouting ability; layering.
Beaked hazelnut (*Corylus cornuta*)	High	Suckering	Very slow (seedlings) to slow (suckers)	+ (partial cut) = (clear-cut)	3 m. Once established, persists for a long time; good sprouting ability.
Raspberry (*Rubus idaeus*)	Low	Buried seeds	Fast	= (partial cut) + (clear-cut)	1.5 m. Inhibits conifer establishment; long seed viability (100 years).
Pin cherry (*Prunus pensylvanica*)	Very low	Seedling (buried seeds)	Fast	– (partial cut) + (clear-cut)	15 m. Good suckering ability; long seed viability (150 years).

[a] + higher, – lower, = similar.

Seeds of other species such as raspberry (*Rubus idaeus* L.) and pin cherry (*Prunus pensylvanica* L.f.) are stored in the soil seedbank [3] and germinate after disturbance. Thus, one big challenge associated with ecological silviculture of mixedwood stands is to favor regeneration of desired tree species, such as yellow birch and red spruce, while mitigating understory shrub competition, especially where herbicides are prohibited (e.g. Quebec's public lands since 2001 [4]).

9.2.2 Structure

Late-successional yellow birch–conifer mixedwoods typically support multi-aged structures, either balanced [5–7] or irregular uneven-aged stands [8, 9]. Historical surveys indicate that mature stands with irregular structures were predominant in preindustrial landscapes [10, 11]. Given the ecological role of irregular stands in yellow birch–conifer mixedwoods, maintaining these structures over a proportion of the territory may help meet ecosystem-based objectives [12]. Late-successional mixedwood stands also have distinctively high levels of spatial heterogeneity and a complex vertical structure (Figure 9.3). Structural legacies include canopy gaps of various sizes and shapes, large veteran trees, and many large snags and coarse woody materials that are rich in microhabitats and contribute to maintaining wildlife and plant diversity (Figure 9.4).

9.2.3 Historical Natural Disturbance Regime

Yellow birch–conifer mixedwoods are usually characterized by a regime of small-scale, moderate disturbances and long fire-return intervals. Preindustrial data indicates that natural fire cycles in Quebec's temperate mixedwood zone vary from a few hundred to more than 2000 years [13, 14], but that average 200–400 years [15]. Effective wildfire suppression during the last decades has reduced natural fire frequency and intensity [13], but the expected impacts of climate change could reverse this trend [16]. After a major disturbance destroys most of the canopy, late-successional temperate mixedwoods require at least 250 years to recover their complex composition and structure [17].

In the absence of stand-replacing disturbances, three main agents drive the natural dynamics by creating canopy gaps (20–2100 m^2 in area; Figure 9.5): (i) natural senescence of overstory trees; (ii) individual or multi-tree windthrow; and (iii) partial insect outbreaks (e.g. spruce budworm [*Choristoneura fumiferana* (Clem.)] [18, 19]). Essentially, the species composing yellow

Figure 9.3 Mature (a) and old-growth (b) yellow birch–conifer mixedwoods showing a complex vertical structure, spatial heterogeneity, and a significant red spruce component. *Source:* (a) Daniel Dumais, (b) Hugo Tremblay.

Figure 9.4 Structural legacies in yellow birch–conifer mixedwoods. (a) Old red spruce veteran. (b) Large yellow birch snag. (c) Coarse woody material. *Source:* (a) Patricia Raymond, (b) Daniel Dumais, (c) Hugo Tremblay.

Figure 9.5 (a) Natural canopy gap in a yellow birch–conifer mixedwood. (b) Small canopy opening created after the fall of a large tree. (c) Released red spruce sapling in a recent small canopy gap. *Source:* Daniel Dumais.

birch–conifer mixedwoods regenerate in these canopy gaps through either establishment of new seedlings or release of advanced regeneration.

9.2.4 Natural Development Model

To understand complex stand dynamics and how this informs ecological silviculture, models that integrate developmental and successional stages are essential (Table 9.2). **Developmental stage** refers to the age class of the main canopy (or time elapsed since the last disturbance).

Table 9.2 Summary of natural dynamics of yellow birch–conifer mixedwoods.

Developmental stage	Attributes	Cause and consequence	Schematic representation (successional stage)
Disturbance and legacy creation	Most of the canopy destroyed; large live trees and snags; downed logs.	Results from a major disturbance (severe insect outbreaks, extended windthrow or fire); establishment of shade-intolerant hardwoods.	**Transition** or **stabilization** 
Young	**Even-aged** 1) Mainly shade-intolerant hardwoods; sparse young fir in the understory.	Results from major disturbance; conifer density initially low.	**Initial** (hardwoods) 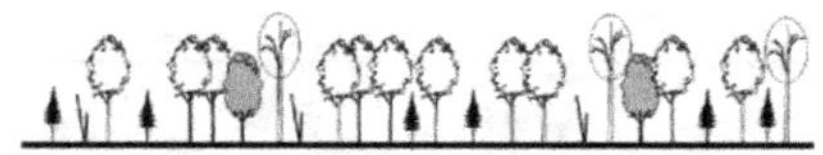
	2) Mix of shade-intolerant and tolerant hardwoods and conifers.	Results from development of shade-intolerant hardwoods and conifers.	**Transition** or **stabilization** (mixedwood) 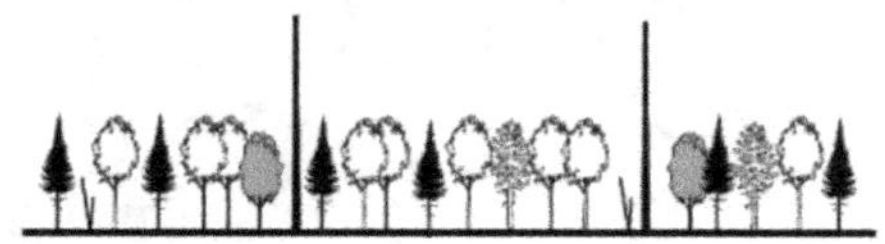
Mature	**Two-aged** 1) Shade-intolerant hardwoods; shade-tolerant conifers in the understory.	Results from growth of paper birch (greater longevity than aspen); conifer advance regeneration develops.	**Initial** (hardwoods) 

(Continued)

Table 9.2 (Continued)

	2) Mix of shade-intolerant and tolerant hardwoods and conifers; shade-tolerant conifers in the understory.	Results from the development of a mixedwood stand; low understory light limits the growth of conifer advance regeneration.	**Transition** or **stabilization** (mixedwood) 
Old	**Multi-aged** 1) Birch dominance persists in canopy; gaps are created.	Gap dynamics created by senescent trees foster growth and recruitment of conifer advance regeneration.	**Initial** (hardwoods) 
	2) Temperate mixedwood canopy with gaps.	Gap dynamics resulting from fir mortality; release of conifer advance regeneration (acclimation, growth).	**Transition** or **stabilization** (mixedwood) 
	3) Temperate mixedwood canopy with complex composition and structure, spatial heterogeneity and canopy gaps, large snags, and downed logs.	Results from long forest development (≥250 years); complex gap dynamics with companion long-lived and shade-tolerant conifer species.	**Equilibrium** (complex mixedwood) 

Symbol legend used in schematic representations (Yellow birch pictogram includes tolerant hardwoods such as red and sugar maple).

Successional stage refers to the forest composition status, from the "initial stage" (early-successional, shade-intolerant species) to the "stabilization stage" (late-successional, shade-tolerant species), and eventually, the "equilibrium stage" (where shade-intolerant species become marginal) [20].

9.2.4.1 Disturbance and Legacy Creation

Although rare, major natural disturbances (e.g. moderate to severe insect outbreaks, extended windthrow, or fire) can destroy most of the canopy but leave important structural legacies (Table 9.2). Long-lived tree species such as yellow birch, cedar, and red and white spruce can generate large live trees with dying crowns and cavities, big snags, and downed logs (Figures 9.4 and 9.6a). These structures help maintain biodiversity and serve as lifeboats for many plant and animal species [21, 22]. In absence of major disturbance, legacy creation can happen at smaller scales in the stand (e.g. at the gap level via gap dynamics).

Figure 9.6 Examples of developmental and successional stages of yellow birch–conifer mixedwoods. (a) Creation of disturbances and legacies: old remnant yellow birch that survived the major past disturbance. (b) Young forest stage: numerous even-aged stems of paper birch, yellow birch, and pin cherry after a major disturbance. (c) Mature forest stage: two-aged aspen–conifer stand. (d, e) Old forest stage: old yellow birch–conifer and red spruce–yellow birch stands. *Source:* (a, b, c, d) Daniel Dumais, (e) Stéphane Tremblay.

9.2.4.2 Young Forest Stage

After a catastrophic fire, shade-intolerant hardwood species form most of the young even-aged forest's cover, along with sparse yellow birch and persistent pin cherry trees [17] (Table 9.2 and Figure 9.6b). Shade-tolerant conifer species (mainly fir and spruces) can appear later in the understory. When the original disturbance is a catastrophic outbreak or windthrow, this stage is characterized by a mix of shade-intolerant hardwoods that share the growing space with shade-tolerant hardwood and conifer species.

9.2.4.3 Mature Forest Stage

Following a catastrophic fire, as a forest develops toward the mature stage, the main canopy is still dominated by shade-intolerant hardwood species, mainly paper birch [17], which can live longer than trembling aspen (*Populus tremuloides Michx.*). In Quebec, trembling aspen generally begins to senescence around 60 years [23]. Understory advanced regeneration is mainly composed of balsam fir [17], but red spruce and white spruce can gradually increase in abundance, particularly on deadwood seedbeds. Understory conifer development results in a two-aged stand structure (Table 9.2 and Figure 9.6c).

If the original disturbance is a catastrophic outbreak or windthrow, this stage is characterized by a mix of shade-intolerant hardwood species that share space with shade-tolerant hardwood and conifer species. Although shade-tolerant conifers are generally present in the understory, low light availability limits their development [24, 25]. In these shady conditions, advanced regeneration adopts an umbrella shape that maximizes light interception [26]. Red spruce is particularly well adapted to such low light conditions [26]; it can resume vigorous growth following release, even after decades of suppression [27].

9.2.4.4 Old Forest Stage

Canopy gaps created by past aspen mortality and birch senescence favor the development of conifer advance regeneration, the establishment of new birch and maple seedlings, and stand structure diversification [28]. Older multi-aged stands are characterized by a temperate mixedwood canopy with scattered natural gaps. Gap dynamics are promoted by the presence of senescent or dead short-lived fir that are prone to insects, decay, and windthrow [19]. Canopy gaps allow shade-tolerant conifer acclimation and regrowth [25], as well as yellow birch and maple seedling establishment [28].

The dynamics of stands aged 250 years or more are mainly governed by small- and moderate-scale disturbances that create complex composition and structures as well as spatial heterogeneity (Table 9.2 and Figure 9.6d, e). At this developmental stage, temperate mixedwood forests possess the greatest biodiversity, due to the variety of microhabitats provided by various ecological attributes (e.g. large live trees with dying crowns and cavities, big snags, and downed logs). Veteran trees are present and provide an invaluable seed source [6, 7].

9.3 An Ecological Silvicultural System for Yellow Birch–Conifer Mixedwoods

9.3.1 Disturbance and Legacy Creation

Given the overabundance of major disturbances caused by forest management (e.g. clear-cuts) and of young forests in modern compared to preindustrial landscapes [10, 11, 14], there is no need to promote this stage in large areas at the landscape scale. Instead, an ecological

Table 9.3 Ecological silvicultural systems for yellow birch–conifer mixedwoods.

Developmental stage	Age[a] (years)	Objective	Potential actions
Young	10–80	Promoting diversity and late-successional species; accelerating the self-thinning process.	• Mixed precommercial thinning • Crop-tree release preferred when species with poor natural pruning (like yellow birch) are present. • Targeted partial harvests with commercial thinning, depending on stand age and maturity
Mature	81–200	Accelerating natural succession by increasing the proportion of conifers and late-successional species.	• Targeted partial harvests with commercial thinning. • Establishment and removal cuts to either establish or release advance regeneration with regular or irregular shelterwood systems. • Retention of legacy trees: seed trees of rare species, habitat trees, and snags. • Ground disturbance to facilitate the germination of long-lived shade-tolerant trees. • Enrichment planting if needed.
Old	>200[b]	Promoting the natural succession and structural complexity.	• Creation of light to moderate disturbances to foster the establishment of new cohorts of regeneration and multi-aged structures with irregular shelterwood or selection systems. • Retention of legacy trees: seed trees of rare species, habitat trees, and snags. • Ground disturbance to facilitate the germination of long-lived shade-tolerant trees. • Enrichment planting if needed.

[a] Approximately.
[b] *Source:* Based on [15].

silvicultural system in managed temperate mixedwoods should aim at accelerating the natural succession by encouraging composition comprised of late-successional species and enhancing structural complexity (Table 9.3). Nevertheless, the disturbance and legacy creation stage happening at gap scales can be incorporated in silvicultural systems in older forest stages as a way to transition structure and composition toward more natural conditions in postindustrial ecosystems.

9.3.2 Young Forest Stage

In young even-aged stands, natural development can be facilitated by accelerating the self-thinning process and by using precommercial thinning to improve the development of understory conifers and other late-successional species (Figure 9.7a). In very young stands (10–20 years) dominated by shade-intolerant species such as aspen, which shelter a conifer

Figure 9.7 Silvicultural treatments that enhance compositional and structural complexity: (a) precommercial thinning in a young aspen-dominated stand; (b) mixed precommercial thinning in a young temperate mixedwood stand; (c) conifer release after a first removal cut in a mature aspen–conifer stand; (d) final removal in a regular shelterwood system in a mature aspen–conifer stand; (e) continuous cover irregular shelterwood in an old balsam fir–yellow birch stand; and (f) hybrid single-tree and small group selection cutting in an old yellow birch–conifer stand. *Source:* (a) Marcel Prévost, (b, e, f) Patricia Raymond, (c, d) Daniel Dumais.

regeneration understory, this tending treatment could improve aspen and conifer growth, as well as spruce survival [29, 30]. The complementary use of growing space by these species could also facilitate the natural pruning of crop trees [31] and decrease browsing damage [30]. Mixed precommercial thinning (Figure 9.7b), a tending treatment aiming at a mixedwood production, can also help increase the proportion and development of late-successional species [32, 33]. As stand ages, commercial thinning can help to further achieve these compositional objectives.

9.3.3 Mature Forest Stage

At this stage, ecological silviculture should aim to accelerate natural succession by increasing the proportion of conifers and late-successional species. This can be achieved with commercial thinning to remove short-lived, shade-intolerant species and increase the proportion of conifers. Shelterwood methods (regular and irregular with retention) can also be appropriate, for example when conifers are present in the understory of stands dominated by shade-intolerant species. This method was experimented with success in eastern Quebec, where establishment cuts (35–50% of basal area harvested) in aspen stands facilitated the development of conifer advance regeneration while limiting aspen suckering (Figure 9.7c) [34, 35]. Ten years later, final removal further released the conifers and contributed to hastening the natural stand succession toward mixedwood production (Figure 9.7d) [36].

9.3.4 Old Forest Stage

In old-growth stands, actions for ecological silviculture will aim at promoting diversity and structural complexity, by ensuring both the establishment of a regenerating cohort after each entry and the retention of biological legacies. A high level of structural complexity will help maintain biodiversity [37]. Conserving seed trees of rare species should help maintain diversity in the long term [8, 26]. Silvicultural systems inspired by natural disturbance dynamics to maintain stand structure, composition, and functions should be promoted. These include partial cuts such as hybrid single-tree and small group selection, and irregular shelterwood methods [12, 28, 38]. For instance, continuous cover irregular shelterwood applied in balsam fir–yellow birch stands can emulate partial disturbances by spruce budworm and serve as an alternative to clear-cut systems [9] (Figure 9.7e). This method promotes the establishment of a diverse cohort of tree species and helps maintain an irregular stand structure, snag abundance, and tree microhabitat abundance and diversity [9, 39].

Both irregular shelterwood and hybrid single-tree and small group selection cutting (Figure 9.7f) generate heterogeneity in understory conditions creating a variety of ecological niches that foster species diversity [7, 28]. This is especially the case when harvested trees create gaps that are 100–300 m^2 in size (no larger than about one tree height) where red spruce can regenerate and grow along with balsam fir and yellow birch [25, 28]. Also, such small harvest gaps limit raspberry development, providing better resource availability to new seedlings of desired tree species [28].

Ground disturbance can also promote yellow birch regeneration, through either fall logging operations before the snow accumulates (passive scarification) [2, 7] or deliberate scarification with machinery when snowless operations are not possible [4]. Nevertheless, red spruce regeneration from seeds is a very long process [40, 41] and with no guarantee of success, even when stands contain reasonably high proportions of seed bearers [7]. For this reason, enrichment planting is becoming a popular means to maintain red spruce and cedar in yellow birch–conifer mixedwoods [42–44].

Finally, any prescription aiming for ecological silviculture in older stands should include retention guidelines for seed bearers of rare species, very large trees, wildlife trees, and standing dead trees, since these elements are highly valuable for the ecosystems [22] and tend to decrease with time in forests managed with partial cuts [45].

9.4 Climate Change Considerations

Ecological silviculture models based on the knowledge of past natural dynamics have the caveat of relying on knowledge from times when the climate was different from today (e.g. preindustrial data [10, 11, 14]). This approach also ignores the confounding effects of anthropic disturbances that occurred during the two last centuries in the temperate mixedwood forest. Indeed, the dynamics that led to current forest conditions likely differ from those that will drive future stand dynamics, in which climate change will also increase variability.

Nevertheless, since ecological silvicultural models aim to increase structural complexity and species diversity to levels that would naturally occur during succession, this approach remains relevant to promote stand resistance and resilience, two strategies recommended to promote adaptation to climate change [46]. In fact, structural and compositional diversity are associated with higher resilience and drive functional diversity [37], a "trait diversity insurance" for forests facing an uncertain future [47]. Ecological silviculture in these ecosystems can increase functional diversity by promoting stand structural complexity and mixing species from different functional groups, such as hardwoods and conifers [37].

In practice, however, late-successional conifers are often difficult to regenerate in temperate mixedwood forests [48]. Silvicultural systems designed to satisfy their ecological requirements can contribute to maintaining the character and diversity of mixedwoods [48]. Enrichment planting may be a valuable tool to maintain late-successional conifers in yellow birch–conifer ecosystems [42–44], while also increasing adaptive capacity in these ecosystems. In particular, assisted migration using seedlings from meridional provenances as part of these plantings could further help to maintain species in current ecosystems and increase the range of climate responses present. If used to incorporate species that are expected to better support future climatic conditions, this approach could also facilitate ecosystem adaptation [46].

9.5 Summary

Since natural dynamics for the mixedwood ecosystem result in an increase over time of conifers, late-successional species, and structural complexity, an ecological silvicultural system should aim to enhance these important attributes. This could be particularly helpful to promote forest resistance and resilience in the context of climate change. In young stands, stand development can be facilitated by accelerating the self-thinning process and increasing the proportion of conifers and late-successional species. This can be achieved with mixed-precommercial thinning, crop-tree release, or commercial thinning, depending on stand composition and age. In mature stands, commercial thinning and shelterwood methods (regular or irregular) with retention can further hasten the natural succession process by increasing the regeneration and the representation of conifers and late-successional species. Ground disturbance can be needed to facilitate the germination of long-lived shade-tolerant trees. In old-growth stands, diversity and structural complexity can be promoted by ensuring that a regenerating cohort becomes established at each entry and that biological legacies of rare seed trees, habitat trees, and snags are retained. Finally, enrichment planting and assisted migration are among the innovative tools that may be used to realize an ecological and adaptive silviculture in yellow birch–conifer ecosystems.

Acknowledgments

We thank Emilie Champagne for her valuable comments on the preliminary version of this chapter, and Denise Tousignant for her assistance with English editing.

References

1 Saucier, J.P., Grondin, P., Brisson, J. et al. (2009). Écologie forestière. In: *Manuel de foresterie, édition revue et augmentée* (ed. Ordre des ingénieurs forestiers du Québec), 165–315. Québec: Éditions Multimondes.

2 Prévost, M. (2008). Effect of cutting intensity on microenvironmental conditions and regeneration dynamics in yellow birch–conifer stands. *Canadian Journal of Forest Research* 38: 317–330.

3 MRN (Ministère des Ressources naturelles) (2013). *Le guide sylvicole du Québec, Tome 1, Les fondements biologiques de la sylviculture, ouvrage collectif sous la supervision de B. Boulet et M. Huot.* Les Publications du Québec, 1044 p.

4 Prévost, M., Raymond, P., and Lussier, J.M. (2010). Regeneration dynamics after patch cutting and scarification in yellow birch–conifer stands. *Canadian Journal of Forest Research* 40: 357–369.

5 Prévost, M. and Dumais, D. (2013). Decennial growth and mortality following uniform partial cutting in yellow birch–conifer stands. *Canadian Journal of Forest Research* 43: 224–233.

6 Duchesne, L. and Prévost, M. (2013). Canopy disturbance and intertree competition: implications for tree growth and recruitment in two yellow birch–conifer stands in Quebec, Canada. *Journal of Forest Research* 18: 168–178.

7 Raymond, P., Royo, A.A., Prévost, M., and Dumais, D. (2018). Assessing the single-tree and small group selection cutting system as intermediate disturbance to promote regeneration and diversity in temperate mixedwood stands. *Forest Ecology and Management* 430: 21–32.

8 Fortin, M., Bégin, J., and Bélanger, L. (2003). Évolution de la structure diamétrale et de la composition des peuplements mixtes de sapin baumier et d'épinette rouge de la forêt primitive après une coupe à diamètre limite sur l'Aire d'observation de la rivière Ouareau. *Canadian Journal of Forest Research* 33: 691–704.

9 Raymond, P. and Bédard, S. (2017). The irregular shelterwood system as an alternative to clearcutting to achieve compositional and structural objectives in temperate mixedwood stands. *Forest Ecology and Management* 398: 91–100.

10 Loo, J. and Ives, N. (2003). The Acadian forest: historical condition and human impacts. *The Forestry Chronicle* 79: 462–474.

11 Barrette, M. and Bélanger, L. (2007). Reconstitution historique du paysage préindustriel de la région écologique des hautes collines du Bas-Saint-Maurice. *Canadian Journal of Forest Research* 37: 1147–1160.

12 Raymond, P., Bédard, S., Roy, V. et al. (2009). The irregular shelterwood system: review, classification, and potential application to forests affected by partial disturbances. *Journal of Forestry* 107: 405–413.

13 Grenier, D.J., Bergeron, Y., Kneeshaw, D., and Gauthier, S. (2005). Fire frequency for the transitional mixedwood forest of Timiskaming, Quebec, Canada. *Canadian Journal of Forest Research* 35: 656–666.

14 Alvarez, E., Bélanger, L., Archambault, L., and Raulier, F. (2011). Portrait préindustriel dans un contexte de grande variabilité naturelle : une étude de cas dans le centre du Québec (Canada). *The Forestry Chronicle* 87: 612–624.

15 Boucher, Y., Bouchard, M., Grondin, P., and Tardif, P. (2011). Le registre des états de référence : intégration des connaissances sur la structure, la composition et la dynamique des paysages forestiers naturels du Québec méridional. Ministère des Ressources naturelles et de la Faune, Direction de la recherche forestière. Mémoire de recherche forestière n° 161. 41 p.

16 Wotton, B.M., Flannigan, M.D., and Marshall, G.A. (2017). Potential climate change impacts on fire intensity and key wildfire suppression thresholds in Canada. *Environmental Research Letters* 12: 095003.

17 Hébert, R. (2003). Are clearcuts appropriate for the mixed forest of Québec? *The Forestry Chronicle* 79: 664–671.

18 Bouchard, M., Kneeshaw, D., and Bergeron, Y. (2006). Tree recruitment pulses and long-term species coexistence in mixed forests of western Québec. *Ecoscience* 13: 82–88.

19 Kneeshaw, D.D. and Prévost, M. (2007). Natural canopy gap disturbances and their role in maintaining mixed-species forests of central Quebec, Canada. *Canadian Journal of Forest Research* 37: 1534–1544.

20 Barrette, M., Dumais, D., Auger, I. et al. (2020). Naturalness assessment performed using forestry maps to validate forest management sustainability. *Ecological Indicators* 119: 106832.

21 Harmon, M.E., Franklin, J.F., Swanson, F.J. et al. (1986). Ecology of coarse woody debris in temperate ecosystems. *Advances in Ecological Research* 15: 133–302.

22 Larrieu, L., Paillet, Y., Winter, S. et al. (2018). Tree related microhabitats in temperate and Mediterranean European forests: a hierarchical typology for inventory standardization. *Ecological Indicators* 84: 194–207.

23 Pothier, D., Raulier, F., and Riopel, M. (2004). Ageing and decline of trembling aspen stands in Quebec. *Canadian Journal of Forest Research* 34: 1251–1258.

24 Dumais, D. and Prévost, M. (2008). Ecophysiology and growth of advance red spruce and balsam fir regeneration after partial cutting in yellow birch–conifer stands. *Tree Physiology* 28: 1221–1229.

25 Dumais, D. and Prévost, M. (2014). Physiology and growth of advance *Picea rubens* and *Abies balsamea* regeneration following different canopy openings. *Tree Physiology* 34: 194–204.

26 Dumais, D. and Prévost, M. (2007). Management for red spruce conservation in Québec: the importance of some physiological and ecological characteristics—a review. *The Forestry Chronicle* 83: 378–392.

27 Wu, X., McCormick, J.F., and Busing, R.T. (1999). Growth pattern of *Picea rubens* prior to canopy recruitment. *Plant Ecology* 140: 245–253.

28 Prévost, M. and Charette, L. (2015). Selection cutting in a yellow birch–conifer stand, in Quebec, Canada: comparing the single-tree and two hybrid methods using different sizes of canopy opening. *Forest Ecology and Management* 357: 195–205.

29 Prévost, M. and Gauthier, M.M. (2012). Precommercial thinning increases growth of overstory aspen and understory balsam fir in a boreal mixedwood stand. *Forest Ecology and Management* 278: 17–26.

30 Prévost, M. and Charette, L. (2017). Precommercial thinning of overtopping aspen to release coniferous regeneration in a boreal mixedwood stand. *The Forestry Chronicle* 93: 259–270.

31 Puhlick, J.J., Kuehne, C., and Kenefic, L.S. (2019). Crop tree growth response and quality after silvicultural rehabilitation of cutover stands. *Canadian Journal of Forest Research* 49: 670–679.

32 Kenefic, L.S., Bataineh, M., Wilson, J.S. et al. (2014). Silvicultural rehabilitation of cutover mixedwood stands. *Journal of Forestry* 112: 261–271.

33 Champagne, E., Dumais, D., Raymond, P. (2023). Precommercial thinning increased diameter growth while maintaining mixedwood stands composition, 15 years after treatment. *Canadian Journal of Forest Research* **53**: 255–270. https://doi.org/10.1139/cjfr-2022-0256.

34 Prévost, M. and DeBlois, J. (2014). Shelterwood cutting to release coniferous advance growth and limit aspen sucker development in a boreal mixedwood stand. *Forest Ecology and Management* 323: 148–157.

35 Prévost, M. and Dumais, D. (2014). Shelterwood cutting in a boreal mixedwood stand: 10-year effects of the establishment cut on growth and mortality of merchantable residual trees. *Forest Ecology and Management* 330: 94–104.

36 Prévost, M. and Charette, L. (2019). Shelterwood cutting in a boreal mixedwood stand: 5-year effects of the final cut on development of aspen suckers and released conifers. *The Forestry Chronicle* 95: 113–123.

37 Thom, D., Taylor, A.R., Seidl, R. et al. (2021). Forest structure, not climate, is the primary driver of functional diversity in northeastern North America. *Science of The Total Environment* 762: 143070.

38 D'Amato, A.W., Raymond, P., and Fraver, S. (2018). Old-growth disturbance dynamics and associated ecological silviculture for forests in Northeastern North America. In: *Ecology and Recovery of Eastern Old-Growth Forests* (ed. A.M. Barton and W.S. Keeton), 99–118. Springer.

39 Martin, M. and Raymond, P. (2019). Assessing tree-related microhabitat retention according to a harvest gradient using tree-defect surveys as proxies in Eastern Canadian mixedwood forests. *The Forestry Chronicle* 95: 157–170.

40 Fraver, S. and White, A.S. (2005). Disturbance dynamics of old-growth *Picea rubens* forests of northern Maine. *Journal of Vegetation Science* 16: 597–610.

41 Dumais, D. and Prévost, M. (2016). Germination and establishment of natural red spruce (*Picea rubens*) seedlings in silvicultural gaps of different sizes. *The Forestry Chronicle* 92: 90–100.

42 Dumais, D., Larouche, C., Raymond, P. et al. (2019). Survival and growth dynamics of red spruce seedlings planted under different forest cover densities and types. *New Forests* 50: 573–592.

43 Dumais, D., Raymond, P., and Prévost, M. (2020). Eight-year ecophysiology and growth dynamics of *Picea rubens* seedlings planted in harvest gaps of partially cut stands. *Forest Ecological Management* 478: 118514.

44 Bourque, C., Dumais, D., Ruel, J.C. et al. (2022). How do advance regeneration and planted seedlings of *Thuja occidentalis* and *Picea rubens* acclimate under a first irregular shelterwood cut? *Canadian Journal of Forest Research* 52: 1412–1422.

45 Angers, V.A., Messier, C., Beaudet, M., and Leduc, A. (2005). Comparing composition and structure in old-growth and harvested (selection and diameter-limit cuts) northern hardwood stands of Quebec. *Forest Ecology and Management* 217: 275–293.

46 Millar, C.I., Stephenson, N.L., and Stephens, S.L. (2007). Climate change and forests of the future: managing in the face of uncertainty. *Ecological Applications* 17: 2145–2151.

47 Messier, C., Bauhus, J., Doyon, F. et al. (2019). The functional complex network approach to foster forest resilience to global changes. *Forest Ecosystems* 6: 1–16.

48 Kenefic, L.S., Kabrick, J.M., Knapp, B.O. et al. (2021). Mixedwood silviculture in North America: the science and art of managing for complex, multi-species temperate forests. *Canadian Journal of Forest Research* 51: 921–934.

10

Ecological Silviculture of Black Spruce in Canadian Boreal Forests

Miguel Montoro Girona[1], Martin Alcala Pajares[2], and Maxence Martin[2]

[1] *Groupe de Recherche en Écologie de la MRC Abitbi (GREMA). Institute of Forest Research, Université du Québec en Abitibi-Témiscamingue, Amos, Quebec, Canada*
[2] *Universidad de Huelva, dr. Cantero Cuadrado 6, Huelva 21004, Spain*

10.1 Introduction

Black spruce (*Picea mariana* (Mill.) BSP) is one of the most important commercial species for the forest industry in the North America because of its wide distribution and excellent wood properties [1]. Black spruce forests cover 41 million ha in Québec, Canada alone, representing 17% of the Canadian boreal forest area. Its high-quality fiber has made black spruce the main species for the Canadian pulp and paper industry and a major wood type used in the United States [2]. This chapter provides an overview of the black spruce forests in eastern Canada and reviews ecological silviculture systems developed through a better understanding of this ecosystem.

10.2 Characteristics of Black Spruce Forests

10.2.1 Distribution and Composition

The native range of black spruce in North America extends from 40° to 70°N and 53° to 160°W [3]. In Canada, it covers approximately 365 million ha, corresponding to 62% of the total forest area, and is present in almost all the territory of the Canadian eastern provinces (i.e. Ontario, Quebec, Newfoundland and Labrador, New Brunswick, Prince Edward Island, and Nova Scotia), finding its northernmost extent in eastern Canada at the tree line located in northern Quebec (Figure 10.1).

Black spruce forests occur as pure or mixed stands (Figure 10.2); for example, black spruce can co-occur with white spruce (*Picea glauca* (Moench) Voss.) in the northern parts of its range near tree line, e.g. in the Yukon Territory and Alaska, where it dominates the north-facing slopes of mountains in areas of permafrost and poor drainage [5]. A typical forest of black spruce on drier, sandy, and rocky upland and well-drained sites tends to be associated with jack pine (*Pinus banksiana* Lamb.) [4]; in contrast, black spruce dominates in lowlands and poorly drained areas in eastern Canada. In eastern Canada, black spruce is often found with balsam fir

Ecological Silvicultural Systems: Exemplary Models for Sustainable Forest Management,
First Edition. Edited by Brian J. Palik and Anthony W. D'Amato.

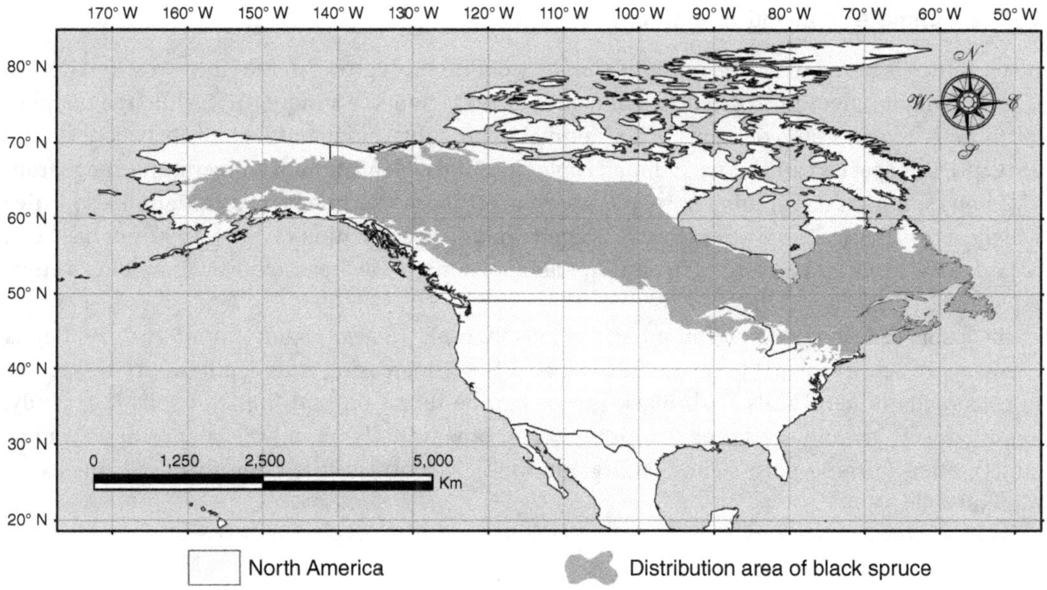

Figure 10.1 Distribution of black spruce forests in North America. *Source:* From [4]/U.S. Department of Agriculture/Public Domain.

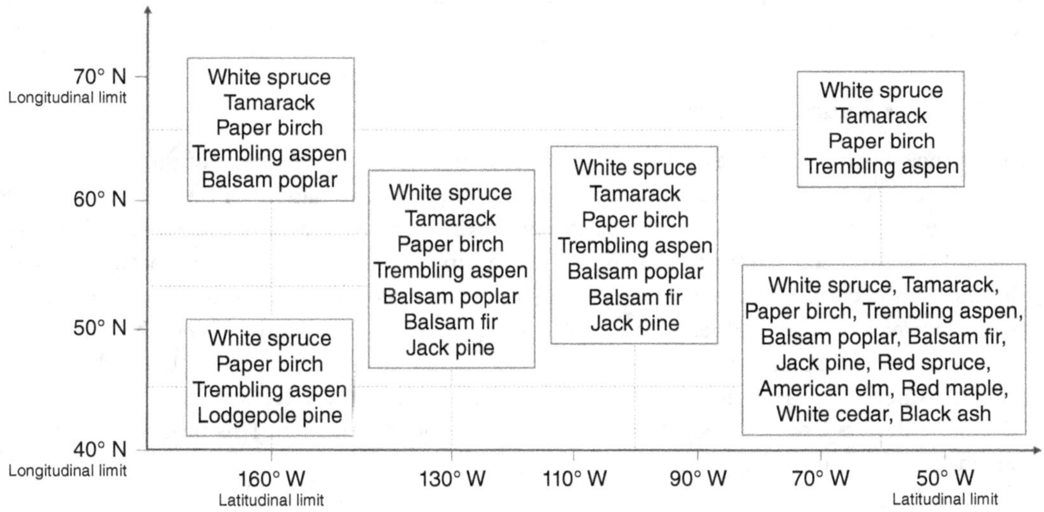

Figure 10.2 Geographic distribution of species associated with black spruce forests. *Source:* National Forest Database of Canada (http://nfdp.ccfm.org).

(*Abies balsamea* Mill.) in late-successional stands on soils having sufficient drainage [6]. It can also grow alongside tamarack (*Larix laricina* (Du Roi) K. Koch) in northern Québec, near the northern tree line. Paper birch (*Betula papyrifera* Marsh.) can also be found in these locations as an associated species, where it tends to replace black spruce after fires. Trembling aspen (*Populus tremuloides* Michx.) and balsam poplar (*Populus balsamifera* L.) are also associated with black spruce forests in northern locations.

10.2.2 Climate, Soil, and Topography

Black spruce is found in a wide range of climatic conditions. It grows in the Northwest Territories, where daylight ranges from 4–5 hours in winter to 19–21 hours in summer [7]. Although net photosynthesis reaches 90% or more of its optimal value when temperatures are between 15 and 25 °C [8], this species can survive in much cooler locations. Mean annual temperatures range from 7 °C in its southern distribution to −11 °C near the northern tree line [9]. Precipitation also varies widely across the black spruce range. Maximum annual precipitation of 1100–1600 mm has been recorded for black spruce forests in eastern Labrador's maritime areas and Newfoundland's afforested areas [10].

Black spruce is generally found on acidic soils with pH values ranging from 5 to 7. Peatlands can be more acidic; soil pH at the surface can be 3.5–4.5, increasing with depth (to pH 6–7) closer to calcareous mineral soils [10]. Black spruce can be found on both mineral soils (i.e. sandy, loamy, clayey, gravelly, and rocky substrates) and organic soils (mainly peatlands and swamplands), often forming pure stands on organic soils. In contrast, it is often in mixed forests on mineral soils.

10.3 Black Spruce Forest Types

Given the wide variety of geomorphological and climatic conditions in which black spruce occurs and the various species that it co-occurs with across its range, black spruce forests have been divided into subtypes (Figure 10.2).

10.3.1 Black Spruce–Feathermoss Forests

Black spruce–feathermoss forests are found in the central and southern parts of the boreal forest region (Figure 10.3). This subtype is characterized by major accumulations of humus and a closed canopy overstory combined with an almost continuous cover of feathermoss species and *Sphagnum* mosses [11]. In poorly drained environments having an extended interval between fires, paludification occurs, which favors the establishment of *Sphagnum* spp. and ericaceous shrubs over feathermoss species and leads to the black spruce–*Sphagnum* subtype [12]. The expansion and coverage of *Sphagnum* mosses reduce water saturation fluctuations in the organic layer and allow the water table to reach the organic layer. Consequently, the soil becomes moister, colder, and more nutrient-poor, and roots can no longer reach the mineral soil. This process leads to an open and less productive forest [13].

10.3.2 Black Spruce–Lichen Forests

Black spruce–lichen forests can occupy former black spruce–feathermoss stands after a natural disturbance, such as repeated wildfires, outbreaks of spruce budworm (*Choristoneura fumiferana* Clemens), or European sawfly (*Gilpinia hercyniae* Hartig), or logging [14]. Following disturbance, the bare soil is first colonized by ericaceous or moss species and then by lichen species, such as *Cladonia* spp. Over time, the expansion of lichen can cover up to 40% of the surface. Black spruce–lichen forests are characterized by scarce tree cover at <25% coverage. Because of their low tree density, black spruce–lichen forests are usually considered to have low productivity and low economic potential and therefore are not the focus of commercial forestry activities.

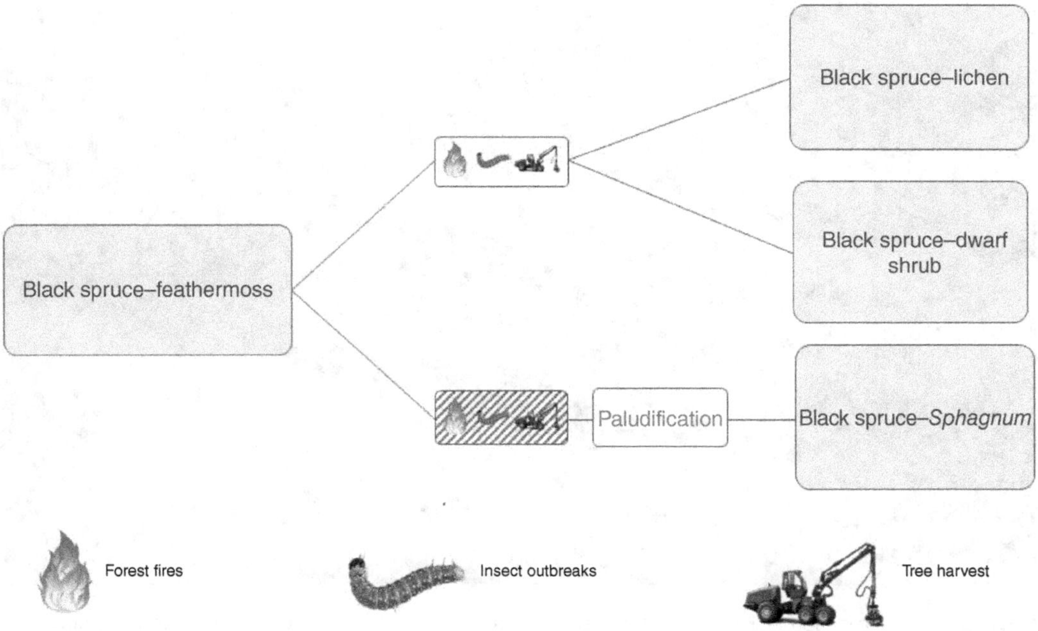

Figure 10.3 Development of the black spruce–feathermoss forest type according to different disturbance scenarios.

10.3.3 Black Spruce–Dwarf Shrub Forests

The black spruce–dwarf shrub subtype is found in the central and southern regions of the boreal forest. In this subtype, natural disturbances favor dominance of Ericaceae and other species, such as *Kalmia* spp., *Callunas* spp., *Vaccinium* spp., and *Gaultheria* spp., [15] because the canopy openings create a more suitable environment for these plants. Ericaceae are considered important nutrient immobilizers during postfire nutrient flushes, and their presence has been linked to increases in the proportion of litter containing phenolic acids, which are considered allelochemicals that can inhibit conifer growth [15].

10.4 Developmental Model for Black Spruce Forests

Black spruce is a fire-adapted species having semi-serotinous cones, but it can also regenerate under its own cover because of its layering ability and shade tolerance. This species is often found in early successional stands (Figure 10.2), but its abundance is highly dependent on the characteristics of the previous stand due to its limited dispersal ability. If the previous stand comprised a mix of black spruce and fire-intolerant species, black spruce is generally the dominant species in the new stand (Figure 10.4). On the other hand, if the previous stand was composed of black spruce with fire-adapted species, such as jack pine or trembling aspen, a similar composition is maintained postfire [16]. However, in the case of short intervals between fires, the proportion of black spruce may decrease in favor of species better adapted to these conditions, notably jack pine but also *Populus* spp. and *Betula* spp. [17]. Two fires occurring in a period insufficient for black spruce to reach sexual maturity can lead to black spruce regeneration failure.

Figure 10.4 Even-aged black spruce stands regenerated after a fire in the Mont-Valin region, Québec, Canada. *Source:* Miguel Montoro Girona.

A wildfire creates a significant amount of deadwood, both standing and downed, at the early succession stage (Figure 10.5). As new black spruce seedlings establish, this deadwood decomposes over the next few decades. Black spruce recruitment occurs within three years of a fire (Figure 10.6) [18]. The next 70 years are followed by successive phases of stem exclusion, stand growth, and self-thinning within an even-aged stand [18]. Black spruce density will gradually decrease during this period. Given the small size of dying black spruce and the decay process of trees killed by fire, the volume of deadwood is very low. After 50 years postfire, black spruce stands often show high-density regeneration in the understory, possibly by early layering [19].

The next stage is canopy breakup resulting from the senescence of the postfire cohort; this marks the beginning of the old-growth stage from an ecological point of view. This *early old-growth forest* can begin at very different ages, typically between 70- and 150-years postfire, depending on stand characteristics and the specific natural disturbances, such as wind or insects. Black spruce regeneration in the understory will progressively start to replace dead trees, and tree height will reach a plateau. As black spruce longevity can be 250–300 years, the complete replacement of the postfire cohort requires more than two centuries after the last fire. Thus, it is possible to observe stands where the canopy remains largely dominated by a low density of large black spruce over 200 years old. If shade-intolerant species are present in a mixture with black spruce, they will gradually disappear, except if sufficiently large gaps occur. Other shade-tolerant species, such as balsam fir, may begin to colonize the stand depending on abiotic characteristics and the presence of seed trees. This stage sees a gradual increase in the volume of standing or downed deadwood and in the density of habitat trees (i.e. living or dead trees that create ecological niches due to structural features, such as cavities, wounds, rotten or dead parts, wood or bark defects, and epiphytes). The peak in habitat tree density and diversity is observed around 200 years after a fire because of the presence of remnant large and old spruce from the postfire cohort [20].

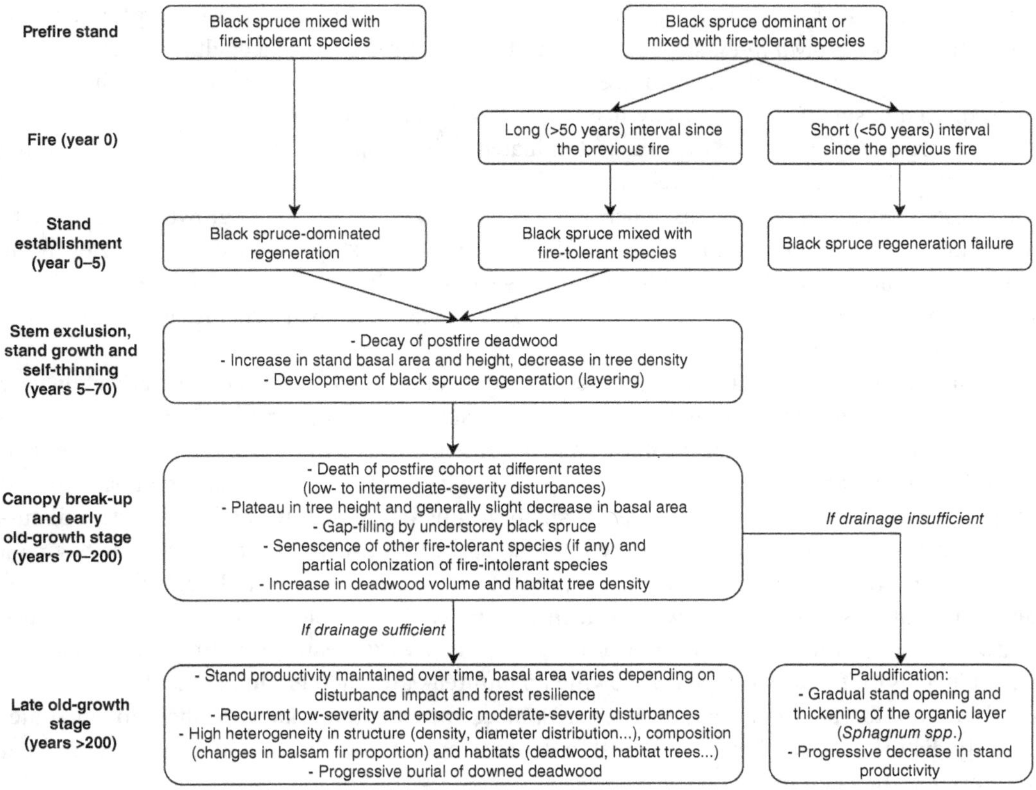

Figure 10.5 Summary of the major developmental stages of black spruce-dominated boreal stands and the observed changes in structure, habitat, and dynamics.

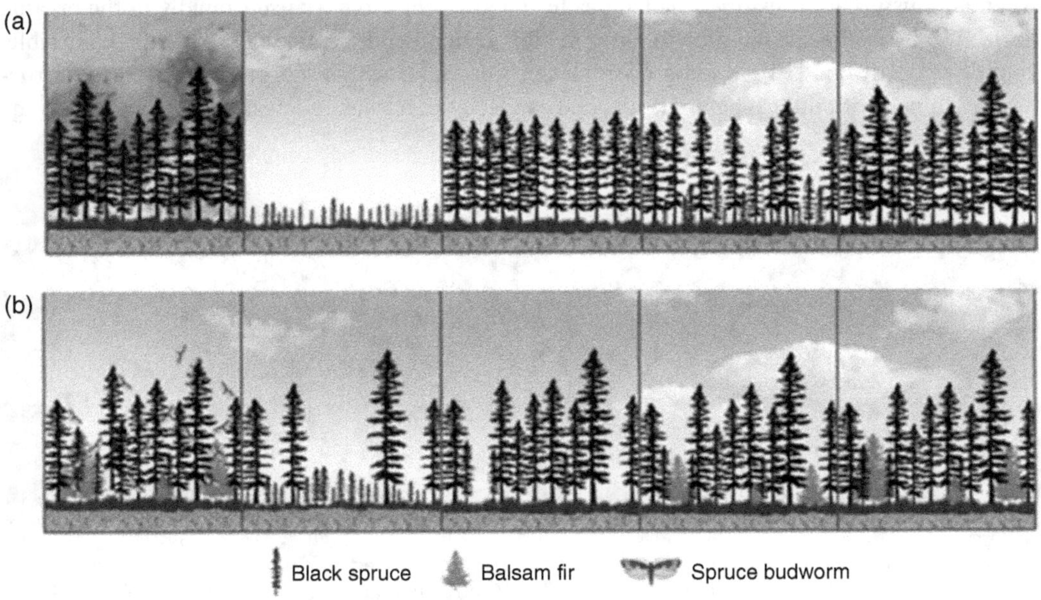

Figure 10.6 Successional stages of black spruce forests: (a) after forest fire and (b) after a spruce budworm outbreak.

Once most of the trees from the postfire cohort have been replaced by new cohort trees, stands enter the *late old-growth forest* stage. These forests are highly uneven-aged, although they do not necessarily show an irregular structure, e.g. with a reverse J diameter distribution, which is generally expected in such forests. Black spruce forests having a higher proportion of balsam fir tend to be more irregularly structured than those dominated by black spruce, possibly because of layering regeneration that inhibits spruce regeneration until the parent trees have died [21]. As the age of the forest exceeds that of the trees, it becomes difficult to study the late old-growth stage, thus in-depth knowledge of these old forests is lacking. However, this stage can last for centuries in the absence of a major disturbance and in soils having sufficient drainage. In cases of soils with insufficient drainage, paludification begins, leading to an increasingly open structure with *Sphagnum*-dominated vegetation.

In sufficiently drained black spruce forests, the late old-growth stage is characterized by very high structural diversity (e.g. heterogeneity of density, canopy openness, diameter structure) because of the cumulated impacts of low- and intermediate-severity secondary disturbances and the subsequent stand resilience. Low-severity disturbances create gaps, resulting in background mortality that allows the regeneration of small groups of trees. Intermediate-severity disturbances are more episodic and generally relate to spruce budworm outbreaks. Stands having a higher proportion of balsam fir are more prone to such disturbances, although black spruce-dominated stands may also show significant mortality. The species-specific nature of spruce budworm means that the proportion of balsam fir can vary greatly over time in a stand; for example, balsam fir can disappear following an outbreak and then gradually recolonize the site until a new outbreak occurs (Figure 10.7) [22]. Black spruce is remarkably resilient to moderate-severity disturbances and can reestablish a dense canopy even when such events are repeated several times in a century.

The high structural diversity of late old-growth forests also results in significant heterogeneity in deadwood volume and habitat tree density among stands. Two late old-growth black spruce forests of similar age can be characterized by different levels of structural diversity because of differing secondary disturbance histories. Fallen deadwood also tends to bury itself rapidly in the organic horizon, causing the late old-growth forest to often contain an exceptionally large, albeit invisible, volume of deadwood [23]. In old-growth black spruce forests, their apparent simplicity of structure combined with little apparent deadwood can lead to a strong underestimation of the true age of these stands.

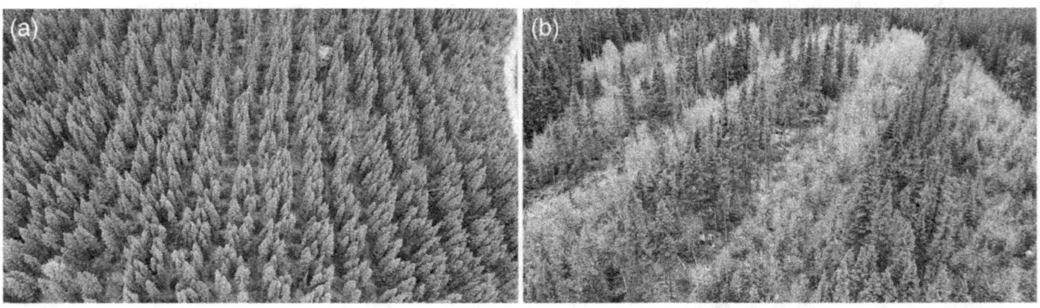

Figure 10.7 Examples of experimental shelterwood treatments undertaken in 2003 by the Canadian Forest Service in black spruce even-aged stands in the Mont-Valin and North Shore regions of Québec, where (a) represent a mini-strip shelterwood and (b) correspond to a seed tree methods. Credit: Ruben Leroy and Miguel Montoro Girona.

10.5 Emulating Natural Dynamics of Black Spurce Forests with Sivliculture

10.5.1 Clear-cutting as Tool to Emulate High-severity Wildfire

Clear-cutting is the traditional method forest managers have applied to black spruce forests over the twentieth century. The historical importance of clear-cutting was related to the need to create large and even-aged stands to increase the efficiency and wood output of mechanized harvesting in response to an increasing demand for pulpwood [24]. Although other types of silvicultural treatments have been applied in the past 20 years, clear-cutting remains the predominant silvicultural method in over 85% of the harvested area in Canada in 2019.

Clear-cutting was considered a fire-emulative treatment, as both forest fires and clear-cuts remove much of the canopy at once. However, fires and clear-cuts differ in several aspects, including their respective frequency, intensity, and severity. Moreover, postfire species composition, stand age distribution, snag abundance, and the presence of species associated with old-growth black spruce forests can also differ between these two disturbances [25]. Overall, the return period for clear-cuts is often much shorter than the historical return period for severe wildfires in the boreal landscapes of eastern Canada that significantly reduces the part of black spruce stands with an age exceeding clear-cut return period. For these reasons, this almost exclusive use of clear-cuts has been increasingly questioned over the past decades; for example, they cause a significant loss and fragmentation of Canada's old-growth boreal forests and, more generally, have favored the increased simplification of managed landscapes.

Clear-cuts remain a relevant option silvicultural option on the condition that they better emulate the impact of fire and that they represent only a minor part of the harvest. The proportion of clear-cuts within harvesting activities should depend on the historical fire cycle; the shorter the fire cycle, the greater the proportion of clear-cuts, and vice versa. Scarification treatments that reduce organic horizon depth in a fire-like manner, and the use of planting or pre-commercial thinning to favors black spruce, will contribute to maintaining black spruce cover [25].

Strip and patch clear-cuts are variations of traditional clear-cuts that were developed to increase the stocking levels of regeneration by supplying seed from the uncut areas surrounding the harvested areas in black spruce forests [26]. They imitate the patches of forests spared or lightly burned within the area of a wildfire. Strip clear-cuts are also carried out in areas characterized by an abundance of *Sphagnum* and low expectations of harvested volume. Stands are divided into different strips, oriented perpendicular to the wind direction. Strip dimensions may vary, with the recommended width twice the height of the seed-supplying trees [27].

10.5.2 Thinning to Restore Complex Structures

Commercial thinning represents 4% of the total partially harvested area of eastern Canada (period 2008–2018). It is mostly applied to even-aged black spruce stands followed by a final harvest 15–20 years after an initial silvicultural intervention. Thinning removes 30–35% of the basal area, mainly affecting lower-diameter trees. Commercial thinning has shown a positive impact on the radial growth of black spruce stems and roots 10 years after harvest without negative effects on the mechanical properties of the wood [28].

From a natural disturbances perspective, commercial thinning can be used to emulate canopy breakup observed in old even-aged stands (120–160 years), where first cohort trees start to die; however, these treatments are generally applied at a younger age (50–70 years). Commercial thinning can also be used to accelerate forest succession in the youngest spruce stands, aiming to prepare the stand for partial cuts (see next points) rather than clear-cuts. Thinning would then focus on second-growth even-aged forests, which are now very abundant in the boreal landscapes of eastern Canada, and can contribute to restoring the complex age structures observed in preindustrial black spruce forests.

10.5.3 Diameter-Limit Cuts to Emulate Secondary Disturbances of High Severity

The two dominant secondary disturbances in the boreal forests of Eastern Canada, spruce budworm outbreaks and windthrows, are both top-down disturbances, killing the tallest trees first. In patches where these disturbances have been severe, it is common that most of the largest trees died. In this context, the use of silvicultural treatments to release preestablished regeneration such as careful logging around small merchantable stems (CLASS) or harvest with regeneration protection (HARP) [29]. This silvicultural treatment is characterized by the removal of 70–90% of trees having a diameter-at-breast height (DBH) of more than 14–15 cm to protect preexisting regeneration and the smallest merchantable stems. CLASS-HARP approaches reduce overstory competition, and their long-term application can produce stands having diversified age structures [29]. The use of diameter-limit cuts is increasing in Canada [30], but similarly to clear-cuts, they are supposed to represent specific and rather minor cases of the natural dynamics of black spruce forests.

10.5.4 Shelterwoods to Emulate Moderate-Severity Secondary Disturbances

The impact of the widespread use of clear-cutting on stand structure, biodiversity, and sustainability is of growing concern to the public, which has affected forest management practices [31] including greater use of partial cutting-based methods.

The shelterwood system can be considered an emulation of the canopy breakup in even-aged stands or intermediate-severity disturbances in uneven-aged stands. In 2020, the shelterwood system was the main partial cutting treatment applied in eastern Canada, but it reached only 8% of the total harvested area (period 2008–2018) (Figure 10.7). The use of partial cuts is therefore still minor in the boreal landscapes of Eastern Canada, underlining the scarcity of treatments emulating secondary disturbances in current management plans. There are several variants of the shelterwood system used in black spruce forests representing different spatial configurations of the harvest, including strip shelterwood, mini-strip shelterwood, distant selection, and close selection (Figure 10.8). All trees are harvested within the strip used by the machinery and the remaining strips can be left unharvested or partially harvested (e.g. harvest concentrated in small patches or homogeneously distributed). Harvest strips can then be scarified or planted to promote black spruce regeneration [31].

Harvest intensity can range from 30% to 60% of the trees [31]. Generally, the shelterwood system has a positive impact on radial growth of black spruce [32] and has promoted regeneration [33], minimized windthrow risk [34], and improved structural and composition heterogeneity. Early phases of the shelterwood system, especially with high retention levels, show positive results in maintaining species associated with old-growth black spruce forests [35]. Since secondary disturbances are spatially variable in severity, promoting a diversity of levels makes it possible to approach the heterogeneity of natural disturbances.

High retention cuts (60–70% of basal area retained) will continue to be favored because of their greater ability to maintain old-growth conditions. Overall, these treatments appear to be the most effective in conciliating timber harvesting and maintaining the functions of preindustrial forests. However, questions remain about the long-term impact of these treatments when applied to

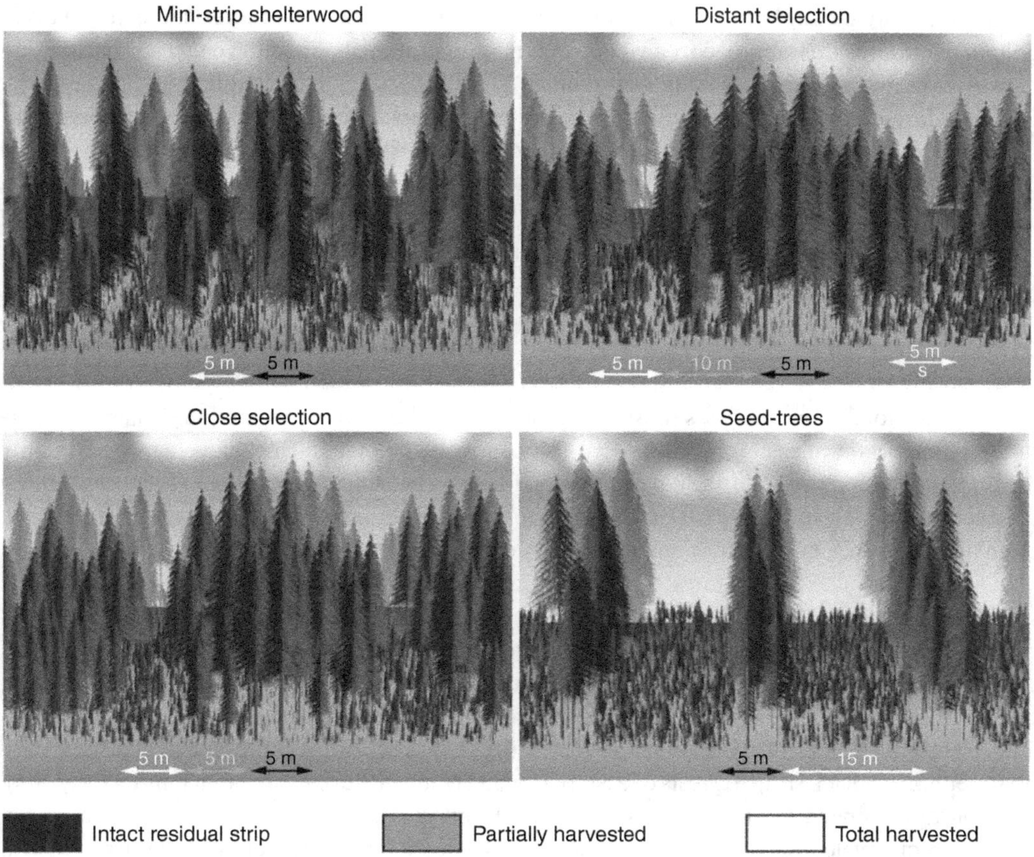

Figure 10.8 Characteristics and spatial patterns of four experimental partial-harvest treatments in boreal forests. The secondary trail is marked *S*. *Source:* From [4]/U.S. Department of Agriculture/Public Domain.

old-growth forests. Successive cuts eventually remove the last strips of old-growth forest, and each strip is harvested every 90–120 years, which may limit the recruitment of attributes found in late old-growth forests (>200 years since the last fire). Retaining a small number of strips out of harvest over the long term, and therefore with old trees and large snags, as well as residual trees in harvested strips is important for maintaining a forest's ecological functionality.

10.5.5 Selection Harvests to Emulate Low-Severity Disturbances and Background Mortality

Selection cutting can be considered as emulative of the low-severity disturbances that drive old-growth forest dynamics. Selection systems comprise 7% of the total partially harvested area of eastern Canada (period 2008–2018). Two subgroups, single-tree selection and group selection are being applied to black spruce stands to develop or maintain uneven-aged stands. Some research has found increased radial growth after applying selection cuts, showing yields higher than observed in stands receiving commercial thinning and surpassed only by the growth response of the shelterwood system. Among both selection cuts, single-tree selection produces slightly better results in radial growth than group selection and a greater diameter growth than patch and strip clear-cuts.

This treatment is certainly the most difficult to apply in black spruce forests due to the low proportion of basal area harvested (<30%) and the small size of boreal trees. While low-severity disturbances are certainly the most common in old-growth black spruce forests, creating a *background mortality*, this approach can be difficult to operationalize effectively. The distribution between low and moderate-severity disturbances in black spruce forests is still poorly understood. It is therefore difficult to determine the proportion of selection cuts that should be favored over shelterwood. At the moment, selection cuts could be favored in the most productive black spruce stands, possibly in mixture with other high-growing species such as white spruce (*Picea glauca* (Moench) Voss). Irregular shelterwoods with a high retention level (60–70% of the basal area remaining) could be then a compromise in less productive stands.

10.6 Summary

Black spruce forests occur across most of the North American boreal biome [36–38] and are particularly important in eastern Canada. Two-thirds of boreal biome in Canada is managed, mainly for timber production [38]. In the past century, forest management's main goal has been timber production, and silvicultural systems designed to optimize timber productivity were put in place, often at the expense of a broader range of ecosystem services.

In most of the black spruce forests in eastern Canada, the clear-cutting system remains the main silvicultural regime. While it may be an appropriate approach in some contexts, it creates fragmented landscapes, promotes young and even-aged stands to the detriment of multi-cohort stand structures, and benefits some commercial species over a more diverse composition. The use of clear-cuts should be based on the historical fire regime, including better emulation of patch sizes and structural legacies of natural disturbance [36]. They would thus represent only a secondary treatment in the black spruce forests of eastern Canada where fire cycles are often long (>150 years).

In contrast, a silvicultural system based on partial cutting is more emulative of the natural model of ecosystem development. Such approaches have been shown to promote residual tree growth, reduce windthrow-related losses, promote regeneration, and help maintain biodiversity. In addition, more diverse silviculture treatments integrating partial cuts in the portfolio of available treatments could increase the adaptive capacity and resilience of black spruce forests to climate change, maintaining a large spectrum of forest compositions and structures at different levels of the landscape [39–41]. The limited knowledge about the secondary disturbance regime in black spruce forests makes it difficult to establish clear guidelines for emulating the natural disturbance regime. Favoring partial cuts (irregular shelterwood and selection cuts), mostly with high retention cuts (60–80%), mixed with a limited area of diameter-limit cuts and clear-cuts currently appears to be the most prudent approach.

References

1 Zhang, S. Y. (tony), & Koubaa, A. 2008. Softwoods of Eastern Canada: Their Silvics, Characteristics, Manufacturing and End-uses. FPInnovations.

2 Zhang, S.Y., Simpson, D., and Kris Morgenstern, E. (1994). *Variation in the Relationship of Wood Density with Growth in 40 Black Spruce (Picea mariana) Families Grown in New Brunswick*. Wood and Fiber Science.

3 Montoro Girona, F. M. (2017). À la recherche de l'aménagement durable en forêt boréale: croissance, mortalité et régénération des pessières noires soumises à différents systèmes sylvicoles (Doctoral dissertation, Université du Québec à Chicoutimi).

4 Heinselman, M.L. (1957). *Silvical Characteristics of Black Spruce (Picea Mariana)*. United States Department of Agriculture.

5 Wirth, C., Lichstein, J.W., Dushoff, J. et al. (2008). White spruce meets black spruce: dispersal, postfire establishment, and growth in a warming climate. *Ecological Monographs* 78 (4): 489–505.

6 Martin, M., Fenton, N., and Morin, H. (2018). Structural diversity and dynamics of boreal old-growth forests case study in Eastern Canada. *Forest Ecology and Management* 44 (15): 125–136.

7 Alexander, M.E., Stocks, B.J., Wotton, M., and Taylor, S.W. (2004). *Characterizing the jack pine-black spruce fuel complex of the International Crown Fire Modelling Experiment (ICFME)*. Canadian Forest Service Publications.

8 Lamhamedi, M.S. and Bernier, P.Y. (1994). Ecophysiology and field performance of black spruce (*Picea mariana*): a review. *Annales Des Sciences Forestières* 51 (6): 517–534.

9 Viereck, L.A. and Johnston, W.F. (1990). *Picea mariana* (Mill.) B.S.P. Black Spruce. In: *Silvics of North America* (ed. R.M. Burns and B.H. Honkala), 227–237. United States Department of Agriculture.

10 Foster, D.R. (1985). Vegetation development following fire in *Picea mariana* (Black Spruce)-Pleurozium Forests of South-Eastern Labrador, Canada. *Journal of Ecology* 73 (2).

11 Bailey, S. (2006). *Humus as an Indicator of Nutrient Availability in a Carefully Logged Boreal Black Spruce-Feathermoss Forest in Northwestern Quebec*. McGill University.

12 Lavoie, M., Paré, D., Fenton, N. et al. (2005). Paludification and management of forested peatlands in Canada: a literature review. *Environmental Reviews* 13 (2): 21–50.

13 Terrier, A., de Groot, W.J., Girardin, M.P., and Bergeron, Y. (2014). Dynamics of moisture content in spruce-feather moss and spruce-Sphagnum organic layers during an extreme fire season and implications for future depths of burn in Clay Belt black spruce forests. *International Journal of Wildland Fire* 23 (4): 490–502.

14 Madec, C., Walsh, D., Lord, D. et al. (2012). Afforestation of black spruce lichen woodlands by natural seeding. *Northern Journal of Applied Forestry* 29 (4): 191–196.

15 Bloom, R.G. and Mallik, A.U. (2006). Relationships between ericaceous vegetation and soil nutrient status in a post-fire Kalmia angustifolia-black spruce chronosequence. *Plant and Soil* 289: 221–226.

16 Danneyrolles, V., Valeria, O., Djerboua, I. et al. (2020). How initial forest cover, site characteristics and fire severity drive the dynamics of the southern boreal forest. *Remote Sensing* 12 (23): 3957.

17 Hart, S.J., Henkelman, J., McLoughlin, P.D. et al. (2018). Examining forest resilience to changing fire frequency in a fire-prone region of boreal forest. *Global Change Biology* 25 (3): 869–884.

18 Harper, K.A., Bergeron, Y., Drapeau, P. et al. (2005). Structural development following fire in black spruce boreal forest. *Forest Ecology and Management* 206 (1–3): 293–306.

19 Martin, M., Leduc, A., Fenton, N.J. et al. (2022). Irregular forest structures originating after fire: an opportunity to promote alternatives to even-aged management in boreal forests. *Journal of Applied Ecology* 59 (7): 1792–1803.

20 Martin, M., Fenton, N.J., and Morin, H. (2021). Tree-related microhabitats and deadwood dynamics form a diverse and constantly changing mosaic of habitats in boreal old-growth forests. *Ecological Indicators* 128: 107813.

21 Martin, M., Girona, M.M., and Morin, H. (2020). Driving factors of conifer regeneration dynamics in eastern Canadian boreal old-growth forests. *PLoS One* 15 (7): e0230221.

22 Pham, A.T., de Grandpré, L., Gauthier, S., and Bergeron, Y. (2004). Gap dynamics and replacement patterns in gaps of the northeastern boreal forest of Quebec. *Canadian Journal of Forest Research* 34 (2): 353–364.

23 Moroni, M.T., Morris, D.M., Shaw, C. et al. (2015). Buried wood: a common yet poorly documented form of deadwood. *Ecosystems* 18: 605–628.

24 Johnston, W.F. (1976). *Black Spruce in the North Central States*. United States Department of Agriculture.

25 Fenton, N.J., Simard, M., and Bergeron, Y. (2009). Emulating natural disturbances: the role of silviculture in creating even-aged and complex structures in the black spruce boreal Forest of eastern North America. *Journal of Forest Research* 14 (5): 258–267.

26 Anderson, B.D., Windmuller-Campione, M.A., Russell, M.B. et al. (2020). Short-and long-term results of alternative silviculture in peatland black spruce in minnesota, USA. *Forest Science* 22 (6): 256–265.

27 Anderson, B.D., Russell, M.B., Windmuller-Campione, M.A. et al. (2018). Development and evaluation of black spruce (*Picea mariana* (Miller) B.S.P.) diameter increment models across silvicultural treatments in northern Minnesota, USA. *Forests* 9 (12): 747.

28 Vincent, M., Krause, C., and Zhang, S.Y. (2009). Radial growth response of black spruce roots and stems to commercial thinning in the boreal forest. *Forestry* 82 (5): 557–571.

29 Lefort, P., Harvey, B., Parton, J., and Smith, G.K.M. (2002). Synthesizing knowledge of the Claybelt to promote sustainable forest management. *The Forestry Chronicle* 78 (5): 665–671.

30 Thorpe, H.C. and Thomas, S.C. (2007). Partial harvesting in the Canadian boreal: Success will depend on stand dynamic responses. *The Forestry Chronicle* 83 (3): 319–325.

31 Montoro, M., Morin, H., Lussier, J.M., and Walsh, D. (2016). Radial growth response of black spruce stands ten years after experimental shelterwoods and seed-tree cuttings in boreal forest. *Forests* 7 (10): 240.

32 Girona, M. M., Moussaoui, L., Morin, H., et al. (2023). Innovative silviculture to achieve sustainable forest management in boreal forests: lessons from two large-scale experiments. In Boreal Forests in the face of climate change: sustainable management (pp. 417–440). Cham: Springer International Publishing.

33 Montoro Girona, M., Lussier, J.M., Morin, H., and Thiffault, N. (2018). Conifer regeneration after experimental shelterwood and seed-tree treatments in boreal forests: finding silvicultural alternatives. *Frontiers in Plant Science* 9: 1145.

34 Montoro Girona, M., Morin, H., Lussier, J.M., and Ruel, J.C. (2019). Post-cutting mortality following experimental silvicultural treatments in unmanaged boreal forest stands. *Frontiers in Forests and Global Change* 2 (4): 1–16.

35 Fenton, N.J., Imbeau, L., Work, T. et al. (2013). Lessons learned from 12 years of ecological research on partial cuts in black spruce forests of northwestern Québec. *The Forestry Chronicle* 89 (3): 350–359.

36 Aakala, T., Remy, C. C., Arseneault, D., et al. (2023). Millennial-scale disturbance history of the boreal zone. In Boreal Forests in the face of climate change: sustainable management (pp. 53–87). Cham: Springer International Publishing.

37 Dunn, A.L., Barford, C.C., Wofsy, S.C. et al. (2007). A long-term record of carbon exchange in a boreal black spruce forest: means, responses to interannual variability, and decadal trends. *Global Change Biology* 13 (3): 577–590.

38 Girona, M. M., Morin, H., Gauthier, S., & Bergeron, Y. (2023). Boreal Forests in the Face of Climate Change: Sustainable Management (p. 837). Springer Nature.

39 Achim, A., Moreau, G., Coops, N.C. et al. (2022). The changing culture of silviculture. *Forestry* 95 (2): 143–152.

40 Girona, M. M., Aakala, T., Aquilué, N., et al. (2023). Challenges for the sustainable management of the boreal forest under climate change. In *Boreal Forests in the face of climate change: sustainable management* (pp. 773–837).
Cham: Springer International Publishing.

41 Gauthier, S., Kuuluvainen, T., Macdonald, S. E., et al. (2023). Ecosystem management of the boreal forest in the era of global change. In *Boreal forests in the face of climate change: Sustainable management* (pp. 3–49). Cham: Springer International Publishing.

11

Ecological Silviculture for Acadian Forests

Robert S. Seymour

School of Forest Resources, University of Maine, Orono ME, USA

11.1 Introduction and Context

The Acadian Forest covers nearly 27.4 million hectares and extends over most of northern New England, northwestern Massachusetts, a section of southern Quebec, and all but the highlands of Nova Scotia and New Brunswick (Figure 11.1). The term was first used by Halliday [1] to describe the Canadian portion of the region, then by Loo and Ives [2] in a detailed ecological characterization. European settlement began during the early seventeenth century; the region is named after early French settlers of Nova Scotia who migrated up the St John River Valley after being displaced by the British ca. 1760. Although large areas were once cleared for subsistence farms, much of the region was never settled by Europeans and remains as a large, virtually unbroken block of contiguous forest stretching from Cape Breton, Nova Scotia, to the Adirondack Mountains of New York. Large areas of publicly owned forest in Canada (crown lands) are leased to the forest industry; forests in the United States are mostly in private ownership by large forest investment companies but also include the White Mountain and Green Mountain National Forests.

11.2 Ecological Characteristics

The signature Acadian species is red spruce (*Picea rubens* Sarg.), a long-lived, shade-tolerant, late-successional conifer that dominated presettlement forests [3, 4] (Figure 11.2). On low-lying landscape positions, red spruce mixes with other northern conifers, primarily balsam fir (*Abies balsamea* (L.) Mill.) and northern white-cedar (*Thuja occidentalis* L.) On better-drained, more fertile uplands, spruce and fir grow in mixedwood stands with various hardwoods [5]. Late-successional associates are sugar maple (*Acer saccharum* Marsh.), yellow birch (*Betula alleghaniensis* Britton), American beech (*Fagus grandifolia* Ehrh.), and sometimes eastern hemlock (*Tsuga canadensis* (L.) Carr.). Except for fir, the longevity of these species generally exceeds 300 years, and most are quite shade-tolerant, typically reproducing by advance regeneration that may persist for many decades in the understory before canopy accession. Although they can grow and develop well in single-cohort stands and are commonly managed this way [6], such structures were uncommon before exploitation began in the eighteenth century. Owing to abundant precipitation throughout the growing

Ecological Silvicultural Systems: Exemplary Models for Sustainable Forest Management, First Edition. Edited by Brian J. Palik and Anthony W. D'Amato.

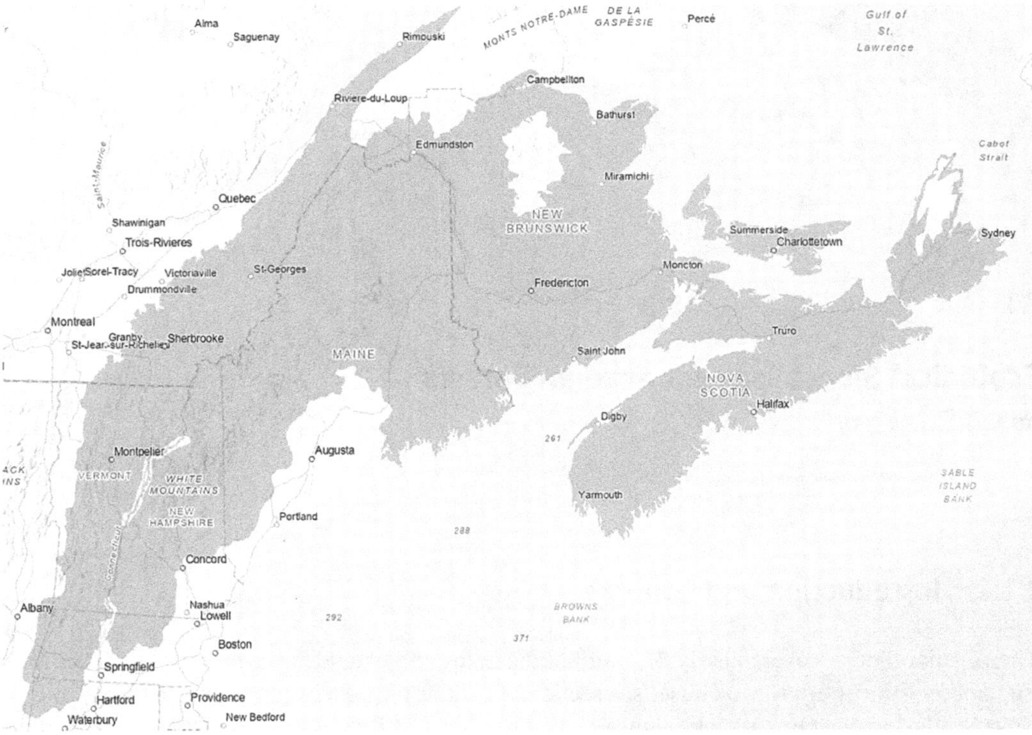

Figure 11.1 Map of the Acadian Forest.

Figure 11.2 Old-growth red spruce stand in northern Maine, USA.

season, stand-replacing fires were virtually unknown, as were stand-replacing windstorms, with estimated return intervals of many centuries to millennia [7]. As a consequence, gap dynamics were the most common natural disturbance, which led naturally to a forest structure dominated by late-successional, multi-aged stands [8, 9]. Extensive outbreaks of the spruce budworm (*Choristoneura fumiferana* Clem.) have an average return interval of 67 years [10] but were stand-replacing only in pure fir stands, a rarity on the presettlement landscape.

Centuries of human exploitation, first for large sawlogs and later (ca. 1900) for smaller-diameter pulpwood, have radically changed the forest structure [11, 12]. Remnants of the primary, old-growth forest are very rare, and most have been reserved from logging and studied intensively by ecologists. The commercial forest landscape is dominated by stands that are much younger and more even-aged than during presettlement. Typical stand compositions have shifted from the slower-growing late-successional species to those that are favored by frequent harvest disturbance, such as red maple (*Acer rubrum* L.), paper birch (*Betula papyrifera* Marsh.), aspen (*Populus* spp.), and balsam fir. Legacies of the presettlement forest, such as large cull trees and small, long-suppressed saplings of late-successional species rare in the overstory, remain in many stands and offer opportunities for restoration with appropriate retention prescriptions.

Acadian forests are inherently diverse in composition; it is not uncommon to have 10 or more commercial species in a single stand on well-drained upland sites (Table 11.1). The exhaustive vegetation classification for Nova Scotia [13] defines three major forest groups (Mixedwood, Spruce-Hemlock, and Tolerant Hardwood) which are subdivided into 23 unique vegetation types (VT). With the exception of northern white cedar, which is very rare in Nova Scotia, this ecological classification applies broadly throughout the Acadian region. These VTs form the basis for an extensive framework of silvicultural keys, the "Silvicultural Guide to the Ecological Matrix" (SGEM) that was recently adopted for Nova Scotia Crown Lands [14]. The SGEM was an outgrowth of the Lahey Report [15] whereby Nova Scotia formally adopted the Triad landscape zoning

Table 11.1 Silvical properties of major Acadian tree species.

Species	Shade tolerance	Regeneration mechanisms	Two-rotation?	LIT species?
Red spruce	Very	Advance	Essential	Yes
Balsam fir	Very	Advance	No	No
Eastern hemlock	Very	Advance	Essential	Yes
Northern white-cedar	Very	Advance, Seed Rain, Layering	Essential	Yes
Eastern white pine	Intermediate	Advance, Seed Rain	Possible	Yes
American beech	Tolerant	Vegetative, Advance	Possible	Yes
Red maple	Tolerant	Advance, Vegetative	Possible	Yes
Sugar maple	Tolerant	Advance	Essential	Yes
Yellow birch	Intermediate	Advance, Seed Rain	Possible	Yes
Paper birch	Intolerant	Seed Rain	No	No
Aspen spp.	Intolerant	Vegetative	No	No

Two-rotation means that the species will lapse into the C stratum beneath more rapidly growing upper strata, and are capable of excellent response if released when the shorter-lived species mature. *Essential* means that the species will not reach sexual maturity and will be lost if managed on short rotations; *Possible* applies to species that can respond to release but also can dominate the canopy from stand origin.
LIT, long-lived, intermediate, or shade-tolerant.

strategy [16, 17] and is meant to apply to all silvicultural operations prescribed for one million hectares of matrix forests that are not allocated to high-production plantations or protected areas. The SGEM has a strong ecological focus on maintaining and restoring late-successional species, using mainly multi-aged variants of the irregular shelterwood system, and can thus be viewed as a model for application throughout the region.

11.3 Models of Disturbance and Stand Development

Using meticulous dendrochronological analysis of tree cores from the 2000-ha old-growth Big Reed Reserve in the heart of the Acadian Forest, Fraver et al. [9] found no evidence of any disturbance exceeding 55% of the canopy killed during any decade on 37 0.15-ha plots spanning 12–18 decades. The mean decadal disturbance rate averaged over five forest types, was 9.6%, approximately 1% per year, confirming estimates in the reviews of Seymour et al. [8] and Lorimer and White [7]. No stand-replacing events were found; the calculated return interval for a 50% canopy removal at the 0.15-ha scale was 1150 years. Disturbance histories for red spruce forests were notably pulsed during decades of severe spruce budworm outbreaks [18] (Figure 11.3); other forest types were remarkably uniform, and virtually no decades escaped some disturbance. Silviculturally, these are true all-aged stands, with somewhat irregular age structures owing to varying severity of disturbances over time. These forests follow the Archetype 3 pattern of Palik et al. [19], except that some gaps result from senescence of single large 300+-year-old trees as well as species-specific pests such as spruce budworm and beech bark disease, in addition to wind. At the stand scale, such a forest is completely dominated by the old-forest stage; earlier stages are not recognizable because young trees are always competing with those of older cohorts owing to their finely patterned age structures. The repeated, exploitative logging of 1860–1930 created millions of acres of young forest over an unprecedented scale over the landscape. These young stands stratify early in development, and by age 20, up to three vertical layers can be recognized. Early successional aspen and paper birch may form an A stratum much taller than the more shade-tolerant species that occupy the lower strata. If present, white pine, red maple, and yellow birch may form a solid B stratum; if not, this layer is composed of densely stocked balsam fir, red spruce, sugar maple, and beech. Eastern hemlock and cedar are even slower growing, forming C or even D strata that appear as an understory even though they are no younger. Because these stands originated mainly from dense, tall advance regeneration, there was no preforest stage.

By 1970, millions of hectares of these artificially even-aged stands reached the mature forest stage, dominating the Acadian landscape. These stands were densely stocked, intensely self-thinning, and in the stem-exclusion stage of stand development; no gaps were present, and any advance regeneration was small and not well established (Figure 11.4).

Superficially these stands appeared to be very even-aged, but closer inspection of the age structure often revealed a 30–60-year range in recruitment age, reaching breast height from 1870 to 1930 [20, 21]. One such example is the Baxter Park Scientific Forest Management Area (SFMA), a 10,000-ha parcel in northcentral Maine that was acquired from industrial landowners in 1955 but not managed until the late 1980s. Tero [21] sampled five mature 10-ha benchmark reserve stands that were in the understory reinitiation stage; small gaps averaging 49.1 m^2 covered 20% of the stand area, and seedlings-sapling density was 89,800/ha dominated by balsam fir, red maple, and red spruce. Age structure was broadly two-aged, with recruitment peaks ca. 1890 and 1930, likely corresponding to heavy spruce sawlog cuttings and the 1913–1919 spruce budworm outbreak, respectively (Figure 11.5). However, some trees recruited during every decade from 1870 to 1960,

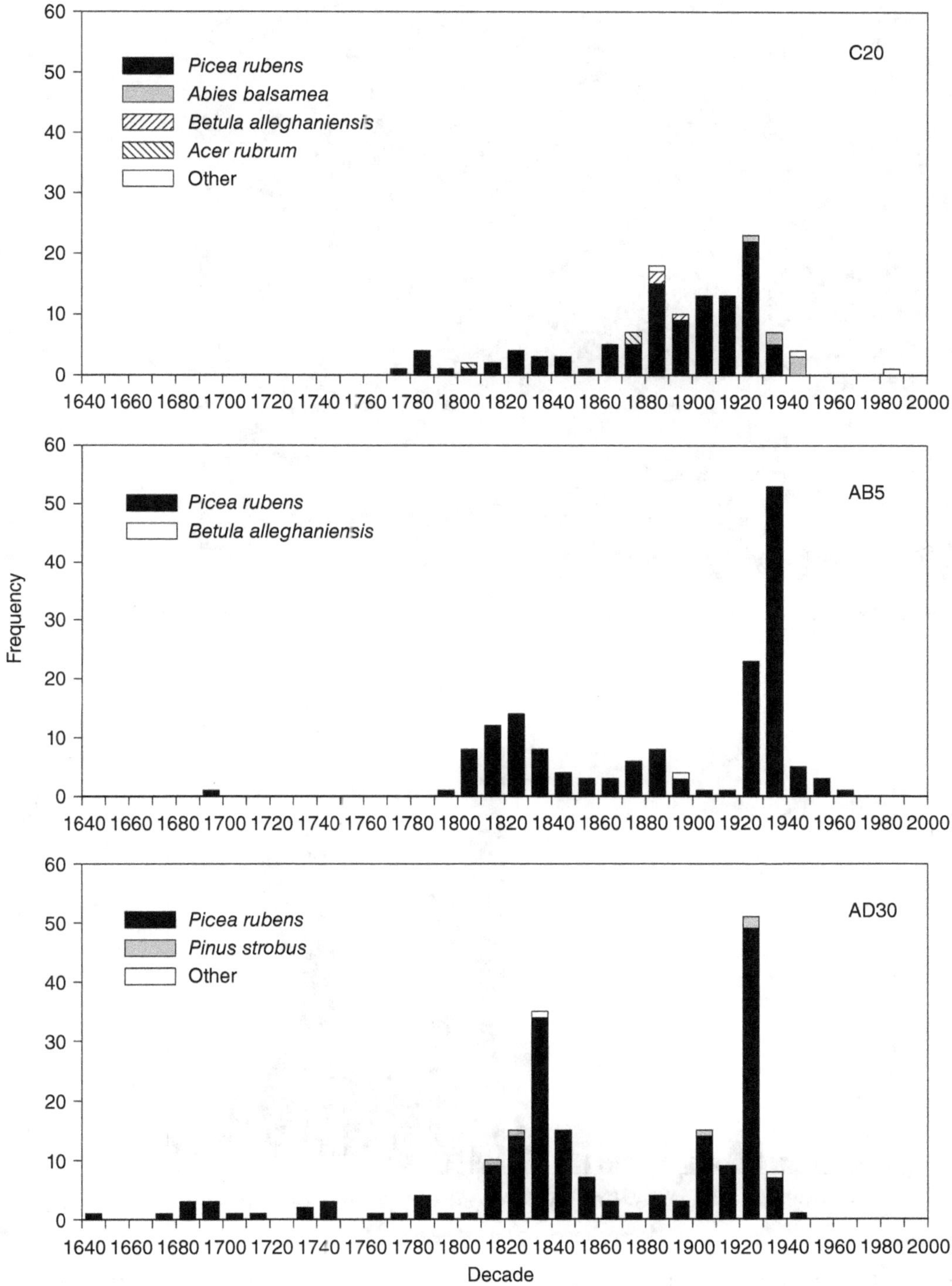

Figure 11.3 Recruitment age-class distributions for three old-growth red spruce stands *Source:* Original courtesy of Shawn Fraver.

Figure 11.4 Densely stocked mature second-growth red spruce stand in western Maine, USA.

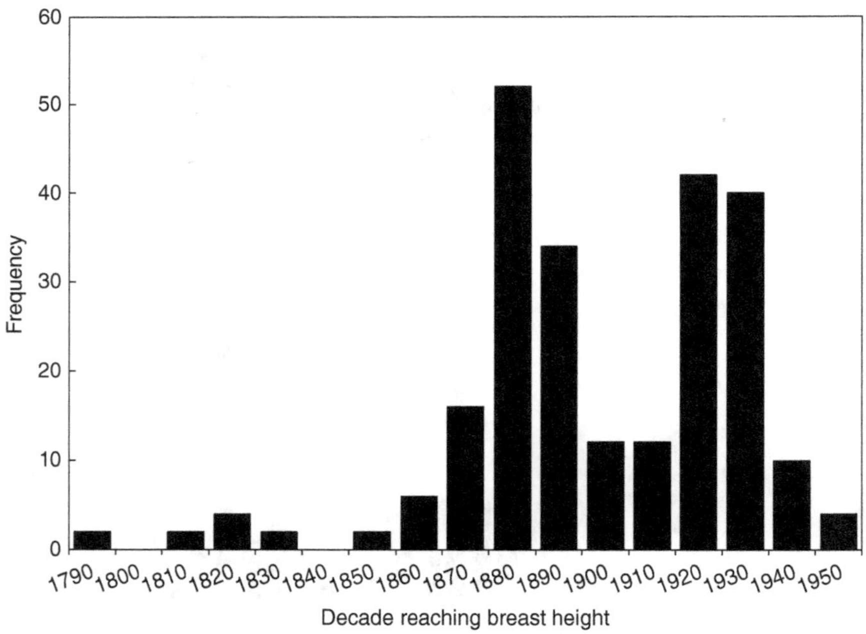

Figure 11.5 Age distribution of red spruce trees from five mature second-growth stands, Baxter Park Scientific Forest Management Area, Maine, USA. *Source:* From [21]/University of Maine.

suggesting that the nineteenth-century harvest history was one of repeated diameter-limit cuttings, first for sawlogs (to 30 cm dbh) and after 1900, pulpwood (down to 15 cm).

Silviculturally this pattern is best characterized as a series of partial removal cuttings, releasing irregularly structured residual stands of old, submerchantable overstory stems and tall advance regeneration. The best modern analog would be an extended irregular shelterwood [22], with two dominant cohorts and several decades where the stand is largely under regeneration of the younger cohort.

From the late 1960s until at least the mid-1990s, the dominant silvicultural paradigm for treating these mature-stage stands was very even-aged. Most industrial landowners simply clear-cut the overstory, under the premise that an attempt at partial cutting in such dense, tall stands was destined to fail as a result of postharvest windthrow. In some cases, small advance regeneration was present, survived the harvest, and was released by aerial herbicide application, leading to third-generation young-forest stands that are very even-aged. Meng and Seymour [23] aged 115 trees in 12 such stands clear-cut in the 1970s, and found only a six-year range in breast-height ages. All precommercially thinned sites in Maine's Commercial Thinning Research Network studied by DeRose and Seymour [24] are similarly uniform (Figure 11.6). By 2000, mature forest stands had mostly been harvested on industrial lands, so within the span of a little over a century, nearly two million acres of Maine's once-diverse Acadian forest dominated by long-lived, intermediate, and shade-tolerant (LIT) species had been converted to single-cohort stands dominated by fir, red maple, and paper birch in the young forest stage.

Some private family ownerships and public owners in the United States resisted the clear-cutting paradigm and worked to implement uniform shelterwood systems, in recognition of the essential importance of advance regeneration for these species. Typical practice involves an establishment cutting to remove balsam fir and short-lived hardwoods, then returning after 10–20 years to remove the overstory of longer-lived LIT species. Although such a system definitely improves regeneration outcomes, it does little to restore the age diversity of the presettlement forest and replicates stand-replacing disturbance at an unnaturally large scale [8]. Beginning in the 1990s, Maine's Bureau of Parks and Lands, the Baxter SFMA, and the United States Fish and Wildlife Service's Umbagog Refuge have adopted longer rotations and reentry cycles by stretching a planned series of light,

Figure 11.6 Young (age 25) even-aged balsam fir stand treated 15 years prior with precommercial thinning.

Figure 11.7 Group selection cutting in a mature red spruce mixedwood stand on the Umbagog National Wildlife Refuge, Maine, USA.

partial overstory removal cuttings over several decades. Some of these have been done by regenerating in distinct groups (Figure 11.7) while others are not spatially clustered. Both systems attempt to add new cohorts at each entry to diversify the age structure and leave large permanent reserve trees as legacies to conserve some of the oldest cohorts, most of which is eventually harvested. These are good examples of the continuous-cover and group variants of irregular shelterwood systems [22] and are at the forefront of practice in the region. Nova Scotians have been slower to embrace multi-aged silviculture until recently: beginning in 2021, all harvesting on Nova Scotia crown lands must follow ecologically based multi-aged silviculture implemented through variants of irregular shelterwood and some group- and single-tree selection [14].

11.4 Restoration Challenges and Possible Pathways

The foregoing makes clear that any serious attempt at ecological forestry (*sensu* 16) in the Acadian region must confront the simplified age structures and altered compositions of repeatedly harvested stands. The goal of such a restoration strategy is to recreate a forest dominated by diverse multi-aged stands, with some having a late-successional component that is most deficient in the commercial forest. In the Acadian region, this problem is arguably more difficult than in regions like much of the Pacific Northwest and elsewhere where the natural stand-development patterns follow heavy partial to stand-replacing disturbance models and the challenge is largely integrating surviving legacies of these events into areas historically managed with even-aged systems using structural retention measures. In the Acadian forest, leaving scattered islands or reserve trees in clear-cuts or uniform shelterwoods of >10 ha, while valuable in some respects, fail to address the more fundamental mismatch of even-aged silviculture with natural processes.

11.5 Regeneration Treatments at the Mature Forest Stage

To convert uniform stands to more irregular, multi-aged structures, foresters must regenerate only a portion of the stand at each entry while keeping the canopy of the surrounding matrix relatively intact and in stem exclusion. Here, foresters define the target structure as an area (not dbh) distribution, defined by [1] the percentage of the stand regenerated at each entry and [2] a distribution of patch (gap) sizes and their locations. An area structure should be patterned after corresponding natural disturbance analogs (frequency, severity, spatial extent) derived from disturbance ecology research [16]. Gap-based multi-aged systems have many advantages, both ecological and operational:

1) Regeneration is managed explicitly, not by assumptions of future ingrowth.
2) Ecological sustainability is guaranteed if the cutting cycle and area regenerated are within natural limits.
3) There is no need to assume a problematic linkage between size and age, as with dbh-based structures.
4) Preharvest layout, logging, artificial regeneration (if needed), and early tending are all concentrated on 10–20% of the stand at a time, with obvious efficiencies.
5) Harvests can be planned and executed with modern spatial information technologies (GIS, GPS); there is no need for a preharvest stand table or keeping a marking tally.
6) Light harvests (removing under 25% of the stocking) are feasible with mechanized equipment because the wood is concentrated. Such light cuttings are impractical operationally if distributed evenly throughout the stand, with too many trees removed for trails and machine access.
7) Perhaps most importantly, advance regeneration in gaps can be largely avoided and thus protected in harvest entries by expanding the gaps outward and locating extraction trails in the matrix.

11.6 Silvicultural Systems Based on Natural Disturbance Parameters – The Acadian Femelschlag

In 1994, a team of scientists from the University of Maine designed a long-term experiment, the Acadian Forest Ecosystem Research Program (AFERP), intended to complement the conventional silvicultural systems on the Penobscot Experimental Forest (PEF) maintained by the U.S. Forest Service since 1950 [25]. Two disturbance-based systems were devised based on the 1% annual disturbance rate, focusing on gaps that are expanded over time, reserving 10% of the growing stock as permanent reserve trees in gaps to serve as biological legacies. Treatments are replicated three times in 10-ha blocks, including untreated controls, occupying 90 ha of mature, conifer-dominated mixedwood forest at the PEF.

The *large-gap* treatment is essentially a multi-aged, irregular group shelterwood system, in which advance regeneration is recruited in defined gaps within the stand, rather than uniformly throughout, leaving reserve trees in the groups after they are regenerated. Instead of stand age structure changing temporarily as in a uniform shelterwood, group shelterwood systems vary spatially, and can contain all stages of the shelterwood sequence: unregenerated matrix awaiting treatment; two-storied patches following establishment cutting; and free-to-grow sapling regeneration after removal of the overstory except reserves. To facilitate efficient harvesting and retain some intolerant species in the regeneration, particularly in light of in-gap retention of reserves, we chose a gap size of 0.2 ha, slightly larger than most natural gaps. Gap cuttings occur as five entries spaced 10 years apart, each comprising 20% of the stand area. After the first entry, which establishes gap

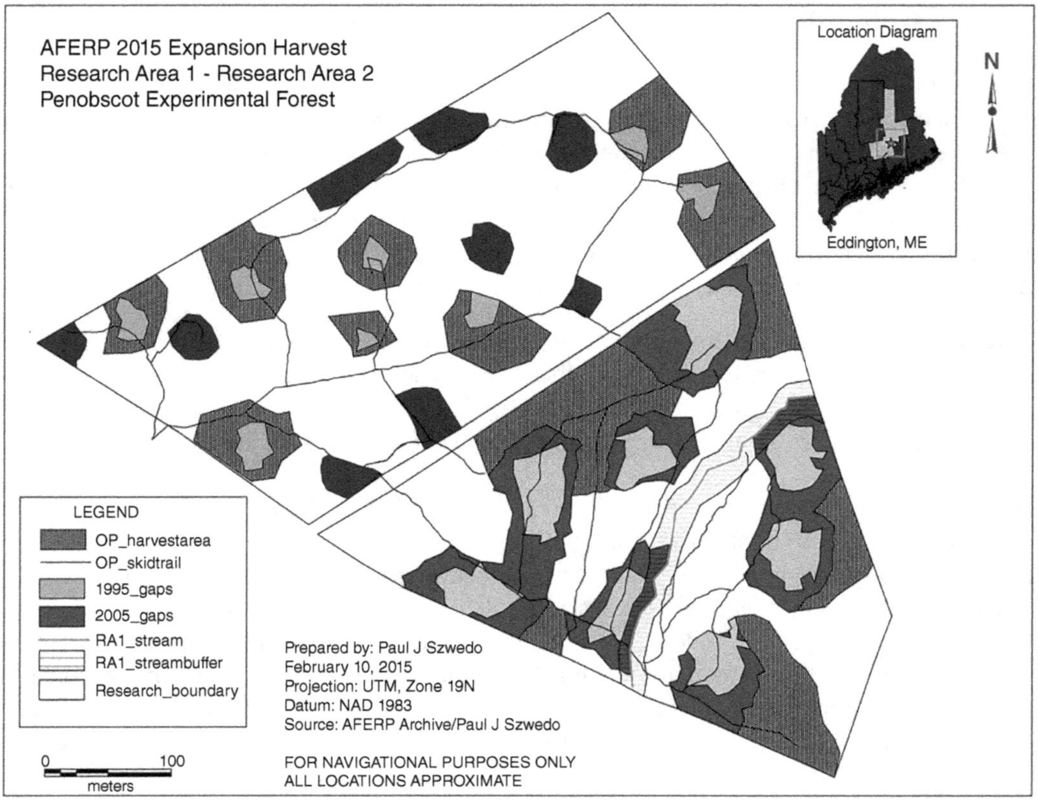

Figure 11.8 Map of gap locations and expansions over three 10-year cutting cycles in one replicate of the AFERP experiment.

locations, further cuttings expand these gaps asymmetrically (Figure 11.8). After five entries, the stand develops with no further regeneration cutting for another 50 years. Regenerating disturbance is thus "front-loaded" during the first 50 years at 2% per year, but still averages 1% annually over the entire 100-year conversion period. There is no attempt to achieve a balanced within-stand age structure, just a diverse, irregular one that nevertheless represents quite a departure from the initial single-cohort structure.

This system is best described as an *irregular group shelterwood with permanent retention of reserve trees within regenerating groups*. European foresters have long applied such a system, known in Germany as the *Bavarian femelschlag*, in which the groups under regeneration are expanded at each entry until they coalesce [26]. In his classic description of European silvicultural systems, Troup [27] describes several regional variants of the femelschlag practiced at that time for converting even-aged stands to more irregular structures, including group and continuous-cover variants [22]. AFERP's combination of prescription elements, notably permanent retention in the gaps, makes this system unique, and it has been named the *Acadian Femelschlag*.

11.6.1 Tending Gap Regeneration – The Young Forest Stage

Gaps in the Femelschlag system are large enough to recognize distinct aggregations of young sapling-sized trees that are not overtopped by scattered reserve trees and thus are amenable to precommercial thinning and cleaning to improve their density and composition. Owing to their highly stratified mixed-species structure, simple spacing prescriptions or even variable-density thinning rules will oversimplify the resulting residual stand relative to ecological benchmarks. Instead, crop-tree selection rules were devised that vary by vertical stratum. Any A-stratum paper birch and aspen are left in a low density, spaced at 14–18 m (30–50 trees ha^{-1}), favoring stems that are not overtopping more valuable (white pine) or rare (red spruce) trees in lower strata. Any white pine, red maple, yellow birch, or red oak in the B stratum are retained on a 6–8 m spacing (160–280 ha^{-1}); hardwood sprouts are sometimes thinned to a single stem. Finally, the lower C stratum of shade-tolerant species is spaced to 2–3 m (1100–2500 ha^{-1}), favoring spruce and sugar maple over balsam fir, beech, and hemlock. Rare species such as northern white-cedar and sometimes red spruce are treated as "invisible," meaning it is neither a crop tree nor competition to one, and thus are just left.

Ideally, these treatments are applied 15–20 years after gap creation. At age 40–50, commercial thinnings can remove most of the short-lived A stratum as well as any balsam fir, leaving a residual stand of the two-rotation species to develop to the mature stage.

11.6.2 Locating Skid Trails, Initial Gaps, and Reserve Trees

Initial gaps were located in two different stand conditions. In patches of well-established advance regeneration resulting from natural canopy breakup in the two decades prior to initiating the experiment, the overstory was removed completely except for the reserve trees. Other gaps were placed in areas in stem exclusion, where we left a gap-level overwood basal area of 14–18 m^2 ha^{-1} to provide shade and seed for new recruitment. This extra residual overwood was removed in the second entry (except permanent reserve trees), as the gaps were expanded.

The matrix between groups was lightly harvested only during the first entry, mainly to pre-salvage balsam fir, paper birch, and aspen that were reaching their natural life span. In the matrix, we did not remove large dominants that would make permanent canopy gaps and thus create unwanted regeneration. Harvest trails connected the gaps, and occasional spur trails were needed to treat the intervening matrix. in order to conserve areas of intact matrix conditions as long as possible, the first two gap-expansion harvests have advanced asymmetrically outward, combining some initial gaps into a single larger one.

Reserve trees were designated when the gaps were marked for cutting. Trees with obvious wildlife usage (e.g. large cavities) were retained; others were selected from the larger dbh classes of long-lived, and sometimes uncommon, species. Target reserve-tree density was 10% of the initial basal area of 30 m^2 ha^{-1}, or 3 m^2; trees were mostly dispersed throughout the gap, not clumped. Many gaps (Figure 11.9) also contain immature, merchantable-size growing stock which is treated as "tall regeneration" of two-rotation LIT species (mainly red spruce, hemlock, and northern white-cedar, Table 11.1), which mature more slowly than their single-rotation faster-growing associates (balsam fir, paper birch, red maple).

11.6.3 Results After 27 Years

The AFERP experiment is nearly 30 years old, and will soon experience the fourth decadal harvest entry. Arseneault et al. [28] reviewed experience from the first decade and reported differences in regeneration patterns between the large and small gaps. Saunders and Arsenault [29] compared AFERP systems against single-tree selection and conventional uniform shelterwood and found that the disturbance-based systems sustained a greater diversity in forest structure and composition while producing yields and financial returns that are competitive with conventional systems.

Figure 11.9 Typical gap in the AFERP experiment, showing permanent legacy trees, immature growing stock, and tall advance regeneration.

Twenty-year changes in vascular plant richness are encouraging. Relative to the controls, which were stable over two decades, species-area curves showed a slight increase of richness in the small-gap treatments, and a near doubling in the large-gap treatments owing to the invasion of early successional plants but no loss of the pretreatment shade-adapted species [30]. Long-term observations of the songbird community showed a decline in total bird abundance after 25 years relative to the initial pre and postharvest survey in the late 1990s, but no difference among treatments [31]. No evidence was found that the large-gap treatment, which had only 40% of its original mature forest left after year 20, had any negative effects relative to the untreated controls.

Survival of reserve trees in the gaps has been very encouraging. Carter et al. [32] recorded the fate of 787 trees of 17 different species and found a 20-year survival of nearly 92%. Even the notoriously non-windfirm, shallow-rooted red spruce lost only 17 of 153 trees.

Operationally, the large-gap treatments have been fully commercial, requiring no harvesting subsidies. Permanent access trails occupy less than 5% of the stand, much lower than comparable uniform systems. This experiment, the first of its kind in the region if not the United States, has created great interest in expanding-gap, irregular group shelterwood systems with permanent reserves in the Acadian region, and has spawned several comparable studies through the eastern United States (e.g. [33]). The author uses variants of irregular shelterwood on over 500 acres of personal woodlot properties in central Maine [34, 35], and its popularity among foresters managing public lands and client's woodlots was an important reason why Nova Scotia recently adopted irregular shelterwood systems to implement ecological silviculture on one million hectares of their public lands [14].

Gap-based irregular shelterwood systems, with multi-aged structures containing just a few cohorts and some old trees, do not emulate their old-growth analogs, nor are they meant to. Although creating finely patterned all-aged stands is theoretically possible with very light, single-tree selection cuttings on short cycles, such a system would never be practical where some level of commodity production is expected from forest management. Indeed, such would be a silvicultural example of the axiom where "the perfect is the enemy of the good enough." Experience has shown that irregular shelterwoods offer the first ecologically based and commercially viable alternative to commodity driven even-aged systems [36], which may be efficient but are highly unnatural and likely have important drawbacks for biodiversity.

11.7 Climate Considerations

Research by paleoecologists [37] suggests that the contemporary Acadian forest is quite a recent assemblage, owing to a slight cooling trend over the last two millennia. The current range of red spruce is only about 1000 years old; earlier in the Holocene, spruce was present only along the immediate Atlantic coast [38]. Species that dominated the warmer and somewhat drier mid-Holocene forest – eastern white pine, oaks, hemlock, birches – are present in today's forest, but under the cold, wet climate of the past millennium, they were at a competitive disadvantage to the more shade-tolerant, long-lived species that thrive in the absence of extensive disturbance, especially fire.

Red spruce is at risk of a major decline in the United States, with refugia persisting in the higher elevations of the White Mountains and northeastern Maine [39]. The *resistance* arm of the three-part climate adaptation strategy [19] should thus focus on conserving extant examples of mature red spruce in these places. A *resilience* strategy should focus on mixed-species silviculture [5], favoring species there were formerly more common several millennia ago (white pine, red maple, oaks), or using enrichment planting where the species is absent.

The most serious effect of a warming climate on Acadian forest vegetation is arguably from introduced pests and pathogens that were formerly excluded by cold winter temperatures. Good examples are two adelgid species, hemlock (*Adelges tsugae*) and balsam fir (*Adelges piceae*) adelgids.

11.8 Summary

1) Historical exploitation of the Acadian forest has simplified composition and age structure, from nearly all-aged during presettlement to single-cohort today. The focus of ecological silviculture is thus on restoring age and species diversity.
2) Disturbance ecology research suggests a gap-based regime on a 100-year return interval, so large-scale even-aged systems, even with retention, are not an appropriate model.
3) Although single-tree and group selection would theoretically be the best options, three decades of experience in an expanding-gap irregular group shelterwood experiment suggests this system also sustains native biodiversity and is more feasible economically.

References

1 Halliday, W.E.D. (1937). *A Forest Classification for Canada*, Forest Service Bull. 89. Ottawa, Canada: Canada Department of Mines Resources. 50 p.

2 Loo, J. and Ives, N. (2003). The Acadian Forest: historical condition and human impacts. *Forestry Chronicle* 79 (3): 462–474.

3 Hosmer, R.S. (1902). A study of the Maine spruce. *4th Rept. Maine Forest Commissioner*, Augusta, ME.

4 Lorimer, C.G. (1977). The presettlement forest and natural disturbance cycle of northeastern Maine. *Ecology* 58 (1): 139–148.

5 Kenefic, L.S., Kabrick, J.M., Knapp, B.O. et al. (2021). Mixedwood silviculture in North America: the science and art of managing for complex, multi-species temperate forests. *Canadian Journal of Forest Research* 51: 921–934.

6 Seymour, R.S. (1995). The Northeastern region. In: *Regional Silviculture of the United States*, 3e (ed. J.W. Barrett), 31–79. New York: Wiley. 643 p.

7 Lorimer, C.G. and White, A.S. (2003). Scale and frequency of natural disturbances in the northeastern US: implications for early successional habitats and regional age distributions. *Forest Ecology and Management* 185: 41–64.

8 Seymour, R.S., White, A.S., and A.S., deMaynadier, P.G. (2002). Natural disturbance regimes in northeastern North America – evaluating silvicultural systems using natural scales and frequencies. *Forest Ecology and Management* 155: 357–367.

9 Fraver, S., White, A.S., and Seymour, R.S. (2009). Patterns of natural disturbance in an old-growth landscape of northern Maine, USA. *Journal of Ecology* 97: 289–211.

10 Fraver, S., Seymour, R.S., Speer, J.H., and White, A.S. (2007). Dendrochronological reconstruction of spruce budworm outbreaks in northern Maine, USA. *Canadian Journal of Forest Research* 37: 523–529.

11 Seymour, R.S. (1992). The red spruce-balsam fir forest of Maine: evolution of silvicultural practice in response to stand development patterns and disturbances. In: *The Ecology and Silviculture of Mixed-species forests* (ed. M.J. Kelty, B.C. Larson, and C.D. Oliver), 217–244. Norwell, MA: Kluwer Publishers. 287 p.

12 Noseworthy, J. and Beckley, T.M. (2020). Borealization of the New England–Acadian Forest: a review of the evidence. *Environmental Reviews* 28 (3): 284–293.

13 Neily, P., Keys, K., Quigley, E. et al. (2013). Forest ecosystem classification for Nova Scotia (2010). Report FOR 2013-1, Renewable Resources Branch, NS DNR. 452 p.

14 McGrath, T., Pulsifer, M., Seymour, R. et al. (2021). *Nova Scotia Silvicultural Guide for the Ecological Matrix*. Nova Scotia Department of Lands and Forestry. 193 p.

15 Lahey, W. (2018). An independent review of forest practices in Nova Scotia. https://novascotia.ca/natr/forestry/Forest_Review/Lahey_FP_Review_Report_ExecSummary.pdf (accessed March 2023).

16 Seymour, R.S. and Hunter, M.L. Jr. (1999). Principles of ecological forestry. In: *Maintaining Biodiversity in Forest Ecosystems* (ed. M.L. Hunter Jr.), 22–61. Cambridge University Press. 698 p.

17 Himes, A., Betts, M., Messier, C., and Seymour, R. (2022). Perspectives: thirty years of triad forestry, a critical clarification of theory and recommendations for implementation and testing. *Forest Ecology and Management* 510: 120103.

18 Fraver, S. and White, A.S. (2005). Disturbance dynamics of old-growth *Picea rubens* forests of northern Maine. *Journal of Vegetation Science* 16: 597–610.

19 Palik, B.J., D'Amato, A.W., Franklin, J.F., and Johnson, K.N. (2021). *Ecological Silviculture: Foundations and Applications*. Waveland Press, 343 p.

20 Fajvan, M.A. and Seymour, R.S. (1993). Canopy stratification, age structure, and development of multicohort stands of eastern white pine, eastern hemlock, and red spruce. *Canadian Journal of Forest Research* 23: 1799–1809.

21 Tero, N.T. (2010). History, composition, and structure in mature second-growth Acadian forest stands managed with irregular shelterwood silviculture. Masters thesis. University of Maine. 97 p.

22 Raymond, P., Bédard, S., Roy, V. et al. (2009). The irregular shelterwood system: review, classification, and potential application to forests affected by partial disturbances. *Journal of Forestry* 107: 405–413.

23 Meng, X. and Seymour, R.S. (1992). Influence of soil drainage on early development and biomass production of young, herbicide-released fir-spruce stands in north-central Maine. *Canadian Journal of Forest Research* 22: 955–967.

24 DeRose, R.J. and Seymour, R.S. (2010). Patterns of leaf area index during stand development in even-aged balsam fir-red spruce stands. *Canadian Journal of Forest Research* 40: 629–637.

25 Seymour, R.S., Guldin, J., Marshall, D., and Palik, B. (2006). Large-scale, long-term silvicultural experiments in the United States: historical overview and contemporary examples. *Allgemeine Forst-und Jagdzeitung* 177: 104–112.

26 Spurr, S.H. (1956). German silvicultural systems. *Forest Science* 2: 75–80.

27 Troup, R.S. (1955). *Silvicultural Systems*, 2e. Oxford University Press. 216 p.

28 Arseneault, J.E., Saunders, M.R., Seymour, R.S., and Wagner, R.G. (2011). First decadal response to treatment in a disturbance-based silviculture experiment in Maine. *Forest Ecology and Management* 262: 402–412.

29 Saunders, M.R. and Arseneault, J.E. (2013). Potential yields and economic returns of natural disturbance-based silviculture: a case study from the Acadian Forest Ecosystem Research Program. *Journal of Forestry* 111 (3): 175–185.

30 Mansfield, M. (2019). Understory response to gap-based, multi-aged silviculture. Masters thesis. University of Maine. 45 p.

31 Pohlman, C., Roth, A.M., Hartley, M.J. et al. (2023). Experimental natural disturbance-based silviculture systems maintain mature forest bird assemblage long-term in Maine (USA). *Forest Ecology and Management* 528: 120630.

32 Carter, D.R., Seymour, R.S., Fraver, S., and Weiskittel, A. (2017). Reserve tree mortality in two expanding-gap silvicultural systems 20 years after establishment in the Acadian Forest of Maine, USA. *Forest Ecology and Management* 389: 149–157.

33 Forrester, J. A., Keyser, T. L., Schnake, D. K. Ecological Silviculture for Southern Appalachian Hardwood Forests. (2024). Pages 98–111 in Palik, B. J., D'Amato A. W. (eds). Ecological Silvicultural Systems: Exemplary Models for Sustainable Forest Management. John Wiley and Sons, LTD, Hoboken New Jersey, USA.

34 Seymour, R.S. (2017). Exploring silvicultural strategies in Northern New England – irregular shelterwood. https://youtu.be/DBOyrDIRwPw (accessed August 2022).

35 Seymour, R.S. (2021). The Acadian Femelschlag - irregular group shelterwood with reserves in Northern Hardwoods. https://youtu.be/vV_ljau2oTc.

36 Raymond, P. and Bedard, S. (2017). The irregular shelterwood system as an alternative to clearcutting to achieve compositional and structural objectives in temperate mixed wood stands. *Forest Ecology and Management* 398: 91–100.

37 Jacobson, G.L. and Dieffenbacher-Krall, A. (1995). White pine and change: insights from the Past. *Journal of Forestry* 93 (7): 39–42.

38 Schauffler, M. and Jacobson, G.L. Jr. (2002). Persistence of coastal spruce refugia during the Holocene in northern New England, USA, detected by stand-scale pollen stratigraphies. *Journal of Ecology* 90: 235–250.

39 Andrews, C., Foster, J.R., Weiskittel, A. et al. (2022). Integrating historical observations alters projections of eastern North American spruce-fir habitat under climate change. *Ecosphere* 13: e4016.

12

Ecological Silviculture for Sierra Nevada Mixed Conifer Forests

Robert A. York

Department of Environmental Science, Policy, and Management, University of California, Berkeley, CA, USA

12.1 Introduction

The Sierra Nevada mixed conifer forest (MCF) extends within a narrow south-to-north oriented band along the western slopes of the Sierra Nevada Mountain range in California, USA. It covers approximately 2.5 million ha between elevations of 900 and 1800 m, generally trending toward lower elevations as latitude increases. Like other conifer forests in western North America, it is defined by a dry summer period that is typical of a Mediterranean climate. It is also characterized by high conifer species diversity, which is often attributed to diverse edaphic conditions. The spatially and temporally complex disturbance regime is dominated by fire and is an important factor in the sustained existence of mixed species (Table 12.1). This fire influence on tree diversity is exemplified by the recent century-long trend toward the fire-sensitive species white fir (*Abies concolor* Gord. & Glend) and incense-cedar (*Calocedrus decurrens* Torr.), which have increased dramatically because of fire suppression. More fire-adapted species such as ponderosa pine (*Pinus ponderosa* Dougl. ex Laws.) and giant sequoia (*Sequoiadendron giganteum* Lindl. Buchholz) have declined in undisturbed stands as a result of fire suppression, but have been maintained with silvicultural systems that have included the application of moderate-severity disturbances (i.e. distinct canopy openings) followed by planting.

Objectives of silvicultural treatments in the MCF are diverse, including sequestering carbon, influencing yields of water for domestic use, and producing timber [1–3]. Timber extraction has influenced the structure of the MCF since the gold rush of the late 1800s, when silviculture as a system of planning for sustained resource benefits was not practiced [3]. Over the last 50 years, the amount of timber harvested from the MCF has declined steadily. Private forestland consisting of both industrial and nonindustrial owners now harvest the majority of timber, and national forests harvest little relative to the amount of forestland owned [4]. Despite the decline in managing for timber, federal forestland is included as an important context for this chapter because of the significant proportion of MCF under federal ownership. Reducing stand densities substantially on federal forestland as a method to improve vigor of large trees in the face of climatic stress will necessarily require the harvest of commercial-size trees in many portions of federal land [5].

To propose a model of ecological silviculture for the MCF, I critically evaluate the current general silvicultural approaches across the three categories of forestland ownership that have

Ecological Silvicultural Systems: Exemplary Models for Sustainable Forest Management,
First Edition. Edited by Brian J. Palik and Anthony W. D'Amato.

Table 12.1 Tree species of the Sierra Nevada mixed conifer forest with shade tolerance, seedbed requirements, max lifespan, and fire tolerance.

Species	Shade tolerance	Capable of germinating on duff/litter?	Maximum lifespan	Fire tolerance
Douglas-fir	Intermediate	Yes	500–1000	Intermediate
Giant sequoia	Intolerant	No	2000–3000	High
Incense cedar	Tolerant	Yes	500	Intermediate
Ponderosa pine	Intolerant	No	300–600	High
Sugar pine	Intermediate	Yes	500	Intermediate
White fir	Tolerant	Yes	300	Low
Black oak	Intolerant	Yes	300–500	Intermediate

conventionally been used: industrial, nonindustrial, and federal. I then describe a silvicultural system that would be applied if using the natural development model as a guide were the primary goal of a landowner in the MCF. Finally, I focus on the potential alterations to existing silvicultural approaches that could be done in order to make current silvicultural systems more in-line with the principles of ecological silviculture as defined in this book.

12.2 Characteristics of Sierra Nevada Mixed Conifer Forests

12.2.1 The Disturbance Regime, Composition, and Structure

Forest composition and structure have changed dramatically over the past century [6] in this ecosystem as a result of disruptions to the disturbance regime and a lack of silvicultural systems designed to replace essential elements of disturbances. Prior to Euro-American colonization (the era of time hereafter referred to as "precolonial"), the disturbance regime was dominated by a high frequency of low-severity fires [7]. Moderate-severity fires were less common but were important for maintaining low overall stocking relative to site potential. Most fire reconstruction studies suggest median fire return intervals at the stand scale between 5 and 10 years. Actual fire return intervals were more frequent since lower intensity fires were not always recorded in tree rings. Widespread communities of people managed the MCF for a variety of objectives before being forcefully removed and persecuted. These people included members of the Nisenan, Miwok, Konkow, Washo, and Maidu Tribes. The descendants of these people are key stakeholders today. During the precolonial era, small fire-created patches (<1.0 ha) were much more common than large fire-created patches, although larger patches did occur at lower frequencies and at higher elevations where the MCF intersects with *Abies*-dominated forest types [8]. Stands were multiaged, with cohorts of regeneration occurring in distinct fire-initiated canopy gaps [9]. During the past century, lighting fires have been actively suppressed [10], and fire has generally been excluded from use. The use of fire is desired by those wanting to restore Indigenous cultural practices [11] and those with general resource management objectives [12]. Consequently, the current disturbance regime has shifted fundamentally from frequent low-severity fires to large, uncontrollable, high-severity wildfires that occur during expanding periods of extreme weather conditions. The resulting profound effects of the altered disturbance regime include: (i) increases in tree density; (ii) reductions of canopy heterogeneity, especially at smaller scales (i.e. 0.1–1.0 ha); (iii) increases in the fire-sensitive species, especially white fir and incense-cedar; (iv) decreases in ponderosa pine and giant sequoia; and (v) increases in the ratio of small to large diameter trees.

12.2.2 Current Silviculture in the Sierra Nevada MCF

Generally, there are three categories of forest structures that exist as a result of silvicultural planning. I describe these below and, as a pretext to describing ecological silviculture, comment on the extent to which they could be considered misaligned with the precolonial disturbance regime and/or the future climatic regime.

12.2.2.1 Planted Forests

Planted forests are single-cohort stands established via planting after either high-severity wildfires or even-aged regeneration harvests, primarily clear-cuts. Disturbances that result in the near-complete mortality of overstory trees across areas greater than 1 ha tend to require artificial regeneration in order to avoid a several-decade or indefinite conversion to shrub-dominated structures [13]. This inability for natural tree regeneration to replace forests reflects the fact that high-severity disturbances are not the dominant disturbance type that initiates new cohorts in the MCF. Forest-to-shrub conversions following high-severity disturbances occur from multiple factors, including a lack of sufficient seed input from wind-dispersal, sprouting shrubs that establish ~100% cover and can preclude seedling establishment, and the tendency for shrubs to reburn with high-severity wildfires that perpetuate shrub dominance. Historically shrubs likely covered ~30% of the forest floor, limited in their cover by competition from a low-density tree canopy that also provided seed input for dense tree regeneration [14]. This overstory-understory dynamic allowed for shrubs and trees to coexist, suppressing shrub cover enough to allow for recruitment of trees. Clear-cut sizes on private lands are limited by state regulations to a maximum of 8 ha when using ground-based yarding. Consequently, 8 ha is the most common patch size occurring on private land managed for timber. Species composition in older planted forests is dominated by two species-ponderosa pine and Douglas-fir [15], although it has recently become more common to plant other species with the objective of promoting mixed species stands that are less vulnerable to host-specific damaging agents and that can yield a diversity of timber products. Reforestation practices commonly include site preparation, planting, and herbicide application to reduce competition with shrubs and grasses. Intermediate treatments may include precommercial thinning and one or more commercial thins prior to another 8-ha regeneration harvest. Planning rotation ages are regulated, requiring a minimum of 50–80 years, depending on site productivity. The vast majority of planted forests in the MCF have not yet reached the minimum rotation age. There have not been attempts to convert single-cohort stands to multiaged structures during mature stand phases, although that option is presumably available. The use of prescribed fire is rare in planted forests managed for timber, but low-intensity prescribed fire has been suggested as an option for reducing risk of loss from wildfires while avoiding significant damage to timber [16].

Related lines of evidence from wildfire behavior modeling, assessments of current forest structural attributes, and postwildfire assessments, suggest that wildfire severity is higher in planted forests compared to mature forests. Wildfire behavior modeling predicts exceptionally high fireline intensities and rates of spread in planted forests, resulting in crown fire behavior during high fire hazard weather conditions [17]. The common physical structures of planted forests when they are young (<40 years old), including homogenous canopies, high shrub cover, low crown bases, and high densities of small trees, reflect structural components that have long been understood to increase vulnerability to fire [18]. Recently, empirical evidence from extensive postwildfire assessments in productive MCF have found increased severities in planted forests compared to mature forests [19].

12.2.2.2 Partial Harvests

Partial harvests involve the periodic, diffuse harvest of individual, commercial-sized trees across mature stands, where a significant proportion of timber yield comes from trees that make up the upper canopy layer. This type of regime has occurred across all landowner types, although it has not been common on federal lands in recent decades. Regionally referred to as "selection" or "single tree selection," advanced silvicultural approaches for allocating growth to multiple age classes have not been well developed. On private lands, there are some protections against this type of harvest leading to high-grading, namely retention requirements of large trees. However, one-cut shelterwood harvests (i.e. overstory removals) are common and also fit in this category of partial harvesting. In this approach to overstory removal, larger trees are harvested, releasing smaller trees assumed to be of younger cohorts. However, because these harvests do not follow a sequence of preparation and cohort-initiation steps (i.e. "seed steps"), the harvested larger trees are often no older than the smaller trees they are releasing. This has been documented as a common form of high-grading that results in the loss of long-term productivity [20] and does not have an analog in the natural development model described below.

The precolonial, fire-dominated disturbance regime maintained structures that, in terms of basal area, were relatively low in overall stocking levels and were dominated by large trees. Partial harvests as practiced in the MCF, in contrast, have typically maintained higher basal areas and can preferentially remove large trees [21]. Where MCF stands are managed for sustained timber yield, stocking densities are maintained at levels considerably higher than what was maintained by frequent low-intensity fire [5]. The suppression of fire has enabled this capacity to manage stands within stocking zones much higher than what would be possible with an intact fire regime. Given the diffuse nature of selection harvesting and high canopy densities, species compositions within this system have trended toward dominance of white fir and incense-cedar, although mid-tolerant species (Douglas-fir (*Pseudotsuga menziesii* (Mirb.) Franco) and sugar pine (*Pinus lambertiana* (Dougl.)) may be more common than in undisturbed forests [22].

Partial harvesting that focuses primarily on large tree removal has resulted in high grading and a corresponding loss of timber productivity and an increase in fire hazard [20]. Partially harvested stands that have been managed for timber yield can retain relatively high proportions of large trees (i.e. not high-graded), but these stands are also vulnerable to high-severity fire, even compared to unharvested reserve stands [17]. This vulnerability has been attributed to an absence of surface and ladder fuel treatments [23]. The shift in species composition away from ponderosa pine and toward white fir and incense-cedar has resulted in a transition from fire-resistant compositions toward fire-vulnerable compositions. The lack of frequent fire or a surrogate that mediates surface fuels has caused a significant departure in surface fuel dynamics, leading to overall increases in surface fuel loads and wildfire severity.

12.2.2.3 Fuel Treatments

Fuel treatments are pursued in stands where, in response to increasing wildfire severity, the objective has been to reduce fire severity via a reduction of surface, ladder, and canopy fuels. A commercial timber harvest that thins canopy fuels can be done either prior to or following the surface and ladder fuel treatment. Harvests are typically a thin from below of the smaller available commercial-size trees to a target basal area or canopy cover. Revenue from commercial harvests may or may not cover the cost of fuel treatments, depending on the amount of timber volume that is available to harvest and given market conditions [24]. Fuel treatments may also occur without any removal of canopy fuels (i.e. noncommercial treatments), albeit at a net cost to the landowner.

Because this system focuses on reducing surface and ladder fuels and retaining large trees, it is closer than the other systems in its alignment with the precolonial disturbance regime and developmental model described below. As a silvicultural system made up of a series of treatments, however, it becomes misaligned over time because of the lack of moderate-severity disturbances that create distinct canopy gaps. While the use of prescribed fire to maintain conditions created by fuel treatments is a critical element that provides a strong qualitative alignment, prescribed fires typically either do not occur because of social constraints or they are not hot enough to create discrete canopy gaps when they do occur [25].

Because fuel treatments focus primarily on reduction of surface and ladder fuels, with canopy reductions as a lower priority [26], they have resulted in a lowering of wildfire severity. However, because fuel treatments are not intended to initiate new cohorts, there is a lack of regeneration of shade-intolerant species [27]. Compared to planted forests and selection harvests, it is more often intended that fuel treatments incorporate prescribed fire in the future. However, rates of fire use on both private and public lands currently lag far behind desired goals [28]. Compared to traditional silvicultural systems, fuel treatments are more aligned with the disturbance regime (especially when they use prescribed fire) and they function well as a method of reducing wildfire severity. However, they overlap with only the low-severity portion of the disturbance regime and therefore are unlikely to sustain all species or structural conditions described in the natural development model below.

12.3 Natural Development Model

The spatial scale at which silvicultural treatments occur in the MCF is 10–100 ha. This range is related to factors of continuity in ecological site conditions (i.e. topography and site productivity) and minimum yields needed to cover costs of operations. The natural development model occurring at this scale is dominated by the process of frequent fires, which generally maintained overall low surface fuels, low stocking levels, and many large trees in terms of relative growing space occupied. Within this 10–100 ha scale over which treatments occur, the natural development model indicates there should be a high complexity in forest structure and species composition. This complexity is a result of frequent fires that create heterogeneity in structure, including distinct canopy gaps, at the 0.1–1.0 ha scale. A cross section of land with a diameter of four mature tree heights (Figure 12.1) provides a visualization of the natural development model for MCF, focusing on how a multiaged structure within a matrix of overall low surface fuels and low stocking is sustained by frequent fires.

12.3.1 Disturbance, Legacy Creation, and Preforest

As with other forest types, disturbances initiate new cohorts of seedlings by releasing sufficient growing space. Of particular importance in MCF, given its dry Mediterranean climate, is the creation of below-ground resources, as well as the creation of bare soil substrates for germination. Disturbances must have sufficient severity (Figure 12.1a) to support the variety of species' germination and establishment requirements (Table 12.1). The spatial pattern over which this can occur is complex, from discrete patches surrounded by mature trees to more diffuse, low-density canopies over larger areas with variable spatial arrangements. Rapid colonization (within 2 years) of disturbed areas via wind-dispersed seeds from surrounding trees (Figure 12.1b) is a critical component of cohort initiation because without rapid seedling establishment sprouting shrubs can exclude tree establishment either indefinitely or for at least several decades [13]. Because the species that occupy postdisturbed areas for long periods in the MCF depend on immediate

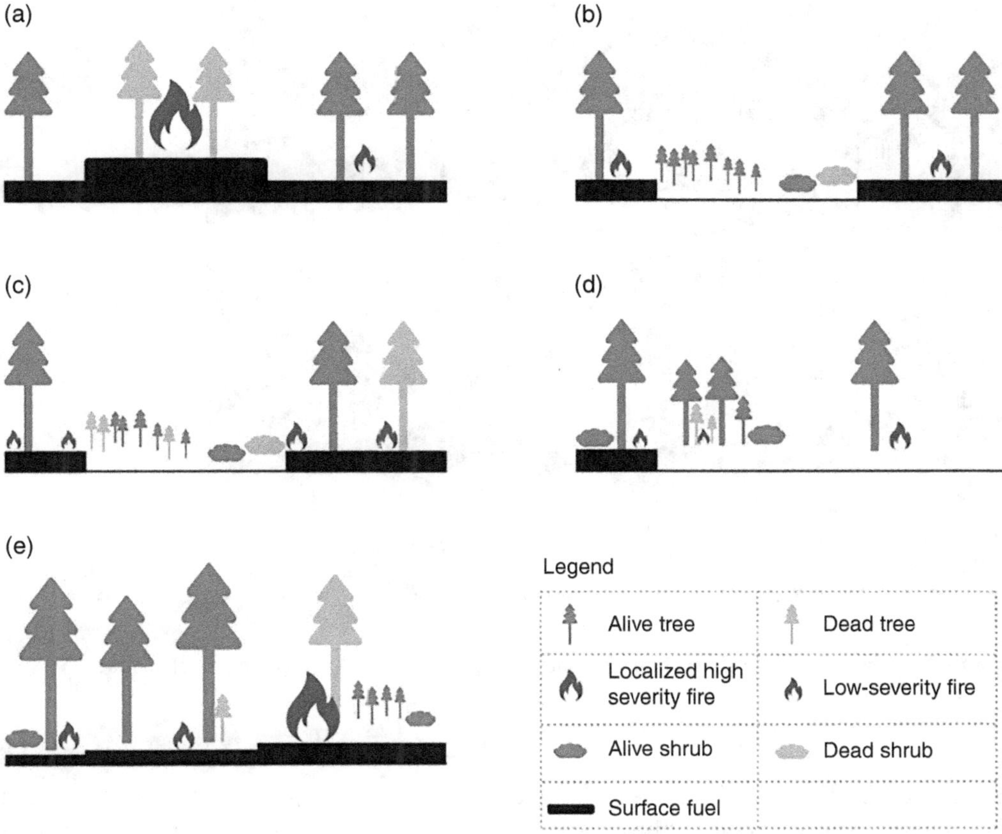

Figure 12.1 A conceptual model of natural development in MCF following a canopy gap created by localized high-severity fire (a), initiation of a dense patch of seedlings (b), thinning of the young cohort, via periodic low-intensity fire (c), canopy recruitment with low density, low surface fuels, and few ladder fuels, maintained by fire (d), and mature forest with high complexity at 1 ha scale maintained with low and moderate-severity fire (e).

establishment following disturbances, I include the preforest stage here as part of the disturbance stage. Species arrangements occupying a given site following disturbances vary substantially, because of highly variable cone production. Large standing dead trees have an ephemeral presence (10–20 years) during cohort initiation, and they are sparse given the overall low basal area that frequent fires maintained. Because fire is the primary disturbance that initiates new cohorts, this phase is characterized by a low amount of surface fuels.

12.3.2 Young Forest

Given the high frequency of low-severity fires following cohort initiation, an important characteristic of the development model for MCF is the rapid initiation of patches of high densities of young trees and shrubs in a young forest stage (Figure 12.1c). High initial densities are reduced over time through a combination of competition-related mortality and especially frequent exposure to low-severity fires (Figure 12.1d). The spatial pattern of tree removal by fire is much more complex than conventional approaches to manual thinning, with fire often removing trees in clumps and along the edges of patches of young trees (Figure 12.2) because of fuel input from surrounding mature

Figure 12.2 A low-severity fire that thinned a patch of young trees with a high degree of spatial patchiness.

trees [29]. Ponderosa pine and giant sequoia typically outgrow other species and can survive fire at higher rates when young [30], leading to increased relative abundance over time. As density is reduced, height to crown base increases during development, from a combination of shading of lower branches and lower crown scorch from surface fires. Crown differentiation, creating nested structural complexity within cohorts, occurs via species' variable responses to competition and frequent fires. Surface fuels during this phase are expected to increase gradually compared to the beginning of the cohort initiation phase as fuel input rates increase from leaf area development and litterfall. Surface fuels increase slowly in a sporadic and ratcheting manner, as they decrease abruptly with low-severity fires, followed by periods of steady recovery between fires.

Within a cohort occurring in a discrete canopy gap, individual trees near the centers are most likely to reach large sizes first because of higher resource availability and because of fire's removal of trees along edges. Recruiting trees thus may be clumped together, with crowns of neighboring trees touching each other but with wide spacing between clumps, eventually forming colonnades between recruiting trees and surrounding mature trees [29]. This process of large tree development may be slow, as growth is curtailed by frequent fire that slows growth rates via leaf area reductions and cambium damage. New cohorts nearby are expected to develop during this phase, sustaining a multiaged structure at larger spatial scales. Frequent fires are expected to maintain overall low surface fuels, with periods of fuel recovery between fires.

12.3.3 Mature and Old Forest

From the original cohort that was established following a locally-severe fire 100+ years earlier, mature trees are now large (>100 cm dbh) and would have persisted through many low-severity fires (Figure 12.1e). Crown bases are high and most mature trees have fire scars at their bases.

Mature trees are developing mast cone crops periodically, colonizing nearby burned patches with wind-dispersed seeds when masting years are coincident with fires. Across the multiaged stand, density is highly variable, with density generally lowering within a cohort as it ages. Mature individuals or clumps of trees eventually die from a complex of fire and biological/abiotic interactions. Mature trees are large enough that the mortality of one particularly large or a few moderately large trees can create sufficient growing space for shade-intolerant species to regenerate and begin the cycle again. Across stand and multistand scales, overall density is much lower than what the site is capable of sustaining when fire is removed.

12.4 An Ecological Silviculture System for Mixed Conifer Forests

The current disturbance regime in the MCF has departed drastically from the one that existed during the precolonial era to the extent that the ecosystem's existence is arguably threatened [31]. The departure is from a frequent, low-severity fire regime to which tree species were adapted toward an infrequent, high-severity regime in which tree species are not capable of surviving. Given the once-omnipresent nature of low and moderate-severity fires that sustained the MCF, of greatest importance are silvicultural systems that strive to achieve the elements of the fire regime that are no longer occurring. Namely, these elements are a ubiquity of low surface fuels, patches of dense regeneration, large trees that are abundant in terms of relative growing space occupied, and overall stocking densities that are low enough to be capable of resisting epidemic mortality from climatic drought stress. The use of prescribed fire when and where feasible is paramount, and foresters should strive for its implementation. Recognizing prescribed fire as a silvicultural treatment that can meet objectives (i.e. "Pyrosilviculture") may help increase its acceptability within silvicultural systems for this forest type [30]. In most current management contexts, however, prescribed fire may be either restricted entirely or at the least continue to be limited in extent and effectiveness. Of critical importance, therefore, are non-fire treatments that can achieve significant structural and compositional changes that fire alone either currently or indefinitely cannot achieve.

The distribution of canopy gap sizes is the structural element that is most clearly missing among the current silvicultural systems (Figure 12.3). An ecological silvicultural system, therefore, is one that relies on gap-based silviculture as the primary regeneration method. A *gap* is defined here as a canopy opening that is large enough to have sufficient growing space to regenerate all species in Table 12.1. They are also small enough so that at least half of the regenerating area has less than full resource availability because of competition with surrounding mature trees (i.e. edge effect). In the mixed conifer forest, this is typically 0.04–1.0 ha, depending on surrounding canopy height and density. Importantly, I include larger areas of regeneration as a gap if retained trees within them are numerous enough to create complex resource gradients. A harvested area larger than 1.0 ha may still be a form of gap-based silviculture if the number of large trees retained within the gap increases with harvested area. These retained trees create micro-sites of edge environment, thus allowing gap areas to increase while still meeting the standard that at least half of the area has less than full resource availability.

The elements of a gap-based silviculture system guided by the natural development model are:

- The creation of many (relative to larger openings) canopy openings <1.0 ha, creating the desired high ratio of smaller to larger gap sizes (Figure 12.3)
- Infrequent larger canopy openings >1.0 ha that represent moderate-severity disturbances and create early seral conditions can be created but with large tree retention to create resource gradients

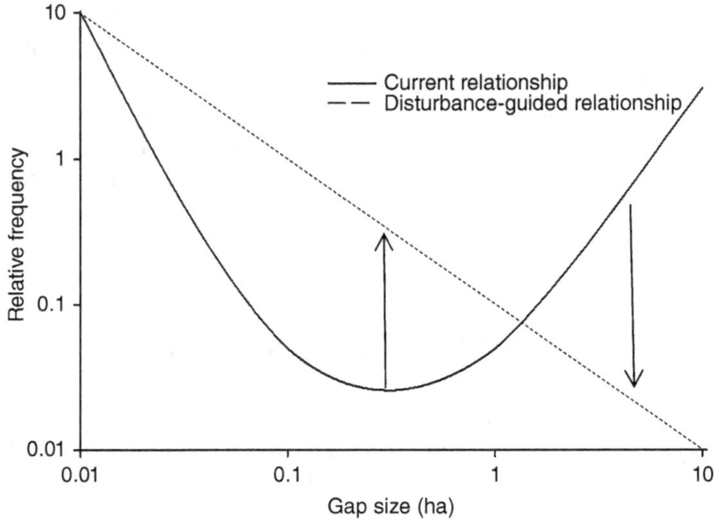

Figure 12.3 Relationship between relative frequency and gap size under the current practices of planted forests, selection harvests, and fuel treatments (solid line). A silvicultural system that used the disturbance regime as a guide would have a downward sloping relationship (dashed line). Gaps greater than 1.0 ha would also have large trees retained within them. Arrows show how the current disturbance regime would need to change for the landscape to be more closely aligned with the precolonial regime.

- Postharvest site preparation of fuels in areas intended for regeneration, establishing low surface fuels during cohort development
- The capacity to plant in order to influence species composition as needed and account for sporadic cone production
- The allowance of shrub development to occur at a level that does not preclude tree establishment and recruitment (e.g. <30% cover)
- Prescribed, low-intensity fires every ~10 years to maintain low surface fuels at the stand scale (10–100 ha) and to lower the density of young cohorts as they develop. Regeneration harvests may occur on alternate entries, every ~20 years.
- Area regulation and planning rotation ages of 100–200 years can be used to ensure long-term sustaining of large, old trees

12.5 Climate Change Considerations

The changing climate of the MCF has revealed and magnified the negative impacts of silvicultural systems that do not sufficiently incorporate key elements of the ecosystem's inherent disturbance and development regime. Simultaneously, it has revealed the negative impacts of the absence of any silvicultural system (i.e. no treatments) in disturbance-dependent forests. While using the past disturbance regime as a guide for designing silvicultural systems may risk being backward-looking when facing a changing climate, it provides a foundation from which to make adjustments in the future. Creating gaps and then planting at regular intervals enables transition silviculture [32], where either new species or new genetic sources of current species that are adapted to the future climate can be regenerated with management experiments. Periodic density management of both regeneration within gaps and mature trees surrounding gaps allows for the probing of new density management regimes that fit within evolving concepts of resistance and resilience. Finding stocking levels that are high enough to support acceptable timber yields but are also low enough to avoid surpassing thresholds of drought-related mortality epidemics [33] is a significant challenge to the continuation of a viable timber ecosystem in the MCF. But finding new approaches that work is critical, because timber harvests are an essential tool for managing density, regenerating desired species, and generating revenue for managing fuels. The loss of this tool is itself a significant threat to the mixed conifer ecosystem.

12.6 Using the Natural Development Model to Alter Existing Systems

Many silvicultural treatments are practiced because they are "known to work" through experience. This traditional professional knowledge is a fundamental component of silviculture that could be lost by attempting to transition to a new system abruptly. I therefore focus below on how existing systems can be modified, rather than replaced. Some existing systems are already closer than others to an idealized ecological silvicultural system, but they would all need to be changed in significant ways (Table 12.2).

12.6.1 Planted Forests

The disturbance-maintained relationship between patch size and frequency is known generally to be negative [8]. That is, more small patches compared to large ones. A silvicultural system that relies primarily upon 8-ha clear-cuts is misaligned but could be brought more into alignment (Figure 12.3) by incorporating a much higher frequency of smaller gap sizes during regeneration harvests. The dominance of one or two species in planted stands is generally misaligned with the composition of MCF. Reforestation in young stands would need to incorporate tree diversity as a higher priority. Given that lifespans (300–600 years) are much longer than planned rotations (50–80 years), age structures will increasingly become misaligned unless rotation ages are lengthened. Managing for rapid individual tree growth via density management can facilitate the

Table 12.2 Areas of misalignment between current practice and ecological silviculture for the three most common silvicultural systems of the Sierra Nevada mixed conifer forest. Adjustments would bring current practices closer to alignment with ecological silviculture.

Existing system	Areas of misalignment	Adjustments
Planted forests	Rotation age, distribution of patch sizes, species composition, lack of low-intensity fire	Longer rotation ages, increase the ratio of small to large gaps, plant shade-tolerant species, use low-intensity fires along with commercial thinning
Selection harvests	Mid-story density, species composition, lack of low-intensity fire	Reduce mid-story density, create discreet canopy gaps, plant shade-intolerant species, experiment with low-intensity fire
Fuel treatments	Canopy heterogeneity, species composition, lack of low-intensity fire as an intermediate treatment	Create discreet canopy gaps, plant shade-intolerant species, commit to low-intensity fires with prescriptions

development of large trees without requiring rotation ages greater than 300 years, which would likely be seen as impractical. The length of time needed to approximate large tree status would likely be 100+ years, depending on site productivity. Incorporating low-intensity prescribed fires into developing planted stands would be a novel but very effective modification that would better align planted forests with ecological silviculture and also reduce the risk of loss from wildfire [16].

12.6.2 Partial Harvests

Modifications to the partial-harvesting-based approaches would need to include the planned removal of precommercial trees simultaneously with commercial tree removal that avoided high-grading. Maintenance of low surface fuel loads, a fundamental element of ecological silviculture, may be challenging to incorporate. Prescribed fire in these stand types have not been explored with the intent of understanding how much regeneration is needed to compensate for losses during prescribed fires. If prescribed fires prove to be too challenging, an alternative could be to transition from diffuse harvesting to discrete, gap-based harvesting approaches. This would allow for fuel reduction (i.e. site preparation) following regeneration harvests of small canopy gaps (<1.0 ha). The surrounding mature forests could be managed similarly to fuel treatments until they are regenerated in the future.

12.6.3 Fuel Treatments

Recent suggestions for altering fuel treatments have focused on introducing some form of variable density thinning [34]. This can create fine-scale heterogeneity but is lacking coarser-scale heterogeneity compared to the natural development model (Figure 12.1a). While maintaining fuel treatments with periodic prescribed fire is often the intent, especially on federal lands, execution lags far behind goals. Pyrosilviculture, specifically altering fuel treatments in order to increase opportunities for conducting prescribed fires in the future, may help increase prescribed fire activity [30]. Given that much of the MCF is still young compared to potential lifespans, existing fuel treatments could continue for several years before regeneration harvests need to occur. Canopy gaps created for regeneration within this thinned matrix could be small, a minimum of 0.04 ha [23], and can be planned with basic tools of rotation ages and harvesting frequencies. Fires may be of low enough intensity that cohorts within gaps can recruit even with frequent burning (Figure 12.2). Alternatively, patches of regeneration, especially those that have shade-intolerant species present, may need to be protected during burns. Planting following gap creation may be needed to create high-enough densities that can absorb periodic fire-related mortality that would be expected from high-frequency fires.

References

1 Stewart, W.C. and Nakamura, G. (2012). Documenting the full climate benefits of harvested wood products in Northern California: linking harvests to the U.S. Greenhouse Gas Inventory. *Forest Products Journal* 62: 340–353.

2 Krogh, S.A., Broxton, P.D., Manley, P.N., and Harpold, A.A. (2020). Using process based snow modeling and lidar to predict the effects of forest thinning on the northern Sierra Nevada snowpack. *Frontiers in Forests and Global Change* 3: 21.

3 Helms, J.A. and Tappeiner, J.C. (1996). Chapter 15: Silviculture in the Sierra. In: *Sierra Nevada Ecosystem Project: Final Report to Congress, Vol. II, Assessments and Scientific Basis for Management Options*, 439–476. Davis: University of California, Centers for Water and Wildland Resources.

4 Stewart, W., Sharma, B., York, R. et al. (2016). Forestry. In: *Ecosystems of California* (ed. H. Mooney and E. Zavaleta), 817–833. University of California Press.

5 North, M.P., Tompkins, R.E., Bernal, A.A. et al. (2022). Operational resilience in western US frequent-fire forests. *Forest Ecology and Management* 507: 120004.

6 Collins, B.M., Bernal, A., York, R.A. et al. (2021). Mixed-conifer forest reference conditions for privately owned timberland in the southern Cascade Range. *Ecological Applications* 31 (7): e02400.

7 Kilgore, B.M. and Taylor, D. (1979). Fire history of a sequoia-mixed conifer forest. *Ecology* 60 (1): 129–142.

8 Collins, B.M. and Stephens, S.L. (2009). Stand-replacing patches within a "mixed severity" fire regime: quantitative characterization using recent fires in a long-established natural fire area. *Landscape Ecology* 25: 927–989.

9 Safford, H.D. and Stevens, J.T. (2017). *Natural Range of Variation for Yellow Pine and Mixed-Conifer Forests in the Sierra Nevada, Southern Cascades, and Modoc and Inyo National Forests, California, USA*, Technical Report PSW-GTR-256. USDA Forest Service General.

10 Stephens, S.L. and Ruth, L.W. (2005). Federal forest-fire policy in the United States. *Ecological Applications* 15 (2): 532–542.

11 Eriksen, C. and Hankins, D.L. (2013). The retention, revival, and subjugation of indigenous fire knowledge through agency fire fighting in Eastern Australia and California. *Society and Natural Resources* 27 (12): 1288–1303.

12 York, R.A., Roughton, A., Tompkins, R., and Kocher, S. (2020). Burn permits need to facilitate- not prevent- "good fire" in California. *California Agriculture* 74 (2): 62–66.

13 Tubbesing, C.L., York, R.A., Stephens, S.L., and Battles, J.J. (2020). Rethinking fire-adapted species in an altered fire regime. *Ecosphere* 11 (3): e03091.

14 Knapp, E.E., Skinner, C.N., North, M.P., and Estes, B.L. (2011). Long-term overstory and understory change following logging and fire exclusion in a Sierra Nevada mixed-conifer forest. *Forest Ecology and Management* 310: 903–914.

15 McDonald, P.M. and Fiddler, G.O. (2010). *Twenty-five years of managing vegetation in conifer plantations in northern and central California: Results, application, principles, and challenges*, USDA Forest Service PSW-GTR-231. Albany, CA: U.S. Department of Agriculture, Forest Service, Pacific Southwest Research Station.

16 York, R.A., Levine, J., Russell, K., and Restaino, J. (2021). Opportunities for winter prescribed burning in mixed conifer plantations of the Sierra Nevada. *Fire Ecology* 17: 33.

17 Stephens, S.L. and Moghaddas, J.M. (2005). Experimental fuel treatment impacts on forest structure, potential fire behavior, and predicted tree mortality in a California mixed conifer forest. *Forest Ecology and Management* 215: 21–36.

18 Kobziar, L.N., McBride, J.R., and Stephens, S. (2009). The efficacy of fire and fuels reduction treatments in a Sierra Nevada pine plantation. *International Journal of Wildland Fire* 18 (7): 791–801.

19 Levine, J.I., Collins, B.C., Steel, Z.L. et al. (2022). High severity fire incidence higher both on and near industrially managed forests. *Frontiers in Ecology and the Environment* https://doi.org/10.1002/fee.2499.

20 York, R.A. (2015). Large-tree removal in a mixed conifer forests halves productivity and increases white fir. *California Agriculture* 69 (1): 27–35.

21 Collins, B.M., Fry, D.L., Lyderson, J.M. et al. (2017). Impacts of different land management histories on forest change. *Ecological Applications* 27 (8): 2475–2486.

22 Levine, C.R., Krivak-Tetley, F., van Doorn, N.S. et al. (2016). Long-term demographic trends in a fire-suppressed mixed-conifer forest. *Canadian Journal of Forest Research* 46: 745–752.

23 York, R.A., Battles, J.J., Wenk, R.C., and In Saah, D. (2011). A gap-based approach for regenerating pine species and reducing surface fuels in multiaged mixed conifer stands in the Sierra Nevada, California. *Forestry* 85 (2): 203–213.

24 Hartsough, B.R., Abrams, S., Barbour, R.J. et al. (2008). The economics of alternative fuel reduction treatments in western United States dry forests: financial and policy implications from the National Fire and Fire Surrogate Study. *Forest Policy and Economics* 10 (6): 344–354.

25 North, M., Innes, J., and Zald, H. (2007). Comparison of thinning and prescribed fire restoration treatments to Sierran mixed-conifer historic conditions. *Canadian Journal of Forest Research* 37 (2): 331–342.

26 Agee, J.K. and Skinner, C.N. (2005). Basic principles of forest fuel reduction treatments. *Forest Ecology and Management* 211: 83–96.

27 Moghaddas, J.J., York, R.A., and Stephens, S.L. (2008). Initial response of conifer and California black oak seedlings following fuel reduction activities in a Sierra Nevada mixed conifer forest. *Forest Ecology and Management* 255: 3141–3150.

28 Clarke, A.K., Tompkins, R.E., Wang, J.A. et al. (2022). Accurate tracking of forest activity key to multi-jurisdictional management goals: a case study in California. *Journal of Environmental Management* 302: 114083.

29 Biswell, H. (1989). *Prescribed Burning in California Wildlands Vegetation Management*. Berkeley, CA: University of California Press.

30 York, R.A., Noble, H., Quinn-Davidson, L.N., and Battles, J.J. (2021). Pyrosilviculture: combining prescribed fire with gap-based silviculture in mixed-conifer forests of the Sierra Nevada. *Canadian Journal of Forest Research* 51: 1–11.

31 Cabiyo, B., Fried, J.S., Collins, B.M. et al. (2021). Innovative wood use can enable carbon-beneficial forest management in California. *Proceedings of the National Academy of Sciences* 118 (49): e2019073118.

32 Nagel, L.M., Palik, B.J., Battaglia, M.A. et al. (2017). Adaptive silviculture for climate change: a national experiment in manager-scientist partnerships to apply an adaptation framework. *Journal of Forestry* 115 (3): 167–178.

33 Bernal, A.A., Stephens, S.L., Collins, B.M., and Battles, J.J. (2022). Biomass stocks in California's fire-prone forests: mismatch in ecology and policy. *Environmental Research Letters* 17: 044047.

34 North, M., Stine, P., O'Hara, K. et al. (2009). *An Ecosystem Management Strategy for Sierran Mixed-Conifer Forests*, Gen Tech. Rep. PSW-GTR-220. USDA Forest Service.

13

Ecological Silviculture for Aspen Mixedwoods in Western Canada

S. Ellen Macdonald, Philip G. Comeau, Charles A. Nock, and Brad D. Pinno

Department of Renewable Resources, University of Alberta, Edmonton, AB, Canada

13.1 Introduction

The boreal mixedwood forest region in western Canada extends across portions of Manitoba, Saskatchewan, Alberta, and British Columbia (Figure 13.1). Forests are established on sedimentary parent material with soils developed from predominantly clay surficial deposits with some till [1]. The region has a relatively dry and cold climate, and the dominant natural disturbance regime is one of relatively frequent, stand-initiating wildfires; drought and insect outbreaks are of secondary importance [2, 3] (Figure 13.2a). Stands targeted for harvesting largely occur on upland mesic sites and are dominated by varying mixtures of broadleaf trees [trembling aspen (*Populus tremuloides* Michx.), balsam poplar (*Populus balsamifera* L.), paper birch (*Betula papyrifera* Marshall)] and conifers [white spruce (*Picea glauca* (Moench) Voss), black spruce (*Picea mariana* (Mill.) Britton, Sterns & Poggenb.), and balsam fir (*Abies balsamea* (L.) Mill.)] (Figure 13.2b, c). Distribution of forest types and successional stages across the landscape reflect disturbance history plus some influence of site and topography [4]. The mixed-species nature of these forests is important for biodiversity (e.g. [5, 6]). Because trembling aspen and white spruce are the main targets of forest management in the region we refer primarily to these two species in this chapter.

Boreal mixedwood forests have supported indigenous people for millennia and are currently valued by local communities, and society at large, for wildlife habitat, recreation, traditional foods and medicines, and aesthetic and spiritual values [7, 8]. They are home to iconic species at risk, such as the woodland caribou (*Rangifer tarandus* L.). The history of industrial forest management is quite short; there was some harvesting of conifers for the past century, but more extensive exploitation of these forests began in the 1980s when trembling aspen became of interest for production of pulp and chips. In addition to natural disturbance and harvesting, exploitation of oil and gas reserves represents a major driver of forest loss and fragmentation [9]. Forest lands in the western boreal are almost exclusively provincially owned public lands, managed to meet multiple objectives with rigorous planning procedures and policies driving most silvicultural decisions. These, in some ways, promote mixedwoods and in other ways make it more challenging to manage in ways that conform to ecological silviculture [10].

Ecological Silvicultural Systems: Exemplary Models for Sustainable Forest Management, First Edition. Edited by Brian J. Palik and Anthony W. D'Amato.

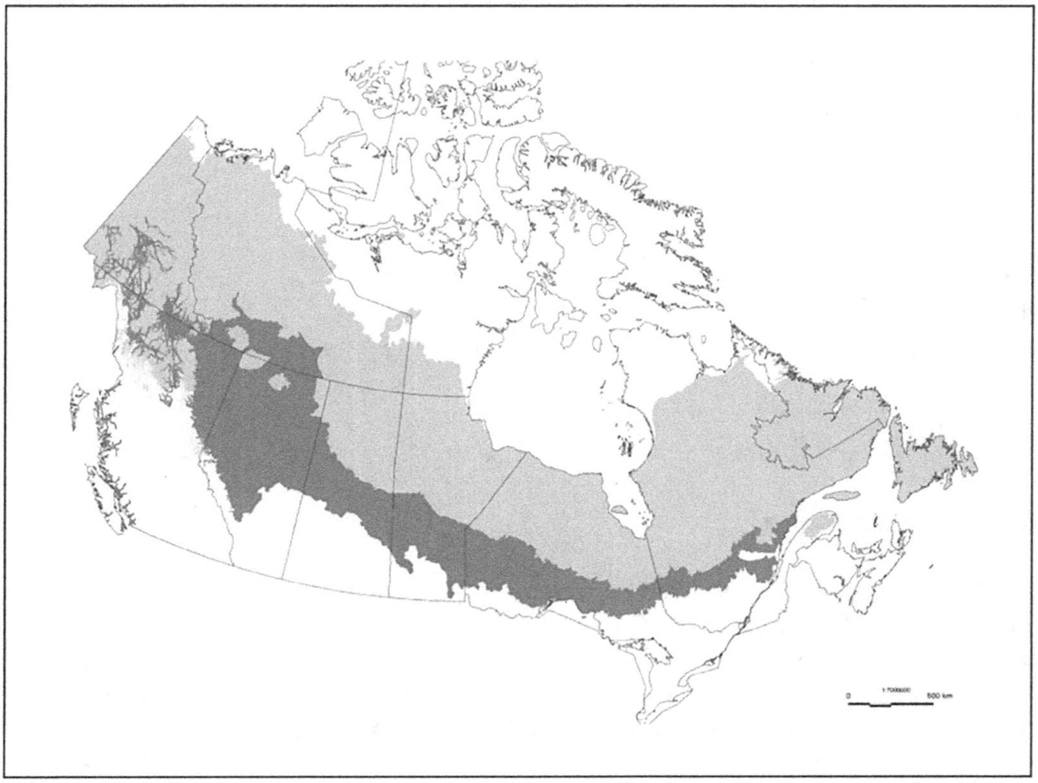

Figure 13.1 The Canadian boreal forest region with the boreal mixedwood shown in dark grey [54, 55]. We focus herein on the western portion. *Source:* Bergeron et al. [11]/Reproduced with permission of Canadian Institute of Forestry.

13.2 Natural Disturbance and Successional Dynamics

The successional dynamics of these forests reflect the life history strategies of the tree species and the natural disturbance regime. A classic transition from broadleaf to mixedwood to conifer is often presumed, but variable disturbance effects and other stochastic factors result in a wide variety of successional pathways and stand conditions [11]. Severe disturbances (i.e. wildfire) kill the majority of mature trees, but only very intense fires severely impact the root systems of aspen, which suckers prolifically even in stands with relatively low pre-disturbance aspen density [12]. With less severe disturbances (e.g. defoliation by forest tent caterpillar, age-related mortality of aspen) canopy openings trigger suckering, leading to multi-aged aspen stands [11, 13].

White spruce is a moderately shade tolerant, masting species that requires a mineral soil or shallow organic seedbed for seedling establishment [14]. Seedbed conditions soon after wildfire or harvesting are good for white spruce and mast years occurring soon after disturbance can result in substantially higher levels of natural regeneration than when disturbances coincide with low seed crops [15]. Delayed regeneration also occurs, notably on substrates such as decayed logs [16, 17]. Because white spruce grows more slowly than aspen, stands often develop as a dominant aspen

Figure 13.2 Unmanaged boreal mixedwoods. (a) Recent wildfire on a mixedwood landscape; (b) Landscape variously dominated by conifer (white spruce) or broadleaf (trembling aspen); (c) Intimate mixture of aspen and white spruce; (d) 90-year-old aspen-dominated stand with white spruce understory; (e) Mixed stand with both aspen and white spruce in the canopy. *Source:* (a–c, e) Jim Witiw/Courtesy of Mercer Peace River Pulp Ltd; (d) Phil Comeau.

canopy with a white spruce subcanopy, even when both species establish simultaneously post-disturbance (Figure 13.2d).

The timing and abundance of white spruce regeneration is a key factor driving subsequent successional development [11]. Common stand types include: pure aspen, aspen with an understory of spruce, intimate mixtures of aspen and spruce, and relatively pure spruce stands (Figure 13.2b–e); any of these could be even-aged or uneven-aged (Table 13.1). Mixedwoods are often multi-aged and can contain veteran (supercanopy) white spruce that are older than the dominant canopy [18]. Recent evidence suggests that some areas in this region experience a mixed fire regime with a large proportion of mixed and spruce-dominated stands being multi-aged [19].

Table 13.1 Naturally occurring mixedwood stand conditions (1 through 7) and the harvesting, regeneration, and tending components of Ecological Silvicultural Systems (A through H) for creating these or transitioning between them. Stand types 3, 4, and 5 could be even-aged or uneven-aged. Once the target stand is created, subsequent management could follow any of the appropriate options for that stand type.

	A.	B.	C.	D.	E.	F.	G.	H.
Initial stand	 (1)	 (1) or (2)	 (1) or (2)	 (3)	 (4)	 (5)	 (5)	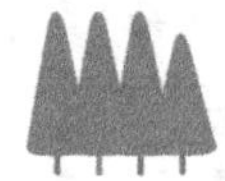 (5) or (6)
Target stand	 (1)	 (2)	 (3)	 (4)	 (3)	 (3)	 (7)	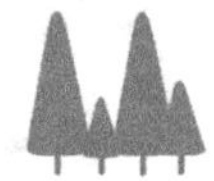 (7)
Harvest	Low retention	Moderate retention		Understory protection	Low retention	Moderate retention	Irregular shelterwood (aspen removal)	Irregular shelterwood
Regeneration	Natural regeneration of aspen	Natural regeneration of aspen	Underplant spruce	Natural regeneration of aspen	Natural regeneration of aspen, plant spruce	Natural regeneration of aspen, and spruce	Natural regeneration of spruce and aspen	Natural regeneration of spruce
Tending	Thinning aspen if desired		Spot tending around spruce if needed		Spot tending around spruce, aspen thinning at age 10	Spot tending around spruce		

1. Even-aged aspen; 2. Uneven-aged aspen; 3. Aspen with spruce sub-canopy; 4. Aspen and spruce mixed; 5. Spruce-dominated mixed; 6. Spruce-dominated (even-aged); 7. Uneven-aged spruce.

Source: Illustration: Bess Callard.

13.3 Current Silvicultural Approaches

The current silvicultural approaches for boreal mixedwoods in western Canada are essentially variations of low-retention harvests: clear-cutting and even-aged regeneration to reproduce the preharvest stand type. Aspen is regenerated naturally from root suckers while spruce is planted immediately after harvest. The result is single cohort stands with the intention of clear-cutting at the next rotation. Due to costs and perceived uncertainties, there is little intentional tending for mixedwoods. The most common tending practice is broadcast aerial herbicide application to boost spruce yield; this is often applied across only portions of a regenerating stand, resulting in spatially segregated mixedwoods [20]. Given the variety of stand conditions arising from natural development, including varying composition, multi-aged condition, and complex vertical structure (Table 13.1), implementing an intentional ecological silvicultural approach requires a change in silvicultural approaches. Recently, there has been increased use of elements of ecological silviculture, including partial harvesting (e.g. of aspen with protection of understory spruce) and retention harvesting (see Chapter 1), but the overall management paradigm of boreal mixedwoods has changed little over the past 30 years.

13.4 Ecological Silvicultural Systems for Boreal Mixedwoods in Western Canada

Lieffers and Beck [21] highlighted several issues with current silvicultural systems for boreal mixedwoods and argued for a seminatural approach to mixedwood management. This was followed by recommendations for *seminatural* silvicultural systems [22], many of which are reflected in the approaches described herein. By implementing ecological silviculture we aim to maintain the wide range of ecological, economic, and social values provided by boreal mixedwood stands and forests, enhance forest resistance and resilience to natural disturbances and climate change, and allow more options for individual stands as conditions (markets, environment, social) change.

Using the naturally occurring range of stand compositions and structures as a foundation, we describe ecological silvicultural systems, including tools for harvesting, regeneration, and tending, to create a variety of naturally occurring stand conditions and facilitate transition between types, as occurs with natural stand development (Figure 13.2, Table 13.1). These systems are based on the following principles:

- ***Maintain varying mixtures spatially and temporally:*** Boreal mixedwood forests exist as a wide variety of compositions and structures arising from natural stand development (Table 13.1). Maintaining this variation and facilitating transitions between conditions that reflect successional stages is key to our recommendations for ecological silviculture. As aspen is shade intolerant, consideration must be given to maintaining it in a mixture while also recognizing that its abundance will change from being dominant following a stand-initiating disturbance, to being a minor component during later successional stages.
- ***Maintain structural heterogeneity (canopy height, age, deadwood, etc.) spatially and temporally:*** Maintaining the variation in size, age, stand condition, and vertical structure that exists across a landscape as a reflection of natural stand development requires working with succession by using partial harvesting rather than forcing ecosystems to move in particular directions (e.g. toward structurally simple spruce-dominated stands) at an accelerated pace.

- ***Accept variability and uncertainty:*** Ecological silviculture approaches must regenerate and maintain a representative balance of stand types and structures within a forest landscape or management unit, accepting that blocks can change between conifer, mixedwood, and deciduous types over time. From a policy perspective, variability should be prioritized over simple homogeneous stand structures, both spatially and temporally. It is essential to move away from locking into a requirement to regenerate to the preharvest stand composition, instead allowing for the stand transitions that occur naturally over time.

13.4.1 Harvesting Options for Boreal Mixedwoods Under Ecological Silvicultural Systems

13.4.1.1 Retention Harvesting

This has emerged as a prominent tool for use in ecological silvicultural systems [23] as it emulates the natural retention of structure and ecological legacies after disturbance (see Chapter 1) and is a treatment that can be applied in clear-cuts or shelterwoods. The key element of retention harvesting is the focus on what is retained (i.e. live trees, large deadwood, and other structural elements), rather than on what is removed. Currently, the vast majority of harvesting in the western boreal mixedwood is a form of low-retention (<15%) harvest, which was thought to emulate stand-initiating fires and to be required to regenerate shade-intolerant aspen [12]. In actuality, a range of retention levels can be emulative of natural disturbance and be used in a silvicultural system, depending on the objective. Low retention can result in an even-aged, aspen-dominated stand, or be followed by planting of spruce to create an even-aged mixture; these emulate natural dynamics following severe disturbance (Table 13.1: Silvicultural System A or E). Moderate or high (as in an irregular shelterwood) levels of retention emulate the effects of less severe disturbance and can be followed by natural regeneration of aspen and spruce, or planting of spruce, to create the naturally existing variety of multi-aged, mixed, structures with varying dominance by aspen and spruce (Table 13.1, System B, F, G, H; Figure 13.3a, b).

13.4.1.2 Understory Protection Harvesting

Mixedwood stands in this western boreal region often have an aspen canopy with a spruce subcanopy (Figure 13.2d; Table 13.1: stand condition 3). Aspen growth peaks at age 40–80, beyond which stem decay increases, but slower-growing spruce generally do not provide much merchantable volume before age 70 [24–26]. Removing some mature aspen by partial harvesting, while protecting understory white spruce, fits well within an ecological silvicultural system because it: facilitates the natural successional transition from aspen- to white spruce dominance; maintains structural diversity; and allows harvest of sound and mature aspen, leaving some to die and decay as an ecological legacy [27]. It also benefits understory spruce by reducing temperature extremes [28], suppressing competitors [29], and reducing white pine weevil (*Pissodes strobi* Peck) damage [30]. In this harvesting system, termed understory protection in western Canada and natural shelterwood in some other regions, all trees are removed in extraction trails, while in removal strips aspen are harvested and most spruce are left (Table 13.1: System D; Figures 13.3c and 13.4). For wind buffering, aspen can be left in strips (3–10 m wide) oriented perpendicular to prevailing winds [25] or by uniform retention of 15–40% of the original aspen basal area. Understory protection harvesting results in stands with three strata: (i) extraction trails, where regeneration is predominantly aspen from root suckers; (ii) removal

Figure 13.3 Elements of ecological silvicultural systems for boreal mixedwoods. Retention harvesting with large patches: (a) two-years post-harvest; (b) 16-years post-harvest; (c) understory protection five-years post-harvest; (d) 20-year-old mixedwood stand with natural (unthinned) aspen density and 500 planted spruce ha^{-1}; (e) underplanted spruce in an aspen stand; (f) 20-year-old mixedwood with aspen thinned at 10-years. *Source:* (a, b) Jim Witiw/Courtesy of Mercer Peace River Pulp Ltd; (c, d) Phil Comeau; (e, f): Brad Pinno.

areas, where advance spruce regeneration dominates with abundant aspen regeneration from root suckers; and (iii) unharvested buffers, which have a mature aspen overstory and a white spruce understory [25]. Release growth of the spruce advance regeneration, plus natural regeneration of aspen, leads to development of an uneven-aged mixture with varying spruce

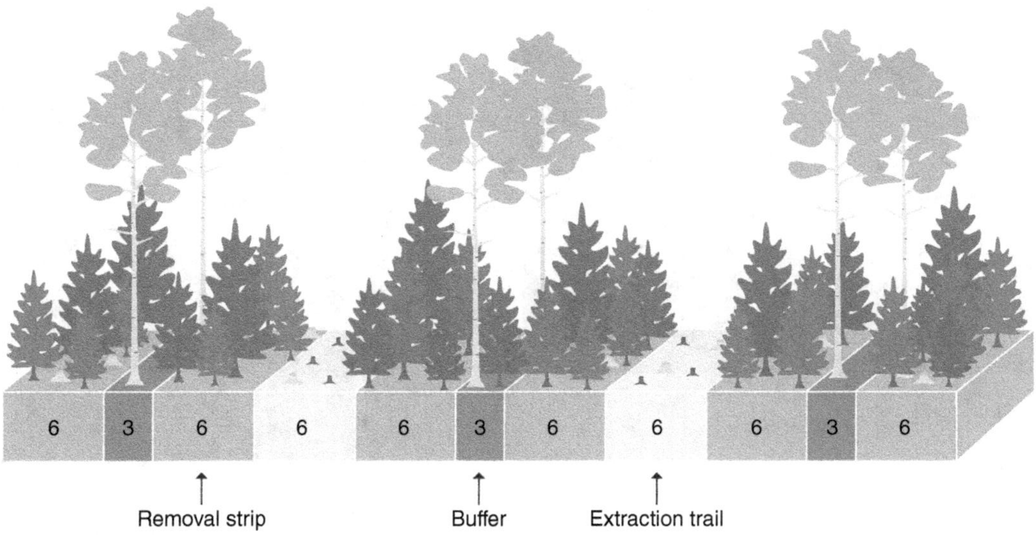

Figure 13.4 Partial harvesting of aspen with protection of understory spruce in a boreal mixedwood. All trees are harvested along 3–6 m wide extraction trails. Aspen is removed, and understory white spruce is retained, in 6–8 m wide removal strips to either side of the extraction trail (width equal to feller-buncher reach). Aspen can be left in 3–10 m wide buffer strips, between two removal strips and oriented perpendicular to prevailing winds. *Source:* Illustration: Bess Callard.

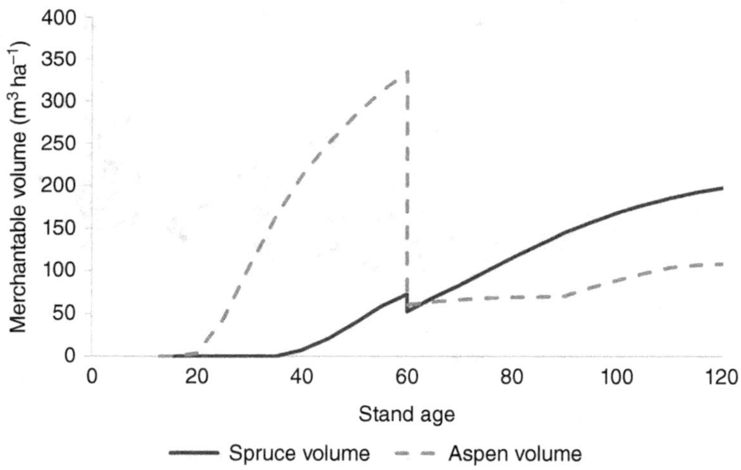

Figure 13.5 General trend in merchantable volume for spruce and aspen for stands treated by partial harvesting of aspen with protection of understory spruce at age 60 (merchantability criteria: 15 cm stump diameter, 10 cm top diameter inside bark (DIB). *Source:* Trends are based on simulations with the Mixedwood Growth Model (MGM21, https://MGM.UALBERTA.CA accessed 6 May 2022).

dominance (Table 13.1: conditions 4–7) 40–80 years later [21, 24, 25, 27]. Understory protection can contribute to maintaining the naturally occurring diversity of mixedwood stand types in a landscape, can reduce silviculture costs [21, 24, 25], and result in good merchantable volume of aspen and spruce (Figure 13.5).

13.4.2 Regeneration Options for Boreal Mixedwoods Under Ecological Silvicultural Systems

Greater reliance on natural regeneration for both aspen and spruce is an important component of ecological silvicultural systems; it can be used following any harvesting treatment and will emulate the natural variability in timing, density, and dispersion of regeneration while reducing reforestation costs and improving public perception. Natural regeneration of aspen from root suckers is generally abundant after retention harvests, similar to postfire densities [12, 31]. Suckering is maximized following winter harvests, when root carbohydrate reserves are high, and is reduced by soil compaction, machine damage to roots [32], and low soil temperatures due to high slash loading, deep organic layers, or leaving log decks on-site over the first growing season [33]. The EMEND experiment (Ecosystem Management Emulating Natural Disturbance) demonstrated successful natural regeneration of both aspen and spruce across a wide range of retention harvesting levels, although leaving >35% aspen canopy suppressed aspen regeneration, due to apical dominance (hormonal suppression of suckering by the retained mature trees) and reduced light [27, 34]. Following understory protection harvesting, aspen regeneration can average >7500 suckers ha^{-1} in extraction trails and >14,000 suckers ha^{-1} in removal areas 10 years after harvest [27]. So even with reduced suckering potential, aspen natural regeneration after partial harvesting will often be substantial (Figure 13.3b, d).

Although spruce is almost always planted after harvesting, natural or artificial regeneration from seed is possible, provided acceptance of variability in stocking, density, and initial growth rates. As with natural regeneration generally, the key requirements for spruce are seed source, seedbed, and environment. Seed source is a challenge given the masting nature of white spruce, but the previous two years' weather may be a reasonable predictor of spruce cone production [35]. Natural spruce recruitment is positively correlated with spruce seed tree density [36], thus spruce seed trees must be maintained throughout harvested areas. Isolated trees can be protected from blowdown by retaining them inside intact aspen patches. If spruce seed trees are not available, aerial seeding may be an option. The best seedbed for spruce germination and establishment, exposed mineral soil, can be created mechanically by removing or inverting the organic forest floor layer. This would be appropriate in low-retention harvests while in partial harvests both exposed mineral soil and raised microsites from decomposing logs will act as suitable seedbeds [14, 37]. Competing vegetation, in particular grasses such as *Calamagrostis canadensis*, can be a major hindrance to spruce seedling establishment [29], so prioritizing natural regeneration to situations with lower levels of competition is advisable.

Planting spruce after low-retention harvests can emulate the natural establishment of even- or uneven-aged mixedwoods (Table 13.1: System E). Planting 1000–1600 seedlings ha^{-1}, often in combination with site preparation such as disc trenching or row mounding, can meet regulatory stocking and productivity requirements. Alternatively, white spruce can be planted underneath an aspen-dominated overstory to proactively create an uneven-aged mixed stand with vertical structural diversity, such as occurs in natural stand development (Table 13.1: System C; Figure 13.3d, e). The aspen overstory reduces risk of frost injury and suppresses competing vegetation [38, 39]; although underplanted spruce grow slower than open-grown seedlings, this approach can set the stage for a future understory protection harvest. Underplanting is likely best conducted when the aspen canopy begins to open at the end of the stem exclusion stage (~age 15–25) with an objective of having spruce that are 3–7 m tall at the time of the future understory protection harvest. Problems with browsing by snowshoe hare can be ameliorated by appropriate timing of planting (i.e. after the hare cycle peak), and avoiding planting in very dense young aspen stands or areas with abundant small black spruce.

13.4.3 Tending Options for Boreal Mixedwoods Under Ecological Silvicultural Systems

Tending young mixedwood stands is a useful tool within ecological silviculture systems because it can: increase structural heterogeneity; accelerate the transition toward increased spruce dominance, such as occurs in natural stand development; and facilitate development of stands with a natural aspen–spruce mixed composition (Table 13.1: Systems A, C, E, F). It can also set the stage for future understory protection harvest of aspen at age 50–70 or retention harvest of both species at age 70–100. Tending can involve competition control by cutting aspen and shrubs using motorized brushsaws or herbicides, although abundant aspen stump sprouting and root suckering following broadcast brushing of aspen suckers can make retreatment necessary [40]. "Spot brushing" aspen 1–1.5 m around planted spruce, while retaining untreated aspen in the surrounding matrix, can alleviate this while enhancing survival and growth of young white spruce regenerating in mixture with aspen [40]. Leaving a matrix of ≥30% of such blocks untreated retains their mixedwood nature while also maintaining sufficient aspen to grow for future application of understory protection harvesting. Spot herbicide treatments in a 2 m radius around planted white spruce showed similar outcomes as spot motor-manual brushing while also reducing damage from snowshoe hare browsing [41]. Broadcast treatments using glyphosate or triclopyr herbicides used to control aspen and other vegetation generally result in spruce-dominated stands when they are effective. An alternative, that can facilitate establishment of natural aspen-spruce mixtures, is to apply herbicides in bands (5–14 m wide) with spot, band, or strip application to reduce the amount of herbicide used [42]. Precommercial thinning of aspen at age 4–15 can emulate the natural process of stand thinning and transition toward an aspen-spruce mixture, enhancing growth of both spruce and the residual aspen (Table 13.1: System A, E; Figure 13.3f) [43–45].

13.5 Policy Challenges

Changing management philosophies and practices to align with implementation of ecological silviculture in boreal mixedwoods will be challenging. The vast majority of boreal mixedwood forest in western Canada is on public land managed by provincial governments. Many regulations and policies govern management activities for these forests, including silviculture, with the land being managed to meet multiple values including economic, social, and environmental goals. Consequently, even though the proposed ecological silviculture practices may bring the system closer to a natural model, there may be conflicts with existing policies that favor more traditional silvicultural approaches. For example, current policies in Alberta focus on even-aged modeling of stands and regenerating stands to the preharvest species composition [46]. These policies, therefore, do not reflect the natural successional processes occurring in boreal mixedwoods, which is the focus of ecological silvicultural systems (Table 13.1). Further, the land tenure system for granting access to timber resources tends to allocate aspen and spruce to different companies, posing challenges for ecological management of mixedwoods at both stand and landscape scales. Current policies and the tenure system thus make it difficult to implement management approaches that could maintain the full range of mixedwood stand conditions across the landscape, and allow for dynamic transitions among them.

13.6 Climate Change Considerations for Boreal Mixedwood Management

Climate change is expected to bring warmer and drier, grassland-like, conditions to the boreal mixedwood region in western Canada, which is already more moisture limited than boreal forests in eastern Canada [11, 47]. This is expected to increase the occurrence of drought, outbreaks of defoliating insects, and the frequency and severity of wildfire [48], with implications for ecological forestry at the tree, stand, and landscape scales. In the near term, drought effects are likely to be important drivers of growth and mortality of trembling aspen and white spruce [49, 50].

Ecological silvicultural approaches that facilitate regeneration of aspen–spruce mixtures, in contrast to pure stands, could enhance resistance and resilience to climate change effects, including drought, winter injury, defoliators, and wildfire. For example, since defoliation of aspen and spruce in the same stand is unlikely to be synchronized, reductions in mixed stand productivity during such events would be less than that of pure stands. However, to date, studies have predominantly focused on climate-growth relationships and drought effects in pure stands, and data from mixed stands is quite limited. Thinning can confer resistance to drought as it provides remaining trees with access to more resources, in particular water. Aspen stands have reduced flammability, relative to spruce, under most weather conditions; therefore, maintaining aspen within regenerated stands can help confer resistance, and promoting aspen stands as firebreaks in strategic locations at a landscape scale could be considered [46]. Because flammability of the forest is connected to multiple values (e.g. timber, human health via smoke) its importance in decision-making may increase; however, the current regulatory framework discourages shifting from the current or even historical composition of the forest [46], presenting challenges to ecological silviculture as applied to climate change adaptation.

Given the long-term predictions for transitions from forest to grassland, some other measures should be considered, which might seem to represent a departure from the principles of ecological silviculture. To address future maladaptation of trees to local climate, potential actions include the planting of local or alternative provenances [51] and/or utilization of improved stock selected for stress resistance [52]. However, recent evidence suggests limited variability in drought resilience in spruce [53], and thus potentially limited benefits from assisted migration (AM). Assisted migration would require quite a different implementation in aspen, given most regeneration is from root suckering. This raises questions about how to establish a new population where an existing clone exists and whether newly establishing clones would grow better than the existing clone, given the established root system.

13.7 How Does This Bring Management Closer to Nature?

The ecological silvicultural systems we describe utilize standard silvicultural tools, sometimes in nonstandard ways, along with less common ones, to manage the diversity of stand compositions, structures, and transitions among them found in unmanaged boreal mixedwood forests of western Canada (Table 13.1). These systems, comprised of harvesting, regeneration, and tending options, are designed to maintain variation in species composition, age, and size structural heterogeneity in stands and to facilitate transitions among stand types, as occurs with natural stand dynamics. Retention harvesting helps maintain critical ecological legacies, such as snags and downed woody material, and increases vertical and horizontal structural heterogeneity; it can be followed by natural regeneration or planting to facilitate transition toward even- or uneven-aged mixed stands (Table 13.1: Systems A, B, E, F). Higher levels of retention can be combined with natural regeneration in an Irregular Shelterwood system to facilitate transitions towards uneven-aged spruce stands (Table 13.1: Systems G, H). This could be important in future as we are faced with increasing numbers of even-aged spruce stands resulting from past silvicultural approaches (Table 13.1: System H). Understory protection followed by natural regeneration can be used to maintain or transition toward, uneven-aged mixed stands (Table 13.1: System D). Underplanting spruce in mature aspen stands emulates the natural transition toward uneven-aged, mixed stands (Table 13.1: System D). The ecological silviculture systems we describe rely heavily on natural regeneration; this will result in greater size, spatial and temporal heterogeneity, such as occurs in unmanaged stands. Given aspen's ability for prolific natural regeneration, some thinning or spot tending may be necessary to maintain structural heterogeneity or to facilitate development of aspen–spruce mixtures (Table 13.1: Systems C, E, F). Through these approaches, we aim to foster adaptive capacity and support the full range of ecological, social, and economic values of these forests.

References

1 Fulton, R.J. (ed.) (1989). *Quaternary Geology of Canada and Greenland*, Geological Survey of Canada, Geology of Canada, No. 1. Geological Society of America.

2 Coogan, S.C.P., Daniels, L.D., Boychuk, D. et al. (2021). Fifty years of wildland fire science in Canada. *Canadian Journal of Forest Research* 51: 283–302.

3 Brandt, J.P., Cerezke, H.F., Mallet, K.I. et al. (2003). Factors affecting trembling aspen (*Populus tremuloides* Michx.) health in the boreal forest of Alberta, Saskatchewan, and Manitoba, Canada. *Forest Ecology and Management* 178: 287–300.

4 Johnson, E.A. (1992). *Fire and Vegetation Dynamics: Studies from the North American Boreal Forest*. Cambridge: Cambridge University Press.

5 Work, T.T., Shorthouse, D.P., Spence, J.R. et al. (2004). Stand composition and structure of the boreal mixedwood and epigaeic arthropods of the Ecosystem Management Emulating Natural Disturbance (EMEND) landbase in northwestern Alberta. *Canadian Journal of Forest Research* 34: 417–430.

6 Chavez, V. and Macdonald, S.E. (2012). Partitioning vascular understory diversity in boreal mixedwood forests: the importance of mixed canopies for diversity conservation. *Forest Ecology and Management* 271: 19–26.

7 Marles, R.J., Clavelle, C., Monteleone, L. et al. (2000). *Aboriginal Plant Use in Canada's Northwest Boreal Forest*. Vancouver, BC: UBC Press. 368 pp.

8 Berkes, F. and Davidson-Hunt, I.J. (2006). Biodiversity, traditional management systems, and cultural landscapes: examples from the boreal forest of Canada. *International Social Science Journal* 58: 35–47.

9 Schneider, R.R., Stelfox, J.B., Boutin, S., and Wasel, S. (2003). Managing the cumulative impacts of land uses in the western Canadian sedimentary basin: a modeling approach. *Conservation Ecology* 7 (1): 8.

10 Lieffers, V.J., Armstrong, G.W., Stadt, K.J., and Marenholtz, E.H. (2008). Forest regeneration standards: are they limiting management options for Alberta's boreal mixedwoods? *Forestry chronicle* 84: 76–92.

11 Bergeron, Y., Chen, H.Y.H., Kenkel, N.C. et al. (2014). Boreal mixedwood stand dynamics: ecological processes underlying multiple pathways. *Forestry chronicle* 90: 202–213.

12 Frey, B.R., Lieffers, V.J., Landhäusser, S.M. et al. (2003). An analysis of sucker regeneration of trembling aspen. *Canadian Journal of Forest Research* 33: 1169–1179.

13 Cumming, S.G., Schmiegelow, F.K.A., and Burton, P.J. (2000). Gap dynamics in boreal aspen stands: is the forest older than we think? *Ecological Applications* 10: 744–759.

14 Gärtner, S.M., Lieffers, V.J., and Macdonald, S.E. (2011). Ecology and management of natural regeneration of white spruce in the boreal forest. *Environmental Reviews* 19: 461–478.

15 Peters, V.S., Macdonald, S.E., and Dale, M.R.T. (2005). The interaction between masting and fire is key to white spruce regeneration. *Ecology* 86: 1744–1750.

16 Lieffers, V.J., Stadt, K.J., and Navratil, S. (1996). Age structure and growth of understory white spruce under aspen. *Canadian Journal of Forest Research* 26: 1002–1007.

17 Peters, V.S., Macdonald, S.E., and Dale, M.R.T. (2006). Patterns of initial versus delayed regeneration of white spruce in boreal mixedwood succession. *Canadian Journal of Forest Research* 36: 1597–1609.

18 Cumming, S., Trindade, M., Greene, D., and Macdonald, S.E. (2009). Canopy and emergent white spruce in "pure" deciduous stands: frequency, predictive models and ecological importance. *Canadian Journal of Forest Research* 39: 1997–2004.

19 Bergeron, J.A.C., Pinzon, J., Odsen, S. et al. (2017). Ecosystem memory of wildfires affects multi-taxa biodiversity resilience in boreal mixedwood forest after retention harvest. *Oikos* 12: 1738–1747.

20 Mihajlovich, M., and Kemball, K. 2017. Alberta Silviculture Guide: Boreal Mixedwood and Lower Foothills Natural Subregions. Version II. Forest Growth Association of Western Canada. https://fgrow.ca/publications/fgrow-alberta-silviculture-guide-boreal-mixedwood-and-lower-foothills-natural-subregions (accessed 12 June 12 2023).

21 Lieffers, V.J. and Beck, J.A. Jr. (1994). A semi-natural approach to mixedwood management in the prairie provinces. *Forestry Chronicle* 70: 260–264.

22 Lieffers, V.J., Macmillan, R.B., MacPherson, D. et al. (1996). Semi-natural and intensive silvicultural systems for the boreal mixedwood forest. *Forestry Chronicle* 72: 286–292.

23 Gustafsson, L., Baker, S.C., Bauhus, J. et al. (2012). Retention forestry to maintain multifunctional forests: a world perspective. *BioScience* 62: 633–645.

24 Comeau, P.G., Kabzems, R., McClarnon, J., and Heineman, J. (2005). Implications of selected approaches for regenerating and managing western boreal mixedwoods. *Forestry Chronicle* 81: 559–574.

25 Grover, B.E., Bokalo, M., and Greenway, K.J. (2014). White spruce understory protection: from planning to growth and yield. *Forestry Chronicle* 90: 35–43.

26 Navratil, S. (1996). Silvicultural systems for managing deciduous and mixedwood stands with white spruce understory. In: *Silviculture of Temperate and Boreal Broadleaf-Conifer Mixtures*, Land Manage. Handb. 36 (ed. P.G. Comeau and K.D. Thomas), 35–46. Victoria, BC: B.C. Ministry of Forests.

27 Bjelanovic, I., Comeau, P., Meredith, S., and Roth, B. (2022). Emulating succession in boreal mixedwood forests in Alberta using understory protection harvesting. *Forests* 13: 533.

28 Filipescu, C.N. and Comeau, P.G. (2011). Influence of *Populus tremuloides* density on air and soil temperature. *Scandinavian Journal of Forest Research* 26: 421–428.

29 Lieffers, V.J., Macdonald, S.E., and Hogg, E.H. (1993). Ecology of and control strategies for *Calamagrostis canadensis* in boreal forest sites. *Canadian Journal of Forest Research* 23: 2070–2077.

30 Taylor, S.P., Alfaro, R.I., DeLong, C., and Rankin, L. (1996). The effects of overstory shading on white pine weevil damage to white spruce and its effect on spruce growth rates. *Canadian Journal of Forest Research* 26: 306–312.

31 Jean, S.A., Pinno, B.C., and Nielsen, S.E. (2020). Early regeneration dynamics of pure black spruce and aspen forests after wildfire in boreal Alberta, Canada. *Forests* 11: 333.

32 Kabzems, R. and Haeussler, S. (2005). Soil properties, aspen, and white spruce responses 5 years after organic matter removal and compaction treatments. *Canadian Journal of Forest Research* 35: 2045–2055.

33 Renkema, K.N., Lieffers, V.J., and Landhaeusser, S.M. (2009). Aspen regeneration on log decking areas as influenced by season and duration of log storage. *New Forests* 38: 323–335.

34 Gradowski, T., Lieffers, V.J., Landhaeusser, S.M. et al. (2009). Regeneration of *Populus* nine years after variable retention harvest in boreal mixedwood forests. *Forest Ecology and Management* 259: 383–389.

35 Krebs, C.J., LaMontagne, J.M., Kenney, A.J., and Boutin, S. (2012). Climatic determinants of white spruce cone crops in the boreal forest of southwestern Yukon. *Botany* 90: 113–119.

36 Solarik, K.A., Lieffers, V.J., Volney, W.J.A. et al. (2010). Seed tree density, variable retention, and stand composition influence recruitments of white spruce in boreal mixedwood forests. *Canadian Journal of Forest Research* 40: 1821–1832.

37 Kokkonen, N.A., Macdonald, S.E., Curran, I. et al. (2018). Effects of substrate availability and competing vegetation on natural regeneration of white spruce on logged boreal mixedwood sites. *Canadian Journal of Forest Research* 48: 324–332.

38 Lieffers, V., and Grover, B. 2004. Alternative Silviculture for Boreal Mixedwood Forests of Alberta. Sustainable Forest Management Network. https://acrre.ualberta.ca/wp-content/uploads/sites/83/2018/09/Alternative-Silviculture-for-Boreal-Mixedwood-Forests-of-Alberta.pdf (accessed 12 June 2023).

39 Kabzems, R., Comeau, P.G., Filipescu, C.N. et al. (2016). Creating boreal mixedwoods by planting spruce under aspen: successful establishment in uncertain future climates. *Canadian Journal of Forest Research* 46: 1217–1223.

40 Comeau, P.G. (2022). Can spot motor-manual brushing treatments be effective for controlling aspen and increasing spruce growth in regenerating mixedwood stands? *Forests* 13 (11): 1784. https://doi.org/10.3390/f13111784.

41 Pitt, D.G., Comeau, P.G., Parker, W.C. et al. (2015). Early vegetation control for regeneration of a single cohort intimate mixture of white spruce and aspen on upland boreal sites - 10th year update. *Forestry Chronicle* 91: 238–252.

42 Comeau, P. (2014). Effects of aerial strip spraying on mixedwood stand structure and tree growth. *Forestry Chronicle* 90: 479–485.

43 Bokalo, M., Comeau, P.G., and Titus, S.J. (2007). Early development of tended mixtures of aspen and spruce in western Canadian boreal forests. *Forest Ecology and Management* 242: 175–184.

44 Kabzems, R., Bokalo, M., Comeau, P.G., and MacIsaac, D.A. (2016). Managed mixtures of aspen and white spruce 21 to 25 years after establishment. *Forests* 7: 5.

45 Bjelanovic, I., Comeau, P., Meredith, S., and Roth, B. (2021). Precommercial thinning increases spruce yields in boreal mixedwoods in Alberta, Canada. *Forests* 12: 412.

46 Lieffers, V.J., Pinno, B.D., Beverly, J.L. et al. (2020). Reforestation policy has constrained options for managing risks on public forests. *Canadian Journal of Forest Research* 50: 855–861.

47 Schneider, R.R. (2013). Alberta's Natural Subregions Under a Changing Climate: Past, Present and Future. Report prepared for the Biodiversity Management and Climate Change Adaptation Project. 86 pages. http://biodiversityandclimate.abmi.ca/wp-content/uploads/2015/01/Schneider_2013_AlbertaNaturalSubregionsUnderaChangingClimate.pdf (accessed 26 May 2022).

48 Gauthier, S., Bernier, P., Kuuluvainen, T. et al. (2015). Boreal forest health and global change. *Science* 349: 819–822.

49 Michaelian, M., Hogg, E.H., Hall, R.J., and Arsenault, E. (2011). Massive mortality of aspen following severe drought along the southern edge of the Canadian boreal forest. *Global Change Biology* 17: 2084–2094.

50 Hogg, E.H., Michaelian, M., Hook, T.I., and Undershultz, M.E. (2017). Recent climatic drying leads to age-independent growth reductions of white spruce stands in western Canada. *Global Change Biology* 23: 5297–5308.

51 Gray, L.K., Hamann, A., John, S. et al. (2016). Climate change risk management in tree improvement programs: selection and movement of genotypes. *Tree Genetics and Genomes* 12: 23.

52 Polle, A., Chen, S.L., Eckert, C., and Harfouche, A. (2019). Engineering drought resistance in forest trees. *Frontiers in Plant Science* 9: 1875.

53 Sang, Z., Sebastian-Azcona, J., Hamann, A. et al. (2019). Adaptive limitations of white spruce populations to drought imply vulnerability to climate change in its western range. *Evolutionary Applications* 12: 1850–1860.

54 Talbot, S.S. and Meades, W.J. (2011). *Circumboreal Vegetation Map (CBVM): Mapping the Green Halo*, Concept Paper. CAFF Strategy Series Report No. 3. Akureyri, Iceland: CAFF Flora Group (CFG), CAFF International Secretariat. ISBN 978-9935-431-05-9.

55 Baldwin, K.A., MacKenzie, W.H., Pfalz, J. et al. (2012). *Level 4 Map, Version 1, Canadian Component of the Circumboreal Vegetation Map (CBVM). Canadian Regional Team of the Circumboreal Vegetation Map Project.* Marie, ON: Natural Resources Canada, Canadian Forest Service, Great Lakes Forestry Centre, Sault Ste. Unpublished map.

14

Ecological Silviculture for Interior Ponderosa Pine and Dry Mixed-Conifer Ecosystems

Andrew J. Larson[1] *and Derek J. Churchill*[2]

[1] *Department of Forest Management, University of Montana, Missoula, MT, USA*
[2] *Forest Resilience Division, Washington Department of Natural Resources, Olympia, WA, USA*

14.1 Introduction

Interior western ponderosa pine (*Pinus ponderosa* Lawson & C. Lawson) and dry mixed-conifer ecosystems with a major ponderosa pine component are the focus of this chapter. These forests are found throughout the interior western United States from Arizona and New Mexico north to southern interior British Columbia, Canada, and extend east into the island ranges of the western Great Plains. Mixed-conifer and mixed-evergreen ecosystems with a minor ponderosa pine component occur in southwestern Oregon, California, and in northwestern Mexico, but they are beyond the scope of this chapter.

This ecosystem was managed for grazing and timber extraction beginning with widespread Euro-American settlement in the middle 1800s [1]. Fire suppression became a management focus after 1905 with the formation of the United States Forest Service. Early twentieth century harvest focused on selective removal of large diameter "overmature" pines (Figure 14.1). In the post-WWII period, this gave way to intensive forestry practices, including even-aged silvicultural systems based on clear-cutting and plantations that were optimized for timber production. This agricultural model of forest management excluded frequent fire, the keystone disturbance process that maintains open, multi-age forests dominated by large, fire-resistant trees.

This chapter establishes the ecological basis and silvicultural strategies to implement ecological silviculture in these ecosystems. The overall objective is to manage conditions that are resilient to wildfire, drought, bark beetles, and other stressors while sustaining native biodiversity and providing for an array of ecosystem services, including wood production. It is a long-term, multi-age silvicultural system that includes considerations for reintroduction of frequent fire, regeneration of fire- and drought-resistant species, and climate change adaptation. Stand-level ecological silviculture in this forest type will be most effective at reducing wildfire risk, enhancing ecosystem resilience and adaptive capacity, and sustaining native biodiversity when implemented as part of a large landscape strategy [2].

Ecological Silvicultural Systems: Exemplary Models for Sustainable Forest Management,
First Edition. Edited by Brian J. Palik and Anthony W. D'Amato.

Figure 14.1 Felling a large diameter ponderosa pine using a crosscut saw Colville Indian Reservation, Washington. Original caption, "Felling crew at work on the Katar unit, ca. 7/1942." *Source:* Harold Weaver/ U.S. National Archives and Records Administration (NARA)/Wikimedia Commons/Public Domain.

14.2 Characteristics of Ponderosa Pine and Dry-Mixed Conifer Ecosystems

14.2.1 Composition and Structure

Large-diameter trees of fire-resistant species, often ranging in age from 120 to >400 years, comprise most of the basal area of active fire regime sites (Figures 14.2 and 14.3). Canopy cover is low, typically ranging from 15% to 35% averaged across entire stands (Figure 14.3). Basal areas are similarly low, ranging from about 5–25 m^2 ha^{-1}. Spatial heterogeneity in tree locations (Figure 14.4) is expressed in the characteristic pattern of widely spaced individual trees, tree clumps, and canopy openings of variable shape and size – a mosaic of small patches (about 0.02–0.5 ha) of different structural conditions [3]. Large-diameter standing snags and downed logs are other important structural features.

Pure or nearly pure ponderosa pine stands are common in the southwestern and Black Hills and Great Plains regions of the United States, while mixed-species stands of ponderosa pine, Douglas-fir (*Pseudotsuga menziesii* var. *glauca* (Beissn.) Mayr), grand fir (*Abies grandis* (Dougl. ex D. Don) Lindl.)

Figure 14.2 A dry mixed-conifer forest dominated by large, old, fire-resistant trees in western Montana managed by The Nature Conservancy using ecological silviculture and prescribed fire. *Source:* Andrew J. Larson.

or white fir (*Abies concolor* (Gord. & Glend.) Lindl. ex Hildebr.), western larch (*Larix occidentalis* Nutt.), and lodgepole pine (*Pinus contorta* Dougl. var. *latifolia* Engelm.) are more common in the northern Rocky Mountains and Inland Pacific Northwest (Table 14.1). Additional species can be present in limited abundance, including broadleaf species such as oaks and aspen that provide important habitat value.

Ponderosa pine and dry mixed-conifer ecosystems are typically embedded in landscapes that include dry woodlands and rangelands at lower elevations and moist or cold conifer forests at middle to higher elevations [2]. In mountainous terrain, these vegetation communities and fire regimes may occur in close juxtaposition, for example, a switch from dry mixed-conifer ecosystems on south and west aspects to moist or cold mixed-conifer ecosystems that burned less frequently but with higher severity on adjacent north and east aspects.

14.2.2 Disturbance Regime

The disturbance regime is characterized by frequent surface fire that resulted from extensive Native American burning and lightning ignitions [1]. Mean fire return intervals range from about 5 to 20 years but on some colder, low-productivity sites return intervals may approach 50 years. Occasional mixed- and high-severity fires did occur but were a minor component of the historical fire regime. Bark beetles, especially the mountain pine beetle, western pine beetle, and Douglas-fir beetle (*Dendroctonous* species), are another important disturbance agent, with periodic outbreaks occasionally causing widespread overstory tree mortality.

Frequent surface fire functions as a stabilizing feedback mechanism, maintaining forest structure through successive fire events [4]. By consuming surface fuels, thinning small regenerating

Figure 14.3 Structural conditions in contemporary active frequent fire regime ponderosa pine/mixed-conifer ecosystems, Bob Marshall Wilderness, Montana (top left), Kaibab National Forest, Arizona (top right), and Selway-Bitterroot Wilderness, Idaho (bottom), USA. *Source:* Photos by Andrew J. Larson (top row) and Mark R. Kreider (bottom).

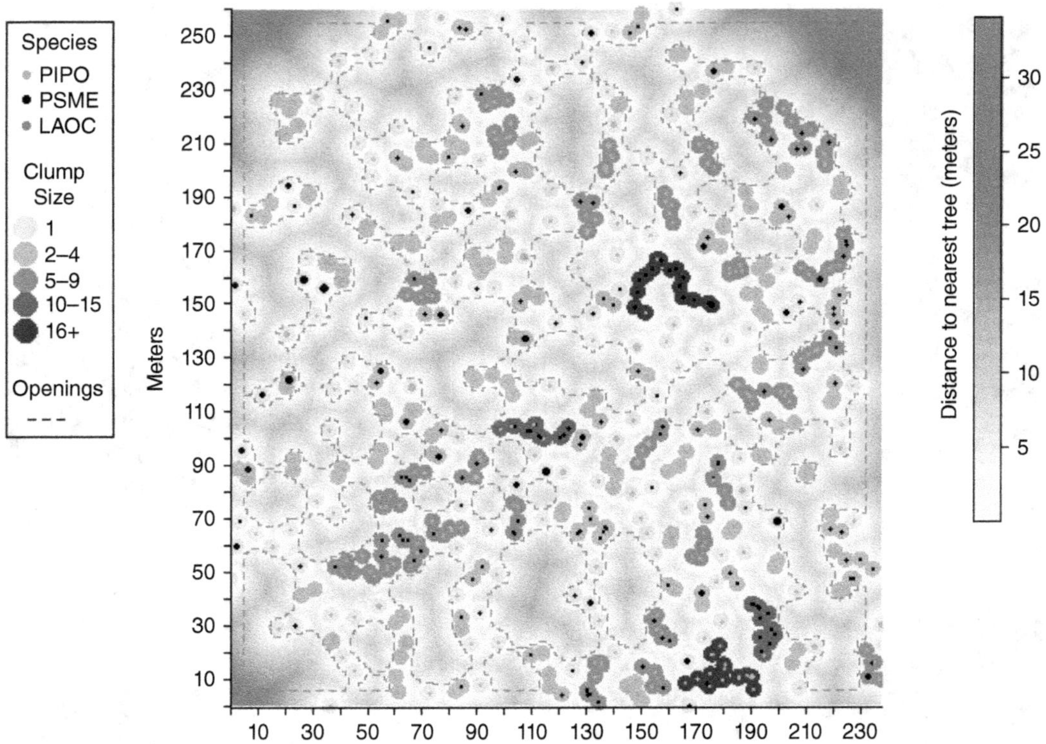

Figure 14.4 Stem map showing the mosaic stand structure of individual trees, tree clumps of various sizes, and openings in a dry ponderosa pine/Douglas-fir ecosystem, eastern Cascade Range, Washington, USA.

Table 14.1 Major conifer tree species and their key characteristics.

Species	Drought resistance	Fire resistance and resilience	Preferred seed bed	Shade tolerance	Max life span (yrs)
Ponderosa pine (*Pinus ponderosa*)	High	Very high: rapid development of thick bark, protected buds, deep roots, resinous wood, high crown base	Mineral	Intolerant	800+
Douglas-fir (*Pseudotsuga menziesii* var. *glauca*)	High	Moderate: develops thick bark with age; intermediate crown base height	Mineral and organic	Moderately tolerant	600+
Western larch (*Larix occidentalis*)	Moderate	High to very high: develops thick bark with age, deciduous, vigorous epicormic sprouter, high crown base	Mineral	Intolerant	700+
Grand fir/white fir (*Abies grandis*/*A. concolor*)	Moderate	Low: thin bark, low crown base, relatively shallow roots	Organic	Tolerant	400+

trees, and pruning lower branches, forest structure is maintained in a low potential energy (i.e. low fuel loading), fire-resistant state. When fires occur, the energy released is insufficient to kill most overstory trees, and crown fire initiation and spread are unlikely [5].

14.2.3 Developmental Model

Ponderosa pine and dry mixed-conifer ecosystems seldom progress through the linear structural development stages common in other forest types (e.g. [6]) due to the rarity of stand-replacement disturbances. Most structural conditions and developmental processes are present, but they occur as a mosaic (Figure 14.4) of small patches (0.02–0.5 ha) instead of more uniform conditions across large areas typical of stand-replacement disturbance regimes. Forest development is presented in [3], with additional detail in [7]. We summarize the key elements here.

The developmental sequence initiates with spatially aggregated tree seedling establishment in canopy openings. Dense thickets of seedlings establish on exposed mineral soil that results from surface fires consuming downed logs that are the result of prior mortality of individual or groups of large trees. This seedbed provides temporary refuge from future surface fires due to scarified soil and consumed fine fuels, as well as reduced competition with understory plants. Tree seedling establishment also occurs at low densities throughout the open forest.

As trees in regeneration clumps grow, they undergo thinning by subsequent fires and density dependent mortality. Over time, tree clumps continue to be thinned by recurring fires, endemic bark beetle attacks, and other agents that cause mortality of individuals or small groups of trees. The developmental cycle continues as individuals and groups of old trees die and eventually fall to the ground, and are consumed in future fires, creating new mineral soil-safe sites for seedlings of fire-resistant species to establish. Repetition of fires, tree regeneration, and mortality of individuals and small groups of trees perpetuates a shifting, fine-scale mosaic of tree clumps and openings in different stages of development [3], with density maintained well below maximum stocking, generally <30% of maximum stand density index [8]. This low-competition environment promotes resilience to fires, drought, and bark beetle attacks.

14.3 An Ecological Silvicultural System for Ponderosa Pine and Dry-Mixed Conifer Ecosystems

Current conditions in ponderosa pine and mixed-conifer forests vary widely based on management history, productivity, and other factors. Foresters implementing this system will often start with stands that are highly departed from the natural developmental model due to past fire exclusion and harvest, where the first order of business is to shift composition and structure to conditions that can be managed sustainably over the long term. In forests that are closer to the natural model (Figure 14.2), multi-age treatments that mimic the natural disturbance regime are utilized.

An ecological silviculture system for ponderosa pine and dry mixed-conifer ecosystems with frequent, low-severity fire regimes is included in detail elsewhere [7]; here we summarize the major elements (Box 14.1). These elements are applicable to all structural conditions and prescriptions, but forest managers may choose to emphasize them to different degrees depending on the initial condition of each management unit (Table 14.2).

Box 14.1 Elements of an Ecological Silviculture System for Ponderosa Pine and Dry Mixed-Conifer Ecosystems

1) **Regeneration of fire-tolerant tree species.** Regeneration of fire-resistant species is critical to the long-term management of frequent-fire forests. This is especially true for shade-intolerant ponderosa pine and western larch, which have a strong affinity for mineral soil seedbeds created by fire. Establishing clumps of regeneration (groups of 10 to >100 trees) in small to medium-sized openings (0.02–0.5 ha) will set stands on a trajectory towards the desired fine-scale, multi-cohort, spatial mosaic. In general, different cohorts should be spatially segregated with smaller trees rarely growing underneath larger trees where they act as ladder fuels. Mechanical scarification during harvests, prescribed fire, or both can be utilized to promote natural regeneration in openings. Planting may be needed to achieve the desired composition, density, and spatial pattern.
2) **Recruitment and perpetuation of large and old trees.** Large diameter trees, which are often old (>150 years), fulfill an array of ecosystem functions and services: they are fire resistant due to thick bark and high crown bases, store most of the aboveground carbon, contribute disproportionately to seed production, and produce large quantities of leaf litter to carry surface fires. Large, old trees develop decadence – dead tops, heart rot, cavities, and complex branches and crown architecture – features that make them especially valuable as wildlife habitats (Figure 14.5). Perpetuation of large tree populations does not preclude harvest or wood production objectives.
3) **Retention of some dying and dead large-diameter trees as snags and logs.** Large-diameter dead wood provides essential habitat for biodiversity (Figure 14.5) and contributes to

Figure 14.5 A decadent, old ponderosa pine tree showing claw marks from repeated climbing by black bears (left) and Great Horned Owl family and nest cavity in a large diameter ponderosa pine snag (right). *Source:* Photos by Andrew J. Larson.

small-scale variation in fire behavior and effects as long-burning, heavy fuels. Leaving some larger trees to die and fulfill these ecological functions is a necessary component of an ecological silviculture system in frequent-fire forests.

4) **Maintenance of low tree density with few ladder fuels.** Low stand density (i.e. low crown bulk density) provides resistance to drought and crown fire spread. Minimizing multistory conditions–trees growing underneath other trees – by thinning understory and midstory trees reduces likelihood of crown fire initiation by maintaining separation between surface fuels and the canopy. Prescribed fire can be an effective tool to reduce ladder fuels and increase crown base height (Figure 14.6). Retaining denser clumps of small trees in canopy openings is important, however.
5) **Perpetuation of a fine-grained spatial mosaic.** Restoring and managing for a multi-cohort patch structure of individual trees, tree clumps, regeneration thickets, and openings perpetuates the natural forest development model in managed forests [2, 3, 9], as well as habitat, snow retention, plant diversity, and other ecosystem functions [10]. Patch sizes should generally range from about 0.02 to 0.5 ha, although patches may be larger in some cases (e.g. 1–2 ha).
6) **Maintenance of surface fuel loads conducive to low-severity fire.** Prescribed fire is the most effective tool for reducing surface fuel loads and keeping them low over time (Figure 14.6). Other methods like piling and burning or mastication can be used when prescribed fire is not feasible. If surface fuel loads are too great even low-density, large tree-dominated forests are at risk of stand replacement fire effects during wildfires.
7) **Maintenance of native understory plant communities and broadleaf tree species.** Understory plants represent the majority of plant species diversity and contribute important ecosystem functions including nitrogen fixation that are not provided by the conifer overstory. Herbaceous plants, especially grasses, are a source of fine fuels that carry surface fires under an active frequent fire regime. In these conifer ecosystems, insect and pollinator communities rely on the flowering understory plants, as do wildlife species, including ungulate grazers and browsers.

Figure 14.6 Conditions during (left) and after (right) prescribed fire treatment in dry ponderosa pine/mixed-conifer ecosystems, western Montana, USA. *Source:* Andrew J. Larson.

Table 14.2 Example structural conditions and potential ecologically based silvicultural treatments in ponderosa pine and dry mixed-conifer ecosystems.

Condition	Potential silvicultural activities[a]
1) High density even-aged	• Commercial and/or non-commercial variable-density thinning (VDT), with explicit guidelines to create spatial variability. • Prescribed (Rx) fire or mechanical fuel treatment to reduce surface fuels and create regeneration opportunities.
2) High-density multi-cohort with fire- and drought-tolerant trees	• Species and size-based commercial and/or noncommercial VDT with explicit guidelines to create spatial variability. • Rx fire or mechanical fuel treatment to reduce surface fuels and create regeneration opportunities.
3) High density with few healthy fire- and drought-resistant trees	• Variable retention harvest to create opportunities for planting, while retaining healthy, fire-resistant trees and large diameter snags. • Rx fire or pile burning to reduce surface fuels. • Plant climate-adapted, fire- and drought-resistant trees to achieve desired species composition and spatial pattern.
4) a. Recently burned by moderate-severity wildfire	• Harvest green and dead trees to further shift composition and structure to large trees of fire- and drought-resistant species and to reduce future surface fuel accumulation while retaining sufficient dead trees for habitat and other ecological values. • Monitor natural regeneration and plant or thin in future if needed.
b. Recently burned by high-severity wildfire	• Harvest dead trees to manage future fuel accumulation while retaining live trees and sufficient snags to provide for wildlife habitat and other ecological values. • Plant climate-adapted, fire- and drought-resistant trees to achieve desired density, species composition, and spatial pattern.
5) Low density, multi-cohort	• Individual and small group selection harvest every 15–30 years combined with Rx fire. • Maintenance Rx fires between harvests when harvest interval is long (>20 years). • Rx fire or managed wildfire every 10–20 years on sites where the landowner does not have timber revenue objectives.

[a] All silvicultural activities should be designed and implemented with careful consideration of the silvicultural system elements listed in Box 14.1.

14.3.1 Incorporating Prescribed Fire in Silvicultural Treatments

A hallmark of ecologically based forest management is working with, not against, the natural disturbance regime. In this forest type, ecologically based silviculture should integrate prescribed fire with mechanical treatments (Figure 14.6), reflecting the central role of frequent surface fires in driving natural forest development and ecosystem function, and because it is the most effective treatment to manage surface fuels [5] and reduce risk of severe wildfires [11]. Use of prescribed fire in the western United States is often challenging due to operational and social factors. It is nevertheless a science-based goal foresters should strive to attain.

Mechanical harvests alone, even when whole tree yarding is utilized, perpetuate or augment surface fuels. Consequently, wildfires in units treated with mechanical methods alone often burn at higher severity compared to sites treated with a combination of harvest and prescribed fire or prescribed fire alone [11]. There are of course costs, as well as risks, associated with using prescribed fire. But forgoing those risks in the short term only increases the risk of severe wildfires in the future. When prescribed fire is not feasible, alternatives such as pile burning and mastication can be used to reduce surface fuels, but they are less effective at reducing fire intensity, facilitating tree regeneration, cycling nutrients, and restoring understory plant communities.

14.4 Example Applications of Ecological Silviculture in Contrasting Initial Conditions

Five structural conditions illustrate the spectrum of current conditions in ponderosa pine and dry mixed-conifer ecosystems (Table 14.2). Conditions 1–3 are characterized by elevated risk of high-severity wildfire and low resilience to stressors like drought and bark beetles. Treatments in these initial conditions prioritize risk reduction in the short term, a prerequisite for sustainable long-term management. Condition 4 addresses the increasingly common problem of how to manage after an uncharacteristically severe wildfire. Condition 5 represents the target condition for frequent fire forests – low-density, fine-grained mosaic structure dominated by large, fire-resistant trees. While these five examples do not capture every condition that exists, the forester can use them to tailor treatments to their specific situation. In all the scenarios described below, a follow-up surface fuel treatment – ideally with prescribed fire (Figure 14.6) – should be used in combination with harvest or noncommercial thinning treatments to reduce surface fuels and create regeneration opportunities for shade-intolerant, fire-resistant species (Table 14.1).

14.4.1 High Density, Even-Aged

This condition results from clear-cut or seed tree harvests and subsequent development of a dense, even-age forest either from natural regeneration, planting, or a combination thereof. This condition can also originate from a high-severity fire followed by dense natural regeneration. Old or large legacy trees are often absent or rare.

Prescriptions are designed to reduce stand density, promote large tree development, shift composition to fire- and drought-tolerant species, and regenerate a new cohort of fire-resistant trees in openings. Restoring a fine-grained spatial mosaic is another key objective and explicit prescription targets for clumps and openings are typically necessary to accomplish this [9, 12, 13].

For management units where commercial thinning is possible, average density targets for these treatments generally range from 50 to 150 trees ha^{-1} or about 9–18 m^2 ha^{-1} of basal area (20–30% of max SDI or stocking). In young plantations, noncommercial, variable-density thinning (VDT) can be used. In some cases, two or three thinning entries can be conducted to gradually reduce density toward these levels. This will shift forest development towards a multi-age structure more slowly but produce revenue periodically over time.

14.4.2 High Density, Multi-Cohort with Abundant Fire- and Drought-Tolerant Species

These forests result from infilling of young trees after fire exclusion and selective logging of some portion of the large and old fire-resistant trees. A similar condition can develop 15–30+ years after fuel reduction and restoration treatments or mixed-severity wildfire. Due to high density and abundant surface and ladder fuels, crown fire risk in these stands is typically very high.

Prescriptions of this type have two primary objectives. First is to decrease fire risk and drought stress for the existing large and old fire-tolerant trees. This is accomplished by removing almost all young trees from within 1–2 times the dripline (about 8–15 m) of large and old trees. The second objective is to restore the desired species composition, density, and spatial arrangement of lower cohorts by thinning the understory and midstory trees in the remainder of the stand. Depending on conditions, this can be accomplished through a combination of commercial (Figure 14.7) and noncommercial thinning, or just a noncommercial treatment of the smaller diameter classes, often called *ladder fuel treatments*. Explicit prescription targets for clumps and openings may be necessary to achieve desired spatial variability targets [13].

14.4.3 High-Density Lacking Healthy Trees of the Desired Species

This condition encompasses a range of stand structures with insufficient healthy trees of climate-adapted, fire- and drought-tolerant species to retain in either a commercial or noncommercial thinning treatment. This condition is most common on mixed-conifer sites where the fire-tolerant component was harvested, and where existing trees of any species have been heavily impacted by insects or pathogens. Another example is plantations where inappropriate species or offsite seed stock were used.

Figure 14.7 Cut-to-length harvester (left) and forwarder (right) system being used to implement an ecological silviculture treatment in a dry mixed-conifer ecosystem, western Montana, USA.
Source: Andrew J. Larson.

A regeneration treatment is generally necessary to address the primary need to establish a new cohort of trees. The amount and spatial location of retention will be a function of the density of existing fire- and drought-tolerant, healthy trees.

In most cases, planting will be necessary to establish a new cohort of climate-adapted, fire- and drought-tolerant species. Planting at lower densities (e.g. 100–250 trees ha^{-1}) in a variable spatial pattern that takes advantage of variable site conditions can establish a pattern of clumps, individuals, and openings [14] and may improve survival and growth of regeneration. Including planting stock from a seed zone that is better adapted to projected future climates should be considered.

14.4.4 Recent Moderate- or High-Severity Wildfire

Due to the increasing extent of large wildfires in the western United States burning under extreme conditions, many areas in conditions 1, 2, and 3 are experiencing wildfires before an initial restoration treatment is conducted. These fires often result in moderate (25–70%) to high (70–100%) mortality of the overstory trees. Treated stands can also experience high mortality in extreme fire weather conditions, especially where prescribed fire was not conducted. After a fire, foresters can evaluate wildfire effects relative to ecological silviculture goals and then develop postfire silvicultural prescriptions that enhance positive wildfire effects and mitigate negative effects [15].

Moderate-severity fire can accomplish many of the goals of ecological silviculture treatments by reducing tree densities, creating openings and regeneration opportunities, shifting species composition, and consuming surface fuels. Moderate-severity fire produces many snags, however, that lead to abundant surface fuels 5–15 years postfire, increasing risk of future high-severity fire [15, 16]. These fires present an opportunity to "finish the job" by harvesting the trees that would have been removed in an ecologically based green tree treatment (conditions 1–3 above), whether they are alive or dead. Appropriate numbers, sizes, and species of snags can be left for wildlife habitat while treating noncommercial-sized trees and activity fuels. If implemented quickly (ideally the winter after the fire in conditions that protect soils), some revenue can be generated to offset costs and reduce the cost and complexity of future entries.

High-severity fire often sets back development of the desired open canopy, multi-age mosaic structure in this forest type. After a high-severity fire managers must first decide whether to remove a portion of the killed trees. The burned forest retains significant ecological values, including the habitat and carbon values of snags and downed logs. On the other hand, high-severity fire often results in uncharacteristically large amounts of dead wood and fuels that make future prescribed fire challenging [16]. Hence, some removal of fire-generated dead trees may be ecologically justified, and treatments can retain live trees, snags, and dead wood based on ecological needs. These benefits must be weighed against potential negative effects from postfire harvest on soils, aquatic systems, and other resources on a case-by-case basis.

The second postfire management decision is whether to replant [15, 16]. Natural tree regeneration, especially of ponderosa pine, can be limited in large patches of high-severity fire due to seed source limitations and unfavorable environmental conditions in the years following the fire. This may be desirable from a landscape perspective as non-forest patches were historically more common in frequently burned landscapes and have many ecological benefits. The decision to replant depends on whether restoring a forest ecosystem on the affected site makes sense given climate change predictions, landscape context, and the objectives of the landowner [15]. If the answer is yes, then planting may be required to achieve the desired density, spatial pattern, and species composition [14], especially in large high-severity patches.

14.4.5 Low Density, Multi-Cohort

This structural condition is the target for frequent fire forests (Figures 14.2 and 14.3). Treatments in conditions 1–4 are designed to move stands toward this condition class. Multiple cohorts exist across the stand in a range of clump sizes but are generally spatially segregated so that tree clumps with multilayered canopies are rare. Overstory density, canopy cover, and surface fuel levels are low; large and old trees of fire-resistant species comprise most of the basal area (Figure 14.2).

Managers have two options for this condition. The first option is to periodically (5–20-year intervals) utilize prescribed fire or managed wildfire to perpetuate stand development processes and resulting conditions typical of the natural model.

The second option is multi-age management designed with the natural model of forest development as a guide (7 pp201). In this forest type, this means a flexible combination of single tree and small group selection with legacy retention that is ideally coupled with prescribed fire or managed wildfire. The natural model of forest disturbance and development is best approximated by area-based silvicultural methods that seek to maintain a shifting mosaic of small patches (about 0.02–0.5 ha) of different structural conditions across a management unit [17]. For example, 20–40% of the area can be in a mixture of openings with and without clumps of regeneration, 10–20% in patches of saplings, 10–30% in medium trees, and 30–40% large trees [10, 12]. Targets for different sizes of tree clumps, as well as the sizes and numbers of openings, can also be used to guide implementation using the Individuals, Clumps, and Openings approach [9, 13, 18]. Specific targets will depend on stand conditions, recent disturbances, landowner objectives, and other factors, and can vary over different entries.

In this system, commercial removal of trees, noncommercial thinning of smaller trees, prescribed fire or other surface fuel treatments, and regeneration are accomplished as needed during the same entry or within a few years of each other. Across a stand, individual tree selection is used to thin clumps or patches of small and medium-sized trees generally from below, although some larger trees may be removed to achieve species composition or other goals. Over several entries, clump sizes are reduced, leaving larger and more vigorous individual trees, as well as some trees with decadent features important for wildlife habitat (e.g. dead tops, cavities).

In other parts of the stand, small group selection is utilized to harvest a portion of the larger size classes during each entry. This creates new openings, ranging in size from 0.02 to 0.5 ha, where aggregated regeneration of fire- and drought-resistant species is actively promoted, perpetuating a fine-grained structural mosaic. Opening sizes may be larger depending on forest conditions (e.g. root rot pockets), as well as the edaphic and topographic template (e.g., microsites with high or low productivity). Maintaining an old tree population in perpetuity is possible with this system. Decisions about the size and abundance of old tree populations will depend upon the balance between ecological, financial, and other goals for the property.

Following harvest entries with prescribed fire treats activity fuels and creates regeneration sites for the desired tree species. The frequency of entries will depend on site productivity and landowner objectives and will usually range from about 15 to 30 years between harvests. More frequent prescribed fires may be warranted to reduce surface and ladder fuels when the interval between harvests is long (>20 years).

14.5 Climate Change Considerations

The simplest climate change adaptation strategy is to treat dense, fire-excluded sites to reduce stand density and surface fuels, and shift species composition to fire- and drought-resistant species. Frequent fire historically maintained tree density far below carrying capacity [8], providing a substantial buffer against drought and resistance to bark beetle attack. Lower fuel moisture levels due to increasing temperatures and vapor pressure deficits mean increased likelihood of extreme wildfire behavior and effects. Thus, a key climate change adaptation strategy is to utilize prescribed fire or managed wildfire (Figure 14.6) to restore and maintain fuel beds resistant to extreme fire behavior and severe effects. These strategies can maintain forest conditions and buy time to implement other climate adaptation strategies like assisted migration.

Proactive treatments, informed by a landscape strategy [2], to transition moist or cold mixed-conifer ecosystems with mixed-severity fire regimes toward conditions conducive to frequent low-severity fire may be desirable to maintain larger landscape connectivity and ecosystem function.

14.6 Summary

An ecologically based silvicultural system for ponderosa pine and dry mixed-conifer ecosystems involves using active management to create and maintain structural conditions that are very different from an agriculturally based timber production system. Key features of this system that follow from the disturbance regime and natural forest development model include:

1) Maintenance of low density, low competition, and open forest conditions for their greater resilience to fire, drought, and bark beetles.
2) Recruitment and perpetuation of large, old trees, and large-diameter snags and logs.
3) Maintenance of low levels of surface and ladder fuels. Prescribed fire is the most effective approach.
4) Regeneration of fire- and drought-resistant tree species in openings using combined harvest and prescribed fire treatments, and planting as needed, to perpetuate a fine-grained, shifting mosaic structure.

References

1 Hessburg, P.F. and Agee, J.K. (2000). An environmental narrative of Inland Northwest United States Forests, 1800-2000. *Forest Ecology and Management* 178: 23–59.

2 Hessburg, P.F., Churchill, D.J., Larson, A.J. et al. (2015). Restoring fire-prone landscapes: seven core principles. *Landscape Ecology* 30: 1805–1835.

3 Larson, A.J. and Churchill, D. (2012). Tree spatial patterns in fire-frequent forests of western North America, including mechanisms of pattern formation and implications for designing fuel reduction and restoration treatments. *Forest Ecology and Management* 267: 74–92.

4 Larson, A.J., Belote, R.T., Cansler, C.A. et al. (2013). Latent resilience in ponderosa pine forest: effects of resumed frequent fire. *Ecological Applications* 23: 1243–1249.

5 Agee, J.K. and Skinner, C.N. (2005). Basic principles of forest fuel reduction treatments. *Forest Ecology and Management* 211: 83–96.

6 Franklin, J.F., Spies, T.A., Van Pelt, R. et al. (2002). Disturbances and structural development of natural forest ecosystems with silvicultural implications, using Douglas-fir forests as an example. *Forest Ecology and Management* 155: 399–423.

7 Palik, B.J., D'Amato, A.W., Franklin, J.F., and Johnson, K.N. (2021). *Ecological Silviculture: Foundations and Applications*. Waveland Press.

8 North, M.P., Tompkins, R.E., Bernal, A.A. et al. (2022). Operational resilience in western US frequent-fire forests. *Forest Ecology and Management* 507: 120004.

9 Churchill, D.J., Larson, A.J., Dahlgreen, M.C. et al. (2013). Restoring forest resilience: from reference spatial patterns to silvicultural prescriptions and monitoring. *Forest Ecology and Management* 291: 442–457.

10 Reynolds, R.T., Sanchez Meador, A.J., Youtz, J.A. et al. (2013). Restoring Composition and Structure in Southwestern Frequent-Fire Forests: A Science-Based Framework for Improving Ecosystem Resiliency. USDA Forest Service Rocky Mountain Research Station General Technical Report RMRS-GTR-310.

11 Cansler, C.A., Kane, V.R., Bartl-Geller, B.N. et al. (2022). Previous wildfires and management treatments moderate subsequent fire severity. *Forest Ecology and Management* 504: 119764.

12 Tuten, M.C., Sánchez Meador, A., and Fulé, P.Z. (2015). Ecological restoration and fine-scale forest structure regulation in southwestern ponderosa pine forests. *Forest Ecology and Management* 348: 57–67.

13 Maher, C.T., Oja, E., Marshall, A. et al. (2019). Real-time monitoring with a tablet app improves implementation of treatments to enhance forest structural diversity. *Journal of Forestry* 117: 280–292.

14 North, M.P., Stevens, J.T., Greene, D.F. et al. (2019). Tamm Review: reforestation for resilience in dry western US forests. *Forest Ecology and Management* 432: 209–224.

15 Larson, A.J., Jeronimo, S.M.A., Hessburg, P.F. et al. (2022). Tamm Review: Ecological principles to guide post-fire forest landscape management in the Inland Pacific and Northern Rocky Mountain regions. *Forest Ecology and Management* 504: 119680.

16 Stevens, J.T., Haffey, C.M., Coop, J.D. et al. (2021). Tamm Review: Postfire landscape management in frequent-fire conifer forests of the southwestern United States. *Forest Ecology and Management* 502: 119678.

17 Graham, R.T. and Jain, T.B. (2005). Application of free selection in mixed forests of the inland northwestern United States. *Forest Ecology and Management* 209: 131–145.

18 Clyatt, K.A., Crotteau, J.S., Schaedel, M.S. et al. (2016). Historical spatial patterns and contemporary tree mortality in dry mixed-conifer forests. *Forest Ecology and Management* 361: 23–37.

15

Ecological Silviculture for North American Pacific Coastal Temperate Rainforests

Justin S. Crotteau[1], William J. Beese[2], John-Pascal Berrill[3], Robert L. Deal[4], Constance A. Harrington[5], and Kellen N. Nelson[6]

[1] USDA Forest Service Rocky Mountain Research Station, Missoula, MT, USA
[2] Vancouver Island University (Retired), Nanaimo, BC, Canada
[3] Cal Poly Humboldt, Arcata, CA, USA
[4] USDA Forest Service Pacific Northwest Research Station (Retired), Portland, OR, USA
[5] USDA Forest Service Pacific Northwest Research Station (Emeritus), Olympia, WA, USA
[6] USDA Forest Service Pacific Northwest Research Station, Juneau, AK, USA

15.1 Introduction

The Pacific coastal temperate rainforest of North America ("temperate rainforest") includes the region adjacent to the Pacific Ocean along the west coast of the United States and Canada, including a narrow fog belt that stretches from northern California through Oregon and Washington, after which it widens to include most of coastal British Columbia (BC) and southeast Alaska, and tapers off in southcentral Alaska (Figure 15.1). The region is characterized by moderate temperatures and abundant precipitation throughout much of the year. In the north, a perhumid climate dominates where extended dry periods are rare. South of BC, seasonal dry periods occur; however, moisture delivery continues in summer via fog. In this unique high-productivity environment, abundant tree regeneration, high growth rates, and infrequent stand-replacing disturbances can lead to extensive forest tracts with large trees and high timber value.

Old-growth forests comprise roughly half of total forest cover in this type in Alaska and BC and under 20% in the southern portion of the region. Frequent gap-to-patch-scale disturbance events such as landslides and windstorms drive natural forest renewal. Clear-cut harvesting was the dominant timber management practice in the region in the twentieth century, resulting in many, even-aged stands. Forest development following stand-replacing disturbances such as clear-cutting and large-scale wind events differs from more typical fine-scale disturbances. Stand-replacing disturbances result in stands of dense closed-canopy conifer regeneration followed by a long-lasting stem exclusion phase that precludes understory vegetation [1]. In contrast, fine-scale disturbances create localized canopy gaps that maintain the multilayered forest structure and diverse vegetation communities characteristic of multi-aged, old-growth stands.

Developing silvicultural approaches and forest management strategies that better approximate natural disturbance and development can help transition conifer plantations into structurally complex forests with abundant understory herbs and shrubs similar to natural forests. This chapter

Ecological Silvicultural Systems: Exemplary Models for Sustainable Forest Management, First Edition. Edited by Brian J. Palik and Anthony W. D'Amato.

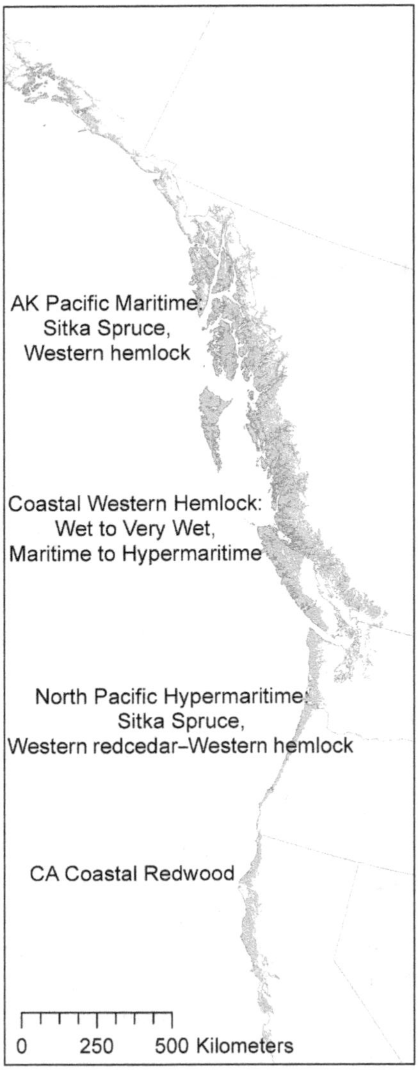

Figure 15.1 Distribution of the temperate rainforest along the Pacific coast of North America. Map shows the variation in forest type nomenclature by geographic subregion.

highlights recent developments and potential applications of ecological silviculture for temperate rainforest ecosystems.

15.2 Characteristics of Temperate Rainforest Ecosystems

15.2.1 Tree Composition, Size, and Age Structure

The temperate rainforest predominantly comprise western hemlock (*Tsuga heterophylla* (Raf.) Sarg.), Sitka spruce *(Picea sitchensis* (Bong.) Carrière), and western redcedar (*Thuja plicata* Donn ex D. Don) from Alaska south through Oregon, transitioning to redwood (*Sequoia sempervirens* (Lamb. Ex D. Don) Endl.) in California. Less common species with economic and cultural value include yellow cedar (*Callitropsis nootkatensis* (D. Don) Oerst. Ex D.P. Little), red alder (*Alnus rubra* Bong.), Pacific silver fir (*Abies amabilis* (Douglas ex Loudon) Douglas ex Forbes), and coastal Douglas-fir (*Pseudotsuga menziesii* (Mirb.) Franco var. *menziesii*).

Western hemlock is the most widespread species across the temperate rainforest occurring as a major component in mixed stands [2]. It is the largest species in the *Tsuga* genus and its high shade tolerance is only exceeded by Pacific silver fir in the region (Table 15.1). Western hemlock is a commercially significant species across the Pacific Northwest and is commonly used in a variety of building material and pulpwood applications including lumber, plywood, pallets and crates, and paper.

Sitka spruce commonly grows in mixed stands with western hemlock; however, pure stands can occur on well-drained sites and in shoreline stands. The species is long-lived and grows larger in stature than western hemlock. It is intermediate in shade tolerance, more so than coastal Douglas-fir but less than western hemlock [3]. Sitka spruce wood has high strength-to-weight ratio and resonance qualities that make it a commercially valuable species for lumber, pulp, and a variety of specialty products including components for musical instruments, light aircraft, and boats [2].

Western redcedar predominantly occurs in mixed species stands from southern Alaska through northern California and forms pure stands to a lesser extent. The species is highly shade tolerant, intermediate in stature, and capable of an exceptional lifespan. Western redcedar wood exhibits high durability and decay resistance with exceptional aesthetic and aromatic qualities. Wood has long been used by indigenous people in this region for cultural lifeways and traditions (e.g. totems, canoes, and long houses). Commercial uses include roofing, siding, utility poles, interior finish carpentry, furniture, and oil and other extractives.

Table 15.1 Shade tolerance, preferred seed bed, mature tree height, and typical lifespan for major tree species in the temperate rainforest.

Tree species	Shade tolerance	Preferred seed bed	Mature height, m (max.)	Typical life-span, years (max.)
Sitka spruce	Tolerant	Mineral soil	60–80 (96)	250–400 (>700)
Yellow cedar	Tolerant	Mineral soil*[3]	25–40 (51)	500–700 (>1800)
Western hemlock	Very tolerant	Organic	50–70 (76)	300–500 (>800)
Red alder	Intolerant	Mineral soil*[2]	25–30 (41)	50–75 (>100)
Western redcedar	Very tolerant	Mineral soil*[3]	35–60 (68)	500–700 (>1400)
Pacific silver fir	Very tolerant	Mineral soil, organic	40–60 (68)	250–400 (>850)
Douglas-fir	Intermediate	Mineral soil	40–80 (100)	400–500 (>1400)
Redwood	Tolerant to very tolerant	Mineral soil*[1,2,3]	60–90 (112)	500–1000 (>2200)
Tanoak	Tolerant	Various*[1,2]	20–45 (63)	150–250 (>400)

* Species also reproduce from root crowns ([1]), stump sprouts ([2]), or branch layering ([3]).
Source: synthesized from Fire Effects Information System https://www.feis-crs.org/feis/ and unpublished data W.J. Beese.

Redwood forms pure stands on alluvial flats but is commonly found intermingled with various conifer and hardwood species including coastal Douglas-fir and tanoak (*Notholithocarpus densiflorus* (Hook. & Arn.) P.S. Manos, C.H. Cannon, & S.H. Oh) [4]. Redwood can grow to exceptional ages and heights, performs well in high light environments, but can also tolerate shade [5]. Regeneration from seed is rare with trees more commonly establishing via vegetative reproduction from stump sprouts that arise after disturbance and grow rapidly in high light or slowly under a partial canopy [6]. Redwood's attractive, decay-resistant heartwood is valuable for exterior construction applications.

15.2.2 Natural Disturbances and Stand Development

The natural disturbance regime in temperate rainforests predominantly includes frequent formation of fine-scale canopy gaps and small patch openings as the result of background mortality of dominant and codominant trees, pockets of insect and disease mortality, or windthrow [1]. Medium-scale stand-replacing events occur less frequently from windstorms and landslides [7, 8]. Landscape-wide, low-severity insect outbreaks also occur infrequently. Broad-scale stand-replacing disturbances are rare. The result is extensive tracts of multi-aged forest with finely patterned, structurally heterogeneous characteristics. Fire regimes across the ecoregion range from all but absent in the perhumid climate zone (i.e. northern BC to southeast Alaska) to greater than 500-year high-severity return intervals where dry seasons occur (southern BC to Oregon; [9]). In the southernmost portion of the ecoregion, redwood forests experienced episodes of flooding, slope failure, and indigenous burning prior to European settlement. These and other disturbances were generally frequent but of low-to-moderate severity and extent, often leading to the initiation or release of different age classes [10].

15.2.3 Developmental Model

Forests in the temperate rainforest most frequently follow a gap- and patch-development model where frequent fine-scale understory infilling and release events give rise to vertically and horizontally heterogeneous forest structure and composition. Here we describe forest development following both frequent (gap- to patch-scale) and infrequent (patch- to stand-scale) disturbances, where infrequent disturbances with higher severity and wider extent are associated with the full suite of developmental stages but lower severity and smaller extent disturbances maintain patches or stands in later stages.

15.2.3.1 Preforest Stage

After canopy disturbance, the short-lived preforest stage is characterized by strong biological legacies that include remnant trees as individuals or in small patches, heavy accumulations of downed wood, abundant understory vegetation, and advanced regeneration released by the disturbance and predominantly composed of shade-tolerant species. Red alder may dominate in locations with disturbed soils and high light availability. In the redwood range, understory vegetation responds rapidly following canopy opening, and sprouting by redwood allows trees to quickly reoccupy disturbed sites.

15.2.3.2 Young Forest Stage

Following patch or stand initiation, young forests form as dense, closed canopy stands in as little as 15–25 years. Low canopy light transmission triggers a rapid decline in understory vegetation and this condition can last for up to 100 years depending on site productivity. This stage terminates when density-dependent stem exclusion disrupts the canopy, spurring a growth release in surviving trees and permitting light to reach the forest floor. Maturing red alder becomes overtopped by competing conifers. In the redwood range, the young forest stage passes more rapidly, and individuals exhibit clumped spatial patterns around parent stumps. Canopy closure is heterogeneous preventing stand-wide elimination of understory vegetation during this development stage.

15.2.3.3 Mature Forest Stage

In the mature forest stage, even-aged patches and stands exhibit increasing horizontal and vertical structural complexity, and large trees become more abundant. Insects and diseases, and wind damage create canopy gaps and stands begin to recruit multiple age cohorts as new trees establish in openings. Understory vegetation reestablishes and diversifies as stand structure becomes increasingly varied. Red alder becomes decadent and declines in dominance, giving way to conifers with longer lifespans. Redwoods overtop shade-tolerant hardwoods and can replace shorter-lived conifers unless larger gaps are created or expanded to promote episodic regeneration and recruitment.

15.2.3.4 Old Forest Stage

Forests in the old forest stage are characterized as structurally complex with canopy gaps of varying sizes and a broad range of tree ages and sizes. Large trees, snags, and downed wood are present and irregularly spaced throughout the stand. Diverse niches support a wide array of understory forbs, ferns, shrubs, bryophytes and fungi.

15.3 An Ecological Silvicultural System for the Temperate Rainforest

The ecological silvicultural system (ESS) presented here attempts to use natural disturbances and developmental processes for this ecosystem and region as a guide; however, we do not attempt to precisely mimic historical disturbance patterns, which may not be feasible or desirable for meeting resource management objectives. We also recognize that there are modifications of traditional systems (e.g. selection or shelterwood with reserves) that could produce similar results. While the recommended ESS is designed at the stand level for management efficiency (with postharvest tactics focusing on management of new cohort patches), it may also be applied to portions of a stand or cutting unit to approximate the scale of typical gap or patch disturbances. Specific treatments for each stage of development may vary depending upon stand origin and characteristics (Table 15.2). A key principle in applying an ESS is to vary the approach to reduce risk and increase

Table 15.2 An ecological silviculture system for the temperate rainforest.

Developmental stage or event	Duration (# of years)	Objectives	Silvicultural activities
Disturbance and legacy creation	Usually less than one year	Provide for continuity of structure and function; retain or create biological legacies; initiate preforest stage	– *Harvest*: VRH (aggregated and dispersed retention), single or multi-pass – *Deadwood*: maintain or create snags (e.g., by topping), downed woody debris
Preforest stage	0–5	Set the stage for tree regeneration; provide short-term forage	– *Site prep/vegetation management*: reduce invasive species, hemlock, or abundant woody shrub species to prioritize desired species; expose mineral soil on microsites to improve germination (e.g. redcedar, redwood, red alder and shrubs)
Young forest stage (early)	20–30	Facilitate/establish new cohort of trees; ensure site occupancy without eliminating understory	– *Thinning*: VDT (precommercial) or thinning to wide spacing to avoid loss of early seral understory vegetation, prioritize underrepresented species – *Browse control*: protect cedars as needed – *Enrichment planting* (of desired underrepresented species): plant spruce, cedars, redwood, hardwoods
Young forest stage (late)	50–70	Maintain or accelerate growth of trees in cohort; develop structural/spatial heterogeneity	– *Thinning*: VDT to release dominant and codominant crop conifers or hardwoods, create spatial heterogeneity and small gaps to maintain understory vegetation
Mature forest stage	70–150	Enhance complexity and spatial heterogeneity; release and accelerate growth on desired trees; enrich species diversity	– *Thinning*: VDT around dominant and codominant crop conifers, with wide spacing around clumps of hardwoods and small gaps to maintain understory vegetation – *Harvest*: VRH as in disturbance stage to initiate new cohort of desired species at stand scale for single pass and patch scale for multi-pass – *Site prep, browse control, and enrichment planting in harvest areas*: as above – *Deadwood creation*: girdle or top trees as part of release to create snags
Old forest stage	150+	Maintain structural complexity and enrich species diversity for resilience to climate and disturbance	– *Thinning*: VDT as above – *Harvest*: VRH as in disturbance stage to initiate new cohort of desired species – *Site prep, browse control, and enrichment planting in harvest areas:* as above – *Decadence/deadwood creation*: as above

options (i.e. avoid doing the same treatments everywhere, [11]). Given the current condition and extent of managed forests in the region, this ESS will often find application in young to mature even-aged forests.

15.3.1 Disturbance and Legacy Creation

Variable retention harvesting (VRH, Chapter 1) is a flexible approach for meeting silvicultural goals at this stage, namely: creating conditions for natural regeneration, retaining live trees and deadwood legacies, and extracting timber value. VRH is typically applied at the stand scale and creates or reinforces within-stand patches for subsequent management. In BC, the "retention silvicultural system" is defined for meeting the goals of VRH, which include (i) maintaining forest influence, and (ii) distributing retention throughout a harvested stand [12]. A practical measure of forest influence is one tree length from standing trees, although many influences extend further into cut areas. This model, and variations of it, are appropriate for implementation across the range of the temperate rainforest.

Retention at the regeneration harvest may include individual trees, aggregates, or varying combinations of both (Figure 15.2). Aggregate locations encompassing specific ecosystem features provide high biodiversity value (e.g. mixed species, bear dens, riparian areas, wetlands, rock outcrops). Large, downed wood may be left along with existing snags, as safety permits. In stands lacking deadwood, these features can be created by leaving logs on site and by girdling or topping live trees. The spatial pattern and retention percentage will vary depending upon specific objectives, including promoting ecological diversity and windfirmness. The high potential for wind disturbance in these ecosystems is an important consideration for cutting boundary location, the size and configuration of aggregates, and the species retained [13]. Other important factors influencing retention are the spread of pathogens such as hemlock dwarf mistletoe (*Arceuthobium tsugense* (Rosendahl)) and root rot, and the potential for decline in yellow cedar.

Based on previous studies, a minimum of 25% retention (by basal area, volume, or area) and an emphasis on aggregates [14–16] supports biodiversity. The spatial pattern of aggregates is often stand-alone groups, ranging from small clumps of a few trees to large groups of 1–2 ha, or strips along riparian corridors. VRH may be a single-pass approach (most typical) or a multi-pass approach. In single-pass VRH, retention patches are established (e.g. 25% of stand), and the matrix (e.g. 75% of stand) is fully harvested in one pass or entry with no additional regeneration harvests scheduled until the end of the matrix rotation (matrix is a single cohort). In multi-pass VRH, retention patches are established (e.g. 25% of stand), but only a portion of the matrix is harvested at each pass or entry, which might range in frequency from 5 to 30 years (e.g. 25% of stand for three managed age cohorts; Figure 15.3). Multi-pass VRH can be used to approximate the spatial and temporal scale of typical disturbances. It is like a modified version of group selection (or patch cutting) with long-term reserves, while traditional regulations for age distribution and stocking control are loosened. Important considerations for multi-pass retention include permanent access roads and layout to facilitate harvesting, hazard tree management, and orientation of patches to optimize light.

15.3.2 Preforest Stage

Because of abundant natural regeneration in temperate rainforests, this stage is typically short (0–5 years) except in cases where competitive shrubs dominate (e.g. *Gaultheria shallon* Pursh or *Rubus spectabilis* Pursh). Specific management objectives may make it desirable to maintain the preforest stage, such as provision of spring forage adjacent to ungulate winter range, or blueberry/

Figure 15.2 Examples of VRH in temperate rainforest ecosystems. (a) Multi-pass retention (53%), N. Vancouver Island, BC Canada; (b) aggregated retention (31%), SW Vancouver Island; (c) dispersed retention after 20 years, N. California USA; (d) aggregated (left) and dispersed (right) retention comparison, southeast Alaska USA; (e) mixed retention (34%), NE Vancouver Island; (f) multi-pass small patch cutting, helicopter yarding, Haida Gwaii BC Canada. Species shown include western hemlock (a, b, d, e, f), western redcedar (a, b, d, e, f), coast redwood (c), Sitka spruce (d, f), and Pacific silver fir (a, e). *Source:* a, b, e, f, (Beese); c, (Berrill); d, (USDA Forest Service).

huckleberry production for traditional cultural use [17]. In other cases, treatment of shrubs may be necessary to set the stage for tree regeneration goals [18]. Mechanical site preparation (MSP) can enhance natural regeneration or create microsites for supplemental planting. MSP emulates soil mixing and bare mineral soil exposure common to windthrow or landslide disturbance. Fire is rare

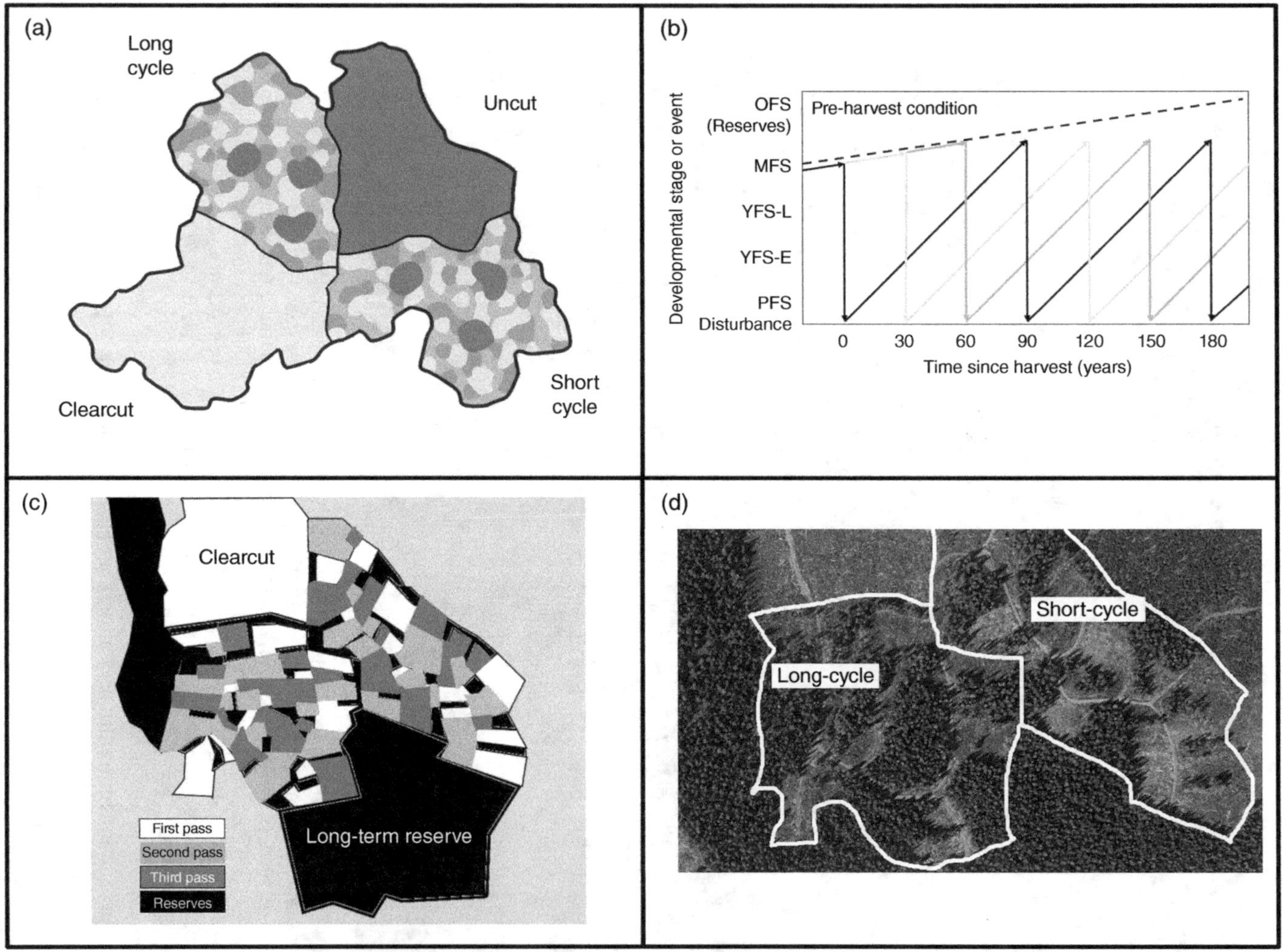

Figure 15.3 Example of multi-pass VRH for maintaining multiple age cohorts surrounding long-term reserves. (a) Conceptual layout of a Variable Retention Adaptive Management (VRAM) comparison of small patch removal using short (5–7 year) vs. long (20–30 year) cutting cycles, with adjacent clear-cut and uncut areas. (b) Diagram of forest developmental stages in one stand under a multi-pass retention system with long-term reserves (dashed line) and three passes or entries (solid lines; black is first pass, light gray second pass, and dark gray third pass) with 30-year cutting cycles, managed as patches within stands. (c) Map of operational layout of the VRAM study on

in this ecosystem, so prescribed fire may not be an appropriate tool for slash disposal or site preparation [9]. Control of non-native, invasive species (e.g. *Cytisus scoparius* (L.) Link) may be done at this stage, preferably through early detection and treatment of small populations.

15.3.3 Young Forest Stage (Early)

The silvicultural goal of this stage is to establish and tend a new cohort of trees, primarily through natural regeneration. Tree density targets are typically low with ESS (e.g. $<700\,\text{ha}^{-1}$) to promote more diverse and enduring understory vegetation. Sitka spruce and western hemlock typically reseed quickly and abundantly into disturbed areas. Supplemental or enrichment planting may be necessary to meet stocking or species composition objectives, e.g. to enhance tree species diversity and maintain desirable but underrepresented species (western redcedar, yellow cedar). Protection from excessive ungulate browsing may be needed in some locations to maintain cedars. Hardwood–conifer mixtures are desirable on ecologically suitable sites (Figure 15.4), predominantly with red alder and bigleaf maple, which are important elements for nutrient cycling and wildlife habitat [19]. As natural regeneration is frequently abundant (especially with advanced regeneration of western hemlock), density control using variable-density precommercial thinning prior to canopy closure is typically appropriate to balance stocking targets and understory vegetation.

15.3.4 Young Forest Stage (Late)

Silvicultural activities in this stage are intended to maintain or accelerate the growth of regenerating trees and to promote greater spatial and structural heterogeneity. These goals can be accomplished through Variable-Density Thinning (VDT) – either precommercial or commercial in either young or older stands. VDT differs from traditional thinning in that inter-tree spacing is

Figure 15.4 Red alder planted to enhance biodiversity in southeast Alaska.

Figure 15.5 Gap centered over residual bigleaf maple (9-m radius) and planted with western redcedar, western Washington. Vexar™ tubes were used for browse protection.

intentionally variable, promoting small gaps to maintain understory vegetation, releasing focal trees, and retaining some clumps for structural complexity ([20]; Figure 15.5; Chapter 1). VDT is also well-suited to emulate structure in riparian ecosystems where natural tree distribution tends to be clumpy due to competition from tall shrubs (e.g. salmonberry, elderberry) or lush herbaceous cover. VDT can be used to alter species composition, especially where western hemlock is overabundant, or to produce understory cover for deer forage [21].

15.3.5 Mature Forest Stage

Primary objectives for this developmental stage include enhancing complexity and heterogeneity, which can be accomplished with divergent pathways. Depending on the management objectives for the stand, mature forests may be harvested on a rotation or cutting cycle basis, or maintained to develop old forest attributes, with or without further treatments.

Regeneration harvests using VRH are appropriate to enhance complexity by maintaining or initiating multi-cohort stand structure with desired species diversity. If a regeneration harvest occurs, retention need not occur at the same level or in the same locations as the previous harvest. It is likely that previously retained areas will have the greatest value for further retention because they will have older trees and more structural diversity than regenerated areas. Current objectives guide the choice of retention level and pattern, including the desired number of age or canopy cohorts. VRH could be used to retain one or more age cohorts while initiating a new cohort throughout the stand with the single- or multi-pass option. As in the Preforest Stage and Young Forest Stage above, site preparation, enrichment planting, and browse control may be necessary to meet species composition objectives in harvested areas.

Stands may be maintained beyond the mature stage to provide for old forest conditions on the landscape, to extend rotation ages to meet wildlife habitat needs for species dependent upon old forests, or to meet specific wood product objectives, particularly for high-value specialty products requiring knot-free, tight grained wood (e.g. musical instruments, cabinetry, furniture, cultural artifacts). Further, stand-tending treatments may be desirable in some cases, such as VDT, but may not be necessary if the desired future condition will be met without intervention. As in the Young Forest Stage (late), VDT might be used to maintain or accelerate growth of focal trees while promoting spatial and structural heterogeneity. Overstory conifers are tall enough at this stage to shade out hardwood and shrub components, so gaps may need to be larger than in previous stages. At this stage in stand development, overstory trees are large enough to contribute to large deadwood pools. Decadent features and deadwood may be enhanced by topping or girdling trees to create snags.

15.3.6 Old Forest Stage

Whether or not to continue harvesting old forests in the temperate rainforest is a controversial topic, even if they have been previously managed. Where these forests have been protected in parks, wilderness, or other special designations, there may be little clear justification for silvicultural intervention; elsewhere on the landscape, there are possible reasons for continued silvicultural activities within this stage. These might include regeneration of early-seral or desired species, releasing new cohorts, creating deadwood, improving nutrient cycling by promoting red alder (a nitrogen-fixing deciduous tree), cultivating berries and wildlife forage, etc. Importantly, targeted species enrichment may be used to improve resilience of this forest stage by promoting native species with greater adaptation to future climate and insect or disease pressures.

As in the Mature Forest Stage, VRH may assist in the maintenance or further development of multi-cohort stand structure with desired species diversity. High levels of retention for portions of the stand with the greatest number of old trees, structural complexity, deadwood, and other old forest attributes may help support biodiversity. As in the Preforest Stage and Young Forest Stage above, site preparation, enrichment planting, and browse control may be necessary to meet species composition objectives in harvested areas.

Although upper canopy trees do not need release, VDT may still be appropriate for enhancing complexity and releasing underrepresented species (e.g. to release or maintain hardwoods in the midstory and valuable forage species in the understory). Deadwood may be enhanced to create habitat for specific organisms.

15.4 Climate Change Considerations

In the temperate rainforest, climate change may disrupt processes that govern tree establishment, growth, and survival and the impacts will vary substantially depending on species and location. Here we present a variety of silvicultural methods to mitigate challenges associated with shifting temperature and precipitation regimes in these ecosystems.

During phases of management that include tree establishment, planting can be used to improve the climate adaptation of individuals within a stand or to improve anticipated post-disturbance outcomes by shifting stand composition. Using assisted population migration, managers might plant a species within its historic range but strategically source seedstock from a climate-adapted region to align seed provenance with anticipated future conditions. Alternatively, assisted species migration or range expansion might include introducing species into areas where they may not have historically occurred because they may be better adapted to future climate conditions or disturbance regimes. For example, it may be appropriate to extend the range of yellow cedar to establish individuals on sites where winter snow will persist under future climates or introduce western redcedar in areas affected by yellow cedar decline [22]. Additionally, drought-tolerant species such as Douglas-fir might be considered for planting whereas Sitka spruce or western hemlock may become less productive due to increasing summer drought. The results from tree improvement or research programs may provide managers with regeneration options such as weevil-resistant spruce, browse-resistant redcedar, Swiss Needle Cast-resistant Douglas-fir, as well as guidance on climate-adapted seed sources for future planting. Finally, planting densities and species mixtures can be varied to create heterogeneous stand structures and enhanced species richness to foster resilience in young stands.

Both VRH and VDT may foster more resilient stand conditions by enhancing structural complexity and age and species diversity. Maintaining stands at lower relative density may enhance tree vigor and drought tolerance. In some cases, enrichment plantings may be suitable in harvest gaps to accommodate species or genotypes that are better adapted to future conditions, with gap sizes designed for the planned regeneration. Forest composition across the temperate rainforest region varies substantially and species will have different responses to shifting climate conditions. Long-term monitoring may help managers assess whether current treatments are effective for maintaining ecosystem productivity or meeting other objectives. In addition, insect and disease activity may respond to shifting climate regimes and new impacts on forests may emerge [22]. Finally, less obvious effects of climate change such as changes in timing of phenology [23] may also have important long-term impacts.

15.5 Summary

An ESS in the temperate rainforest emphasizes diversity of forest structure and composition for improved ecological function over a timber-oriented silvicultural system. Key attributes of this model include:

1) Legacy trees, snags, and downed wood are retained as lifeboat structures for late-successional forest biota. Retention of these structures better aligns posttreatment stands with typical features following predominant natural disturbances.
2) Both postharvest and throughout the ESS, emphasis is placed on creating and maintaining openings for forbs and shrubs to thrive. While rapid forest growth often makes this a short-lived ecological condition at any one location in a stand, harvesting and tending entries enable this condition to persist in either static locations or shifting mosaics, as might occur with frequent wind disturbance.
3) Western hemlock typically regenerates naturally and prolifically. This ESS encourages diverse natural and artificial regeneration of Sitka spruce, cedars, redwood, and hardwoods.
4) Culturing high heterogeneity within stands is practiced in every phase of management. From irregular retention patches to VDT, this system is designed to emulate the complexity associated with a gap-dynamics disturbance regime.
5) Harvest timing is flexible. Managers can select single-pass VRH to limit entries or could opt for multi-pass VRH to create multiple age classes at gap and patch scales. With retention patches providing continuation of old forest values, managers are free to choose short or long rotations and cutting cycles.

6) There are opportunities to diversify composition and age classes for climate adaptation at each harvest entry, and opportunities to tend future-adapted species during each thinning entry. Using this system, managers can integrate climate adaptation strategies in every developmental phase.

Acknowledgments

This research was supported by the USDA Forest Service, Rocky Mountain Research Station. The findings and conclusions in this publication are those of the authors and should not be construed to represent any official USDA or U.S. Government determination or policy. Any use of trade names is for descriptive purposes only and does not imply endorsement by the U.S. Government.

References

1 Alaback, P.B. (1982). Dynamics of understory biomass in Sitka spruce-western hemlock forests of southeast Alaska. *Ecology* 63: 1932–1948.

2 Harris, A.S. (1990). *Picea sitchensis* (Bong.) Carr. Sitka spruce. In: *Silvics of North America*, Conifers, vol. 1 (ed. R.M. Burns and B.H. Honkala), 260–267. Washington, DC: USDA For. Serv.

3 Minore, D. (1979). *Comparative Autecological Characteristics of Northwestern Tree Species - A Literature Review*, Gen. Tech. Rep. PNW-87. Portland, OR: USDA For. Serv., Pacific Northwest Forest and Range Experiment Station.

4 Berrill, J.-P., Beal, C.B., LaFever, D.H., and Dagley, C.M. (2013). Modeling young stand development towards the old-growth reference condition in evergreen mixed-conifer stands at Headwaters Forest Reserve, California. *Forests* 4 (2): 455–470.

5 Olson, D.F., Roy, D.F., and Walters, G.A. (1990). *Sequoia sempervirens* (D. Don) Endl. Redwood. *Silvics of North America* 1: 541–551.

6 O'Hara, K.L., Cox, L.E., Nikolaeva, S. et al. (2017). Regeneration dynamics of coast redwood, a sprouting conifer species: a review with implications for management and restoration. *Forests* 8: 144.

7 Buma, B. and Johnson, A.C. (2015). The role of windstorm exposure and yellow cedar decline on landslide susceptibility in southeast Alaskan temperate rainforests. *Geomorphology* 228: 504–511.

8 Kramer, M.G., Hansen, A.J., Taper, M.L., and Kissinger, E.J. (2001). Abiotic controls on long-term windthrow disturbance and temperate rain forest dynamics in southeast Alaska. *Ecology* 82: 2749–2768.

9 LANDFIRE: LANDFIRE Biophysical Settings (BPS) Layer. 2016. USDI, Geological Survey, and USDA. http://landfire.cr.usgs.gov/viewer/ (accessed 1 January 2022).

10 Lorimer, C.G., Porter, D.J., Madej, M.A. et al. (2009). Pre-settlement and modern disturbance regimes in coast redwood forests: implications for the conservation of old-growth stands. *Forest Ecology and Management* 258 (7): 1038–1054.

11 Bunnell, F.L., Kremsater, L.L., and Wind, E. (1999). Managing to sustain vertebrate diversity in forests of the Pacific Northwest: relationships within stands. *Environmental Reviews* 7: 97–146.

12 Mitchell, S.J. and Beese, W.J. (2002). The retention system: reconciling variable retention with the principles of silvicultural systems. *Forestry Chronicle* 78: 397–403.

13 Beese, W.J., Rollerson, T.P., and Peters, C.M. (2019). Quantifying wind damage associated with variable retention harvesting in coastal British Columbia. *Forest Ecology and Management* 443: 117–131.

14 Deal, R.L. and Tappeiner, J.C. (2002). The effects of partial cutting on stand structure and growth of western hemlock-Sitka spruce stands in southeast Alaska. *Forest Ecology and Management* 159: 173–186.

15 Halpern, C.B., Halaj, J., Evans, S.A., and Dovčiak, M. (2012). Level and pattern of overstory retention interact to shape long-term responses of understories to timber harvest. *Ecological Applications* 22: 2049–2064.

16 Beese, W.J., Deal, J., Dunsworth, B.G. et al. (2019). Two decades of variable retention in British Columbia: a review of its implementation and effectiveness for biodiversity conservation. *Ecological Processes* 8: 33.

17 Deur, D. and Turner, N.J. (2005). *Keeping It Living: Traditions of Plant Use and Cultivation on the Northwest Coast of North America*. University of Washington Press.

18 Mallik, A. and Prescott, C.E. (2001). Growth inhibitory effects of salal on western hemlock and western redcedar: allelopathy and competition. *Agronomy Journal* 93: 85–92.

19 Deal, R.L., Orlikowska, E.H., D'Amore, D.V., and Hennon, P.E. (2017). Red alder-conifer stands in Alaska: an example of mixed species management to enhance structural and biological complexity. *Forests* 8 (4): 131.

20 Willis, J.L., Roberts, S.D., and Harrington, C.A. (2018). Variable density thinning promotes variable structural responses 14 years after treatment in the Pacific Northwest. *Forest Ecology and Management* 410: 114–125.

21 Crotteau, J.S., Rue-Johns, A.Z., and Barnard, J.C. (2020). Effects on understory biomass and forage 8-10 years after precommercial thinning of Sitka spruce – western hemlock stands in southeast Alaska. *Canadian Journal of Forest Research* 50: 215–225.

22 Hennon, P.E., McKenzie, C.M., D'Amore, D. et al. (2016). *A Climate Adaptation Strategy for Conservation and Management of Yellow-Cedar in Alaska*, Gen. Tech. Rep. PNW-GTR-917. Portland, OR: USDA, Forest Service, PNW Research Station. 382 p.

23 Ford, K.R., Harrington, C.A., Bansal, S. et al. (2016). Will changes in phenology track climate change? A study of growth initiation timing in coast Douglas-fir. *Global Change Biology* 22 (11): 3712–3723.

16

Ecological Silviculture for Oak Ecosystems of the Central Hardwoods Region, USA

Miranda T. Curzon[1] *and John M. Kabrick*[2]

[1] *Department of Natural Resource Ecology and Management, Iowa State University, Ames, IA, USA*
[2] *Northern Research Station, USDA Forest Service, Columbia, MO, USA*

16.1 Introduction

This chapter focuses on the largely oak (*Quercus*)-dominated forest, woodland, and savanna ecosystems that characterize the Central Hardwoods Region. These ecosystems occur across much of the eastern United States, including five ecoregion provinces (Figure 16.1) [1, 2].

The Central Hardwoods Region contains productive agricultural land and a high proportion of the population of the United States. Approximately 75% of the forest land is privately owned [3], exceeding 90% in many states. Fragmentation continues to decrease the size of privately owned forests and woodlands with average area ranging from less than 8 ha in Ohio, Pennsylvania, and New York to 12–16 ha in Missouri, Illinois, and Virginia [4].

Ownership patterns strongly influence management. Passive management favored by nonindustrial private forest owners has led oak-hickory (*Carya*) ecosystems to succeed in more mesic mixed hardwood forests region-wide [5]. High grading is also common and has degraded forest quality. Where active, planned management occurs, even-aged regeneration methods (e.g. clear-cutting, shelterwood) are most commonly applied using rotations of 70–120 years. On poor sites in the southwestern-most Central Hardwoods where oak competitors are of less concern, uneven-aged methods (e.g. selection) can successfully regenerate oak.

Applying knowledge of natural disturbance and historical range of variability to management of these ecosystems is challenging, considering the substantial change that has already occurred (e.g. loss of American chestnut, land conversion, fire suppression) and the way multiple disturbances of varying type, frequency, and severity interact [6]. Yet, there is interest in silvicultural approaches that better reflect the dynamics and structure associated with a natural model (i.e. ecological silviculture), particularly given private landowner concerns over maintaining mature canopy and habitat, often at the expense of successful regeneration [7]. Incorporating ecological principles into oak silviculture has been discussed previously [3, 6, 8, 9], and the growing use of prescribed burning reflects a understanding of the role disturbance played in maintaining oak species historically. However, the application of ecological silviculture remains limited. This chapter provides a general framework for designing ecological silvicultural systems that can be applied in oak ecosystems across the Central Hardwoods Region.

Ecological Silvicultural Systems: Exemplary Models for Sustainable Forest Management, First Edition. Edited by Brian J. Palik and Anthony W. D'Amato.

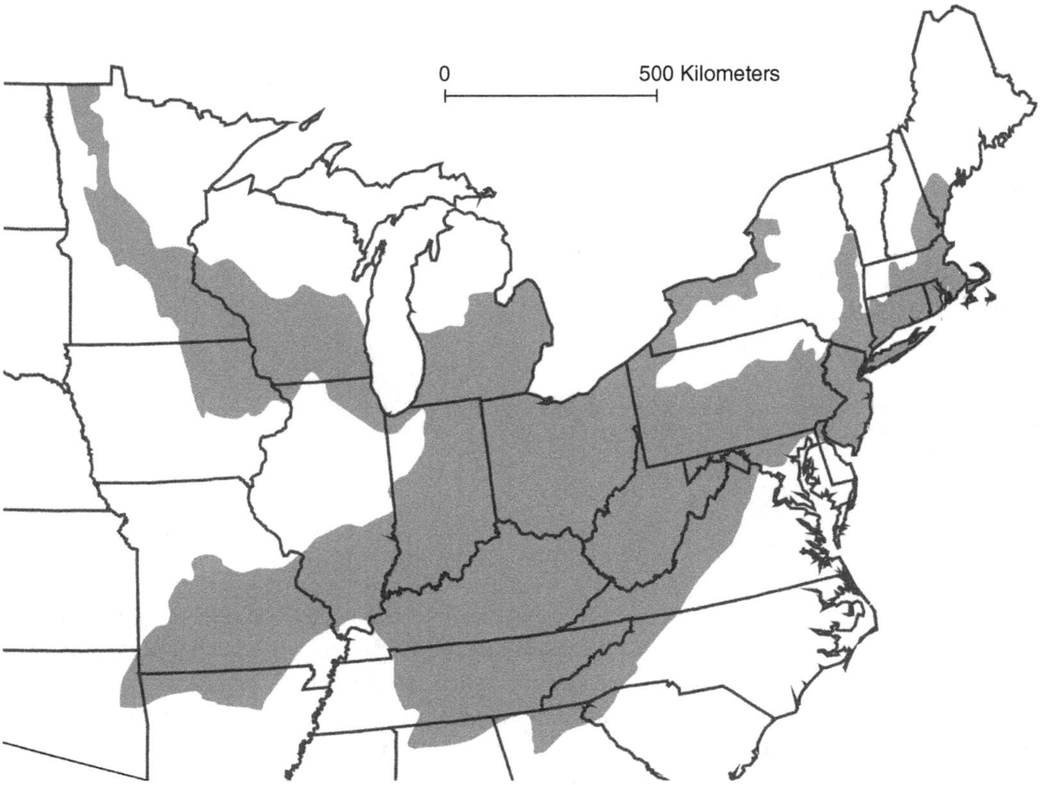

Figure 16.1 Central Hardwoods Region of the United States. *Source:* Adapted from Hanberry et al. [1].

16.2 Characteristics of Central Hardwood Forests and Woodlands

16.2.1 Composition and Structure

While the Central Hardwoods region contains a wide range of environmental conditions and over 100 native tree species, oak, and hickory species are most common (Table 16.1), and the forest type is broadly classified as oak hickory [12]. Oak ecosystems can be organized into three general types with distinctive structures and compositions: savannas, woodlands, and forests (Figure 16.2). These communities develop in predictable patterns across landscapes in association with specific soils, topography, and disturbance patterns. In reality, structure and composition of each occur along a continuum [17].

Savannas have the lowest tree density [16] with widely spaced or small clusters of trees each with a dominant crown class, large diameter, and spreading crown indicative of open growth (Figure 16.2). Associated species flourish in the higher temperatures and light levels present in open conditions. While the structure can appear simple, multiple cohorts may be present [18].

Woodlands have widely spaced trees with spreading crowns, but sufficient tree-to-tree competition to encourage crown differentiation (development of crown classes). There may be multiple cohorts present. Many herbaceous species flower and produce seed during the summer, a result of past disturbances that reduced stand density sufficiently to support a light-demanding understory.

Table 16.1 Silvical characteristics for common species in Central Hardwoods ecosystems [10, 11].

Species	Shade tolerance	Mast frequency (years)	Maximum longevity (years)	Fire response
Bitternut hickory (*Carya cordiformis* (Wangenh.) K. Koch)	Intolerant (to intermediate in bottomlands)	3–5	200	Easily wounded and top killed; sprouts and suckers following fire
Black oak (*Quercus velutina* Lam.)	Intermediate	2–3	250	Susceptible to wounding and decay; stump sprouts
Bur oak (*Q. macrocarpa* Michx.)	Intermediate	2–3	400	Mature trees resistant; stump sprouts
Mockernut hickory (*C. tomentosa* (Poir.) Nutt.)	Intolerant (varies from tolerant to intolerant)	2–3	300	Easily top-killed; stump sprouts
Northern red oak (*Q. rubra* L.)	Intermediate	2–5	400	Susceptible to wounding and decay; stump sprouts
Pignut hickory (*C. glabra* Mill.)	Intolerant (Northeast) to tolerant (Southeast)	1–2	325	Somewhat resistant, seedlings typically top-killed; stump sprouts
Post oak (*Q. stellata* Wangenh.)	Intolerant	2–3	400	Frequently top-killed; smaller trees readily stump sprout
Shagbark hickory (*C. ovata* (Mill.) K. Koch)	Intermediate	1–3	350	Stump sprouts; root suckers more common as parent trees age
White oak (*Q. alba* L.)	Intermediate	4–10	600	Stump sprouts
Scarlet oak (*Q. coccinea* Muenchh.)	Very intolerant	3–5	250	Very susceptible to fire damage and mortality; stump sprouts
Chestnut oak (*Q. montana* Willd.)	Intermediate	4–5	400	Stump sprouts
Red maple (*Acer rubrum* L.)	Tolerant	1–2	150	Easily top-killed, stump sprouts
Black cherry (*Prunus serotina* Ehrh.)	Intolerant	1–5	250	Easily top-killed, stump sprouts

Forests have higher tree densities and closed, multitiered canopies. On dry sites, white oak and hickory can occur in the midstory and understory whereas maples (*Acer*), beech (*Fagus*), and shade-tolerant shrubs frequently dominate the woody understory community on mesic sites. Multiple cohorts are usually present. Spring ephemerals are important herbaecous indicators; they flower and produce seed prior to spring leaf out by trees and suggest a recent absence of fire.

16.2.2 Natural Disturbance

Low-intensity, frequent fire is a dominant disturbance in oak savannas, woodlands, and forests [19]. Historically, most fire in the Central Hardwoods Region was set by humans [20].

Central Hardwoods Oak Ecosystems	A	B	C
Overstory	Bur oak, post oak, white oak, chinkapin oak, blackjack oak	Bur oak, black oak, post oak, white oak, black hickory, mockernut hickory, shagbark hickory	Northern red oak, white oak, bitternut hickory, mockernut hickory, pignut hickory, shagbark hickory, sugar maple
Subcanopy	Eastern redbud, slippery elm, ironwood, paw paw, musclewood	Fragrant sumac, green briar, Carolina buckthorn, virginia creeper, summer grape, *vaccinium* spp.	American hazelnut, wild plum, rough-leaf dogwood
Ground flora	Big bluestem, switchgrass, Indian grass, sideoats grama	Bristly sunflower, nodding fescue, panic grasses, river oats, sedges, asters, milkweed, horsemint, beard-tongue, dittany	May apple, bloodroot, wild geranium, phlox, false Solomon's seal, Dutchman's breaches, trilliums, trout lily, wild comfrey
Structure	Structure equivalent to 10–30% stocking, 1–3 cohorts; Widely spaced or small clusters of trees each with a dominant crown class, large diameter, and spreading crown indicative of open growth	Structure equivalent to 30–75% stocking, 1–3 cohorts; Widely spaced trees with a simple structure including dominant, codominant, and some intermediate trees with spreading crowns. Bell-shaped diameter distributions are common. Open understory.	Structure equivalent to >75% stocking, 1–4 cohorts; Closely spaced trees with a multi-tiered canopy; a reverse-J diameter or irregular diameter distribution is common, semi-open understory

Figure 16.2 Representative plant species and structural conditions associated with the three, general oak-dominated ecosystems that occur in the Central Hardwoods Region: (A) savanna; (B) woodland; (C) forest [13–15]. Species composition varies across the large region. Densities that characterize each ecosystem type have been presented in terms of stocking [16] (see Figure 16.3). *Source:* Photos: USDA Forest Service.

Lightning is also a source of ignition, but it does not occur frequently enough to generate the fire return intervals reconstructed across the region [21]. Fire frequency largely drives community type with mean return intervals ranging from 10–20 years in forests to 1–5 years in savannas. The variability in recovery periods following fire interacts with mean fire interval to influence composition and structure [22]. For instance, variable recovery periods between fires in an ecosystem with short mean fire intervals favor the germination and establishment of oak, whereas regular fires with consistent, equally short recovery periods and the same mean interval encourage dominance by grasses, as found in prairies and savannas. Historically, large, stand-replacing fires occurred infrequently with return intervals of 100–500 years [23], varying across the region [20, 24]. Other important disturbances include wind, insect and disease outbreaks, drought, ice damage, and grazing and browsing [25]. Each of these disturbance types, particularly in combination, contributes to greater heterogeneity in oak ecosystem structure at both stand and landscape scales than created by traditional, timber-focused silvicultural systems.

16.3 Natural Developmental Model

The natural development model for oak-dominated ecosystems is summarized in Table 16.2. This model begins with one (or more) moderate- to high-intensity disturbances that release oak advance regeneration and retain some live and deadwood legacies. It proceeds through young, mature, and old stages of development. Throughout, repeated, moderate-intensity surface fires under a sparse

Table 16.2 A developmental model for oak-dominated savannas, woodlands, and forests.

Stage	Attribute	Cause or Consequence
Disturbance and Legacy Creation	Live tree and deadwood legacies retained; aggregated (savanna or woodland) to dispersed (forest)	Near stand-replacing to partial-canopy disturbance
Preforest stage	Oak and hickory sprout dominated (woodlands or forests) with herb/shrub/grass occurring (savanna); absent to lasting <5 years (woodland, forest) to 15 years (savanna)	Hardwood sprouting shortens this process where sprouts are exceptionally abundant
Young stage	Recruitment and development of oak sprouts in mixtures of hickories; thinning from reintroduction of fire (savanna, woodland) or self-thinning from competition (forest)	Recruitment into overstory may take up to 50 years; stems of sprout origin may be locally abundant; density is reduced due to competition-induced mortality and/or to low-intensity surface fire and browsing
Mature stage	Oaks with wide crowns dominate the overstory; dead wood increases in abundance; greater species diversity occurs in the understory and ground layer (savanna, woodland) or in the seedling layer (forest)	Surface fire creates canopy openings commensurate with intensity; low-intensity fires retain closed canopies of dominant trees (closed woodland) and high-intensity fires create partial openings (open woodland); oak and hickory seedling sprouts accumulate in the understory during fire-free periods (forest)
Old stage	Dominated by oaks with wide crowns with two or more ages classes present; deadwood abundant; patchy canopy with abundant snags; diverse species composition in the ground layer (savanna, woodland); some presence of more mesic, shade-tolerant tree species (forest)	Overstory structure becomes more complex due to tree mortality; low-intensity fire continues to facilitate establishment and early growth of oak stump sprouts (savanna, woodland); less frequent fire (forest)

(savanna) or open canopy (open woodland) or low-intensity surface fires under a closed canopy (closed woodland, forest) are required to maintain characteristic understory species. Fires also encourage development of oak advance regeneration. Stages are described at the stand scale; horizontal complexity across stands results from inherent variability in soil properties such as depth, nutrient, and water supply, and microtopography as well as the influence of legacy trees on the understory microenvironment.

16.3.1 Disturbance and Legacy Creation

Canopy disturbances are needed to increase light availability and release oak advance regeneration. Due to variations in fire frequency and regularity, recruitment of regeneration in oak ecosystems is generally episodic rather than continuous, occurring during fire-free intervals in each community type [22]. Mature legacy trees serve a minor role in stand regeneration; rather, new cohorts typically recruit from advanced regeneration. The spatial distribution of legacy trees following disturbance varies, depending on topography and soil conditions. Legacy trees tend to occur in aggregated patterns near and along streams or other natural fire breaks and in a more dispersed pattern where natural fire breaks are absent.

16.3.2 Preforest Stage

The structure and composition of the preforest stage vary widely across the region and within a landscape depending on disturbance type, disturbance severity, and site characteristics. To varying degrees, shrubs, forbs, graminoids, vines, and vegetative sprouts from top-killed oak and hickory advance regeneration can dominate the preforest stage [26]. Regardless of specific circumstances, the herbaceous and woody understory is well developed [26], and the greater abundance of fruit, quality browse, small prey, and other foods benefit wildlife [27]. The length of this stage varies between savannas, woodlands, and forests. Frequent fire can maintain elements of the preforest stage for extended periods in savannas and woodlands, but eventually, a fire-free period is necessary to allow recruitment. In forests, the preforest stage can last as few as 10 years if ample advance regeneration leads to fast recruitment [28].

16.3.3 Young Stage

In all three ecosystems, the Young Stage may last as many as 50 years as the newest cohort grows enough to resist top-kill by fire and recruits to the canopy. In forests and woodlands, the understory diminishes as the canopy approaches closure and competition leads to crown differentiation and self-thinning (forests). This can occur across an entire stand or in small patches within a stand [29]. On dry and nutrient poor (e.g. the Ozark Highlands) and/or frequently disturbed sites, oaks and hickories emerge as the dominant species, eventually overtopping other tree species less tolerant of stressful conditions.

16.3.4 Mature Stage

The mature stage in savannas and woodlands, both open forest ecosystems [13], is characterized by an abundance of large-diameter, fire-tolerant oaks. Forests in this stage have a closed canopy with an abundance of black oak, white oak, and northern red oak (Figure 16.2). During this stage, the species composition and structure depend on fire frequency and severity. Low-intensity, infrequent surface fires top kill only some of the trees <12 cm dbh, leaving the canopy closed in forests and diverse. High-intensity or frequent surface fires top kill greater numbers of trees <12 cm dbh and can shift composition over time, removing all but the most fire-tolerant oaks (e.g. bur oak [*Q. macrocarpa* Michx.] or post oak [*Q. stellata* Wangenh.]) in stands with partial openings and an overall lower stocking [30]. During fire-free periods, the small-diameter oaks and hickories top-killed by fire resprout, and a rich ground flora develops comprising spring ephemerals that flower and set seed in the spring prior to leaf out (forests) or legumes, forbs, and C3 grasses that flower or set seed in the summer (woodland). In savannas, openings created and maintained by fire lead to abundant, light-demanding C4 grasses.

16.3.5 Old Stage

The old stage begins when 20–30% or more of the overstory exceeds an age of at least 90 years, depending on the region (ranging from >90 years in Minnesota to >150 years in the central Appalachians) [3, 22]. Old savannas, woodlands, and forests exhibit greater structural complexity than earlier stages as well as abundant deadwood. Infrequent, but moderate-intensity fire provides fuel for low-intensity surface fires that move quickly, remove herbaceous biomass, and top-kill accumulating oak sprouts. In woodlands, white oaks and hickories become dominant in the stand as shorter-lived oaks (e.g. black oak [*Quercus velutina* Lam.]) die, which provide snags and downed wood. Resulting canopy gaps favor the accumulation of seedling sprouts in isolated patches; these are kept in check by infrequent low-intensity surface fire. In oak forests, white oaks and northern red oaks typically dominate alongside hickories, red maple, and other species more or less abundant depending on location (e.g. sassafras [*Sassafras albidum* (Nutt.) Nees], hackberry [*Celtis occidentalis* L.]), and understory shrubs. Canopy heterogeneity develops as shorter-lived tree species die and gap-scale disturbances occur, allowing the development of a distinct midstory and understory. Variability in the light environment creates heterogeneity in the understory plant community. Downed wood volumes frequently exceed 35.9 m^3 ha^{-1} and vary across the region [31–34]. Older trees that have experienced wind, ice, or other damage develop features such as cavities that provide habitat and contribute to greater structural complexity. The section that follows illustrates how knowledge of the natural model can be applied to management of these ecosystems through an ecological silviculture system.

16.4 Ecological Silvicultural Systems for Central Hardwoods Ecosystems

16.4.1 Disturbance and Legacy Creation

Variable retention harvesting (VRH, similar to shelterwood with reserves or clear-cut with reserves) or an irregular shelterwood method can be applied in a way that models natural disturbance and maintains or restores key ecosystem attributes and functions in oak ecosystems (Table 16.3). Regeneration harvests enhance structural complexity by creating biological legacies as well as space for a new cohort.

If the desired future condition includes a multi-cohort age structure, continuous cover irregular shelterwood [35] can be used to create openings (gaps) across a stand such that one-third or less of the stand is harvested in any entry (Figure 16.4). The size of the openings necessary for successfully regenerating oak (generally 0.08–0.2 ha) [3] varies with tree height and slope as well as region, climate, and site quality. Larger gaps may be necessary to provide oak and hickory enough light to compete successfully with shade-tolerant species, particularly on richer soils and where prescribed fire is less frequently applied. Reducing stocking in the matrix to less than 75% and maintaining the midstory, as needed, will also increase light availability in gaps (Figure 16.3). Alternatively, VRH can be applied to achieve a two-cohort structure (Figure 16.4). The level and spatial distribution of retention should be informed by the natural disturbance regime as well as stand and site objectives.

Many historical woodlands and savannas now resemble forests, following decades of fire suppression and exclusion. VRH is recommended for creating legacies and initiating the preforest stage [9, 36]. To achieve woodland structure, postharvest stocking should not exceed 30% (Figure 16.3). Assuming overstory trees are of merchantable size and age (>40 cm DBH), retention should include 50–75 trees ha^{-1}. Retained, live tree legacies can be arranged in clumps (aggregates), dispersed across a stand, or some combination of both (Figure 16.4). As with woodlands, VRH can be used to initiate the preforest stage in savannas, but with fewer live trees retained.

Across all three Central Hardwoods ecosystems, existing snags and large downed dead wood should be retained to the extent possible. It may also be desirable to supplement large dead wood pools at the time of harvesting, i.e. use harvesting equipment or girdling to create dead wood.

Table 16.3 An ecological silvicultural system for oak-dominated savannas, woodlands, and forests.

		Outcome			Potential silvicultural actions		
Stage	**Years**	**Forest**	**Woodland**	**Savanna**	**Forest**	**Woodland**	**Savanna**
Disturbance/ Legacy Creation	0	Retention of legacy trees and deadwood; initiation of preforest stage			First establishment cutting for irregular shelterwood; harvest 30% of stand area in gaps >0.2 ha; residual stocking in remainder of stand <75%; alternatively, apply variable retention harvest (VRH, clear-cut with reserves)	VRH; dispersed and aggregated retention (stocking 20–30%)	VRH; dispersed retention (stocking 20–30%)
					Post-sale slashing operations (where there is no pine component); lopping slash and removing it from the vicinity of legacy trees to reduce severity of future prescribed fires		
					Plant shortleaf pine or pitch pine seedlings where these species are a desired component		
Preforest stage	1–10	Promote ground flora in all ecosystems, particularly woodlands and savannas. Establish new cohort (primarily released advance regeneration) and reduce understory competition from mesophytic tree species			Conduct post-sale slashing operations if delayed to favor pine regeneration		
					Moderate-intensity ground fire applied shortly after harvest to reduce competition by mesophytic tree species		
					If natural regeneration is understocked, supplement with seedlings		
Young stage	10–40	Recruit new cohort			Fire-free recruitment period; free thinning of youngest cohort (crop-tree release)		
	40	Retention of legacy trees and deadwood; initiation of patch-scale preforest stage	Promote ground flora, increase tree growth, and enhance crown expansion		Establishment cutting; harvest 30% of stand area in gaps >0.5 acres, thin remainder of stand to <75% stocking	Thin to maintain stocking <60%; periodic prescribed fire; treat invasive species as needed	Thin to maintain stocking <30%; periodic prescribed fire; treat invasive species as needed
	40–50	Establish new cohort and enhance structural heterogeneity			Where pine component is desired, conduct delayed post-sale slashing operations; moderate-intensity ground fire applied shortly years after harvest to reduce competition by mesophytic tree species; where natural regeneration is understocked, plant seedlings		
	50–80	Recruit new cohort			Fire-free recruitment period; free thinning of youngest cohort (e.g.		

Mature stage	80	Retention of legacy trees and deadwood; initiation of patch-scale preforest stage	Promote ground flora, increase tree growth, and enhance crown expansion	Establishment cutting; harvest 30% of stand area in gaps >0.5 acre, thin remainder of stand to <75% stocking; OR VRH can be applied to initiate the preforest stage across the stand	Thin to maintain stocking <60%; periodic prescribed fire; treat invasive species as needed	Thin to maintain stocking <30%; periodic prescribed fire; treat invasive species as needed
	80–90	Establish new cohort, enhance structural heterogeneity		Where pine component is desired, conduct delayed post-sale slashing operations; Moderate-intensity ground fire applied five to seven years after harvest to reduce competition by mesophytic tree species; where natural regeneration is understocked, plant seedlings		
	90–120	Recruit new cohort		Fire-free recruitment period; free thinning of youngest cohort (e.g. crop-tree release)		
Old stage	120+	Can initiate preforest stage in patches. Enhance structural heterogeneity, including addition of deadwood. Increase tree growth and encourage crown expansion. Discourage potential mesophytic species in the understory. Promote ground flora in woodland and savanna.		Establishment cutting; harvest 30% of stand area in new gaps >0.5 acre with 1 retained tree; residual stocking in remainder of stand <75%	Thin to maintain stocking <60%; periodic prescribed fire; treat invasive species as needed	Thin to maintain stocking <30%; periodic prescribed fire; treat invasive species as needed
		Establish new cohort		Where pine component is desired, conduct delayed post-sale slashing operations; moderate-intensity ground fire applied shortly years after harvest to reduce competition by mesophytic tree species		
		Recruit new cohort		Fire-free recruitment period; free thinning of youngest cohort (croptree release)		

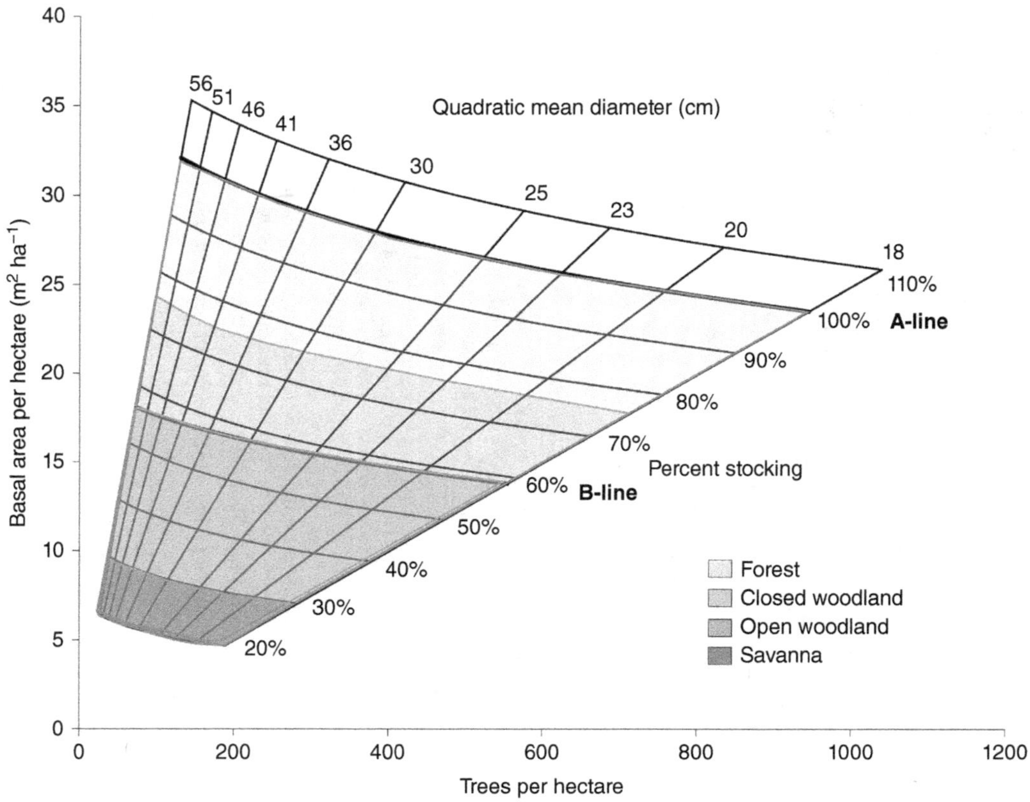

Figure 16.3 Stocking guide [16] that illustrates thresholds for managing savannas and woodlands compared to forests. Woodlands can be further divided into those with closed (lighter grey shading) or open (darker grey shading) canopies [3].

16.4.2 Preforest Stage

Across all three ecosystem types, moderate-intensity surface fire should be applied shortly after harvest to reduce competition from shrubs and mesophytic tree species. Timing depends on fuel loading and advance regeneration. Assuming oak advance regeneration is present, seedlings will be most likely to resprout following fire if the root collar diameter exceeds 0.6 cm [37]. If advance regeneration is not adequately stocked, periodic fire can be applied to reduce competition and leaf litter in anticipation of a future mast year, particularly on richer sites. If artificial regeneration is required, delay planting as necessary to allow an appropriate period of dominance by herbaceous species and shrubs.

16.4.3 Young Stage

During this stage, treatments appropriate for forests differ from those recommended for managing woodlands and savannas. Frequent, low-intensity fires are needed to maintain the understory in woodlands and savannas. Thinning during this stage should maintain stocking at less than 60% for woodlands and less than 30% for savannas, enhancing spatial variability, encouraging overstory tree growth, and allowing ample light to reach the understory.

Fire is excluded during most of the young stage in forests to allow established regeneration to recruit to the canopy. Once the most recently recruited cohort has developed fire-resistant bark,

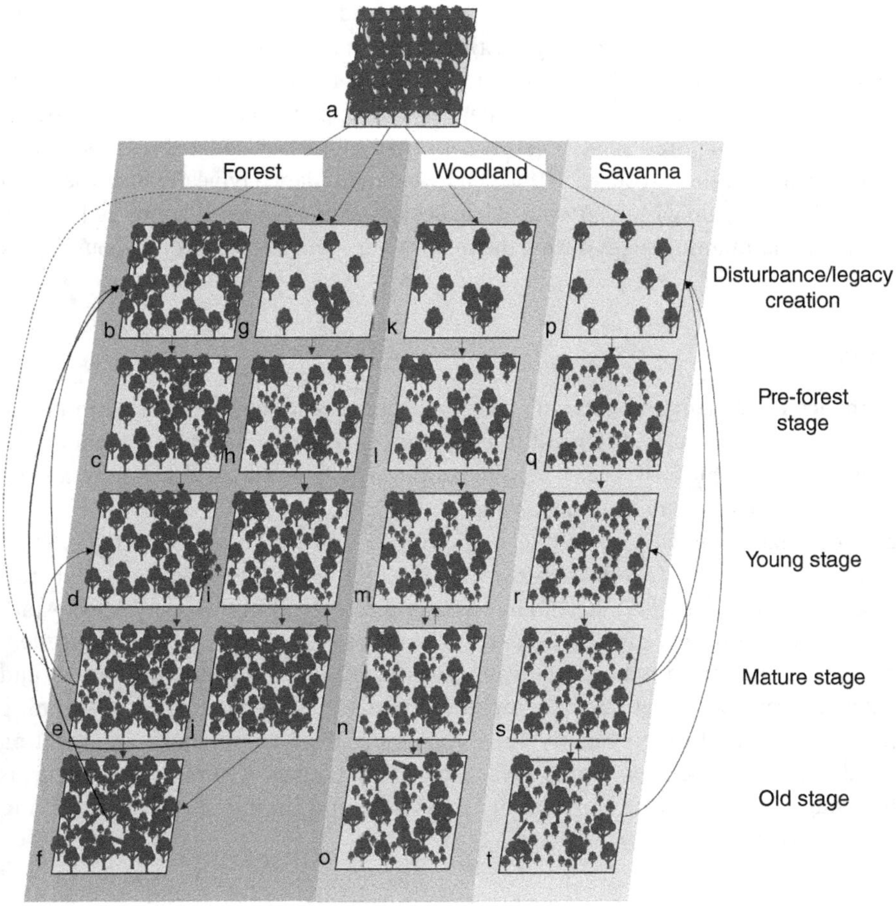

Figure 16.4 Stages of development for Central Hardwoods forests managed with a continuous cover irregular shelterwood system (b–f) and forests (g–j), woodlands (k–o), and savannas (p–t) managed with VRH. The presentation of a single, forest-like starting point (a) across ecosystem types assumes an absence of recent disturbance as is common across the region. Arrows indicate different potential pathways. The dashed line illustrates the option to move mature forest stands initially managed within an irregular shelterwood system back to an earlier developmental stage using VRH.

prescribed burns can be reintroduced to maintain the understory flora, reduce competition, and encourage advanced regeneration. A second establishment cutting will make space for a new cohort in patches across the stand as well as create new legacies. Thinning will encourage diameter growth and crown expansion. This can be accomplished through a thinning from below, but free thinning methods such as crop-tree release are better suited for encouraging growth of specific trees for either ecological or timber objectives, so long as stocking is also maintained. Once a new cohort is established, fire should be excluded for a period to allow regeneration to grow until it develops fire resistance (and recruits to the canopy).

16.4.4 Mature Stage

At this stage, two to three cohorts should be present and well-established. Silvicultural actions may appear similar between the young and mature stages, but the overall structure will differ as legacy trees continue to grow, develop cavities and similar features, and die (contributing to dead wood pools). Management can direct further development via one of multiple pathways. Forests, woodlands, and savannas can be maintained in the mature stage indefinitely with prescribed fire and periodic, partial harvests as described above. Alternatively, the application of VRH can return all three ecosystem types to the preforest stage (Figure 16.4). Lastly, forests, woodlands, and savannas can be allowed to advance to the old stage. While dead wood abundance typically increases after harvest, it begins to decline again within 20 years due to decay and an absence of continued recruitment [38]. Creating or maintaining snags during the mature stage will ensure a continuous supply of large dead wood.

16.4.5 Old Stage

Given the disturbance-dependent and dynamic nature of these ecosystems, strictly defining the old stage and developing an associated silvicultural approach presents a challenge. Except on the poorest, driest sites, oak, and hickory species require periodic fire to recruit to the canopy, and an absence of disturbance leads to compositional change.

Silviculture in old-stage stands is most likely to advance non-timber objectives [3]. Gap-scale disturbances (e.g. fire, mild to moderate wind, insects) will create openings that may recruit or release younger cohorts [29]. Partial harvest, as described above with repeated establishment cuttings in a continuous cover irregular shelterwood system, can also be continued. Prescribed fire can be applied to maintain advanced regeneration if dominance by oak and hickory species is desired. Other actions may include invasive species removal, free thinning to create additional structural heterogeneity and supplement dead wood pools, and treating the understory to reduce competition. Some presence of mesic, shade-tolerant tree species is expected at this stage in oak forests, so treatments will depend on site characteristics and desired future conditions.

16.5 Climate Change Considerations

Projected changes in future climate, as well as emerging pests, pathogens, and invasive species, will interact with the legacy of extensive changes to disturbance regimes, composition, and structure that have already occurred across the Central Hardwoods Region. The region is large and highly variable, so climate change projections vary, and acquiring up-to-date information about specific locations is key to effective climate adaptation management. Broadly, annual precipitation is expected to increase across the region [39, 40], but higher temperatures, longer-growing seasons, and greater seasonal variation in precipitation may increase water stress [41] and shift composition toward more southern and/or xeric species [42].

Despite these expected changes, oak and hickory species of the Central Hardwoods Region are well-adapted to disturbance and stressful conditions [8], traits that may buffer impacts from projected climate change. Managing oak ecosystems in a changing climate has been addressed in greater detail elsewhere [3, 40], and we provide broad guidelines below.

As proposed for other ecosystems in this text, a practical framework for adaptation includes resistance, resilience, and transition strategies [43, 44]. Resistance tactics for maintaining current species composition and structure largely center on thinning treatments that reduce tree-to-tree competition and prioritize release of species that are particularly valued for products, habitat, or other qualities. By increasing resource availability, thinning lessens vulnerability to disease, insects, and other stressors, at least in the short term.

The resilience strategy focuses on allowing some change within the range of historical variability and an eventual return to prior composition and structure (or at least function). Tactics for encouraging resilience also include thinning to reduce competition but with greater emphasis on encouraging more future-adapted native species, increasing age- and structural diversity through the application of two- or multi-cohort silvicultural systems like those described in this chapter, and maintenance of the understory with fire or other tools to minimize establishment of invasive species and ensure regeneration potential. Many of the most common species (e.g. bur oak, white oak, mockernut hickory, post oak, and northern red oak) are considered less vulnerable to climate change [3] and might be encouraged.

The transition strategy embraces the likelihood of future change and modifies forest structure as suggested for resistance (lower stocking) and resilience (two or more cohorts), as well as species composition, to facilitate adaptation. In addition to thinning and applying uneven-aged silvicultural systems, potential tactics include enrichment planting with stock selected to increase the genetic diversity of species already present, such as through seed sources from zones that currently resemble projected future conditions for a given area. Assisted range expansion, planting species projected to have suitable habitat but not currently present, may also assist with transition, particularly in the northern part of the Central Hardwoods Region.

16.6 Summary

Ecological silviculture for oak ecosystems differs from traditional timber-focused approaches that prioritize timber production, efficiency, and wood quality.

Keypoints

1) Open forest ecosystems (e.g. savannas and woodlands) are important components of the Central Hardwoods Region, both culturally and ecologically, having been maintained largely through prescribed fires applied by indigenous peoples. They should be managed accordingly with greater focus on ground flora than tree recruitment.
2) Regeneration harvesting should prioritize biological legacies and the maintenance of a healthy and diverse understory, creating structures that reflect the dynamic and disturbance-dependent nature of these ecosystems.
3) The preforest stage provides valuable early successional habitat that can be enriched and encouraged through fire while also encouraging desired oak and hickory regeneration. In woodlands and savannas, characteristics of this stage are maintained throughout development.
4) Application of treatments should reflect the best available knowledge of the natural and historic disturbance regime. Specifically, such knowledge should inform recovery periods (between prescribed fires and/or harvest entries) and disturbance severity (e.g. severity of surface fire, proportion of canopy removed, abundance of dead wood created).
5) Diversity and variability in structure, timing (e.g. recovery periods), and community composition in the understory (if not trees) are prioritized in these ecological silvicultural systems to promote native biodiversity and maintain adaptive capacity.

Acknowledgments

We thank Brian Palik, Anthony D'Amato, Benjamin Knapp, Greg Edge, Jessica Flatt, and Joe Herring for providing thoughtful comments that helped improve this chapter.

References

1 Hanberry, B.B., Kabrick, J.M., Dunwiddie, P.W. et al. (2017). Restoration of temperate savannas and woodlands. In: *Routledge Handbook of Ecological and Environmental Restoration* (ed. S.K. Allison and S.D. Murphy), 142–157. Routledge.

2 Fralish, J.S. (2003). The central hardwood forest: its boundaries and physiographic provinces. In: *Proceedings of the 13th Central Hardwood Forest Conference*, Gen Tech Rep NC-234 (ed. J.W. Van Sambeek, J.O. Dawson, F. Ponder Jr., et al.), 1–20. US Department of Agriculture, Forest Service, Nor.: St Paul, MN.

3 Johnson, P.S., Shifley, S.R., Rogers, R. et al. (2019). *The Ecology and Silviculture of Oaks*, 3e. Boston, MA: Cabi.

4 Butler, B.J., Caputo, J., Robillard, A.L. et al. (2021). One size does not fit all: relationships between size of family forest holdings and owner attitudes and behaviors. *Journal of Forestry* 119 (1): 28–44.

5 Dey, D.C. (2014). Sustaining oak forests in Eastern North America: regeneration and recruitment, the pillars of sustainability. *Forest Science* 60 (5): 926–942.

6 Zenner, E.K. (2016). Managing Central Hardwood forests within the context of the historic range of variability (HRV): challenges and opportunities. In: *Natural Disturbances and Historic Range of Variation* (ed. C.H. Greenberg and B.S. Collins), 371–391. New York: Springer.

7 Knoot, T.G., Schulte, L.A., Grudens-Schuck, N., and Rickenbach, M. (2009). The changing social landscape in the midwest: a boon for forestry and bust for oak? *Journal of Forestry* 107: 260–266.

8 Dey, D. (2002). The ecological basis for oak silviculture in eastern North America. In: *Oak Forest Ecosystems* (ed. W.J. McShea and W.M. Healy), 60–79. Baltimore: Johns Hopkins University Press.

9 Schweitzer, C.J., Janzen, G., and Dey, D. (2016). Regenerating oak stands the "natural" way. In: *Managing Oak Forests in the Eastern United States* (ed. B. Collins and C.H. Greenberg), 75–84. Boca Raton, FL: CRC Press, Taylor & Francis Group.

10 Burns, R.M. and Honkala, B.H. (1990). *Silvics of North America: volume 1: conifers. USDA, Forest Service. Agriculture Handbook*, 654–675.

11 Iverson, L.R. (1999). *Atlas of Current and Potential Future Distributions of Common Trees of the Eastern United States*, vol. 265. USDA, US Department of Agriculture, Forest Service, Northeastern Research ... .

12 Barrett, J.W. (1995). *Regional Silviculture of the United States*. Wiley.

13 Hanberry, B.B., Jones-Farrand, D.T., and Kabrick, J.M. (2014). Historical open forest ecosystems in the Missouri Ozarks: reconstruction and restoration targets. *Ecological Restoration* 32 (4): 407–416.

14 Nelson, P.W. (2010). *The Terrestrial Natural Communities of Missouri*. Jefferson City, MO: Missouri Natural Areas Committee.

15 WDNR (2022). Wisconsin's Natural Communities. https://dnr.wi.gov/topic/endangeredresources/communities.asp (accessed 15 July 2022).

16 Gingrich, S.F. (1967). Measuring and evaluating stocking and stand density in upland hardwood forests in the Central States. *Forest Science* 13 (1): 38–53.

17 Hanberry, B.B., Bragg, D.C., and Hutchinson, T.F. (2018). A reconceptualization of open oak and pine ecosystems of eastern North America using a forest structure spectrum. *Ecosphere* 9 (10): e02431.

18 Dettman, C.L., Mabry, C.M., and Schulte, L.A. (2009). Restoration of Midwestern U.S. savannas: one size does not fit all. *Restoration Ecology* 17 (6): 772–783.

19 Arthur, M.A., Bray, S.R., Kuchle, C.R., and McEwan, R.W. (2012). The influence of the invasive shrub, *Lonicera maackii*, on leaf decomposition and microbial community dynamics. *Plant Ecology* 213 (10): 1571–1582.

20 Stambaugh, M.C., Knapp, B.O., and Dey, D.C. (2021). Fire ecology and management of forest ecosystems in the Western Central Hardwoods and Prairie-Forest Border. In: *Fire Ecology and Management: Past, Present, and Future of US Forested Ecosystems*, Greenberg C.H., Collins B., eds. 149–199. Springer.

21 Guyette, R.P., Muzika, R.M., and Dey, D.C. (2002). Dynamics of an anthropogenic fire regime. *Ecosystems* 5 (5): 472–486.

22 Frelich, L.E., Reich, P.B., and Peterson, D.W. (2015). *Fire in Upper Midwestern Oak Forest Ecosystems: An Oak Forest Restoration and Management Handbook*, Gen Tech Rep PNW-GTR-914. Portland: United States Department of Agriculture, Forest Service, Pacific Northwest Research Station, 64 P; 914.

23 Guyette, R.P. and Kabrick, J.M. (2002). The legacy and continuity of forest disturbance, succession, and species at the MOFEP sites. In: *Proceedings of the Second Missouri Ozark Forest Ecosystem Project Symposium: Post-treatment Results of the Landscape Experiment. Gen. Tech. Rep. NC-227* (ed. S.R. Shifley and J.M. Kabrick), 26–44. St. Paul, MN: U.S. Dept. of Agriculture, Forest Service, North Central Forest Experiment Station.

24 Abrams, M.D. (1992). Fire and the development of oak forests. *BioScience* 42 (5): 346–353.

25 Dey, D.C. and Kabrick, J.M. (2015). *Restoration of Midwestern Oak Woodlands and Savannas*, Stanturf JA Ed Restor Boreal Temp For, 2e, 401–428 Chapter 20. Boca Raton, FL: CRC Press.

26 Greenberg, C.H., Collins, B., Thompson, F.R., and McNab, W.H. (2011). Introduction: what are early successional habitats, why are they important, and how can they be sustained? In: *Sustaining Young Forest Communities* (ed. C.H. Greenberg, B.S. Collins, and F.R. Thompson III), 1–10. New York, NY: Springer Dordrecht.

27 Greenberg, C.H., Perry, R.W., Harper, C.A. et al. (2011). The role of young, recently disturbed upland hardwood forest as high quality food patches. In: *Sustaining Young Forest Communities*, 121–141. Springer.

28 White, P.S., Collins, B., and Wein, G.R. (2011). Natural disturbances and early successional habitats. In: *Sustaining Young Forest Communities* (ed. P.S. Keyser, T. Fearer, and C.A. Harper), 27–40. Springer.

29 Hart, J.L. (2016). Gap-scale disturbances in central hardwood forests with implications for management. In: *Natural Disturbances and Historic Range of Variation*, 33–47. Springer.

30 Knapp, B.O., Hullinger, M.A., and Kabrick, J.M. (2017). Effects of fire frequency on long-term development of an oak-hickory forest in Missouri, U.S.A. *Forest Ecology and Management* 387: 19–29.

31 Shifley, S.R., Brookshire, B.L., Larsen, D.R., and Herbeck, L.A. (1997). Snags and down wood in Missouri old-growth and mature second-growth forests. *Northern Journal of Applied Forestry* 14 (4): 165–172.

32 Goebel, P.C. and Hix, D.M. (1996). Development of mixed-oak forests in southeastern Ohio: a comparison of second-growth and old-growth forests. *Forest Ecology and Management* 84 (1–3): 1–21.

33 Rubino, D.L. and McCarthy, B.C. (2003). Evaluation of coarse woody debris and forest vegetation across topographic gradients in a southern Ohio forest. *Forest Ecology and Management* 183 (1–3): 221–238.

34 Spetich, M.A., Jenkins, M.A., Shifley, S.R. et al. (2022). Characteristics of dry-mesic old-growth oak forests in the Eastern United States. *Earth* 13 (3): 975–1009.

35 Raymond, P., Bédard, S., Roy, V. et al. (2009). The irregular shelterwood system: review, classification, and potential application to forests affected by partial disturbances. *Journal of Forestry* 107 (8): 405–413.

36 Kabrick, J.M., Dey, D.C., Kinkead, C.O. et al. (2014). Silvicultural considerations for managing fire-dependent oak woodland ecosystems. In: *Proceedings, 19th Central Hardwood Forest Conference: Held in Carbondale, IL. 10–12 March 2014*; General Technical Report NRS-P-142 (ed. J.W. Groninger, E.J. Holzmueller, C.K. Nielsen, and D.C. Dey), 2–15. Newtown Square, PA: U.S. Department of Agriculture, Forest Service, Northern Research Station.

37 Brose, P.H. and Van Lear, D.H. (1998). Responses of hardwood advance regeneration to seasonal prescribed fires in oak-dominated shelterwood stands. *Canadian Journal of Forest Research* 28 (3): 331–339.

38 Wang, S., Knapp, B.O., Ehlers, S. et al. (2021). Forest management effects on downed dead wood at stand and landscape scales in a temperate forest of the central United States. *Forest Ecology and Management* 482: 118905–118905.

39 Dale, V.H., Hughes, M.J., and Hayes, D.J. (2016). Climate change and the future of natural disturbances in the central hardwood region. In: *Natural Disturbances and Historic Range of Variation: Type, Frequency, Severity, and Post-disturbance Structure in Central Hardwood Forests USA* (ed. C.H. Greenberg and B.S. Collins), 355–369. New York: Springer.

40 Brandt, L., He, H., Iverson, L. et al. (2014). *Central Hardwoods ecosystem vulnerability assessment and synthesis: a report from the Central Hardwoods Climate Change Response Framework project*, Gen Tech Rep NRS-124, 1–254. Newtown Square, PA: United States Department of Agriculture, Forest Service, Pacific Northwest Research Station 254 P;124.

41 Dietze, M.C. and Moorcroft, P.R. (2011). Tree mortality in the eastern and central United States: patterns and drivers. *Global Change Biology* 17 (11): 3312–3326.

42 Wang, W.J., He, H.S., Thompson, F.R. et al. (2016). Landscape- and regional-scale shifts in forest composition under climate change in the Central Hardwood Region of the United States. *Landscape Ecology* 31 (1): 149–163.

43 Millar, C.I., Stephenson, N.L., and Stephens, S.L. (2007). Climate change and forests of the future: managing in the face of uncertainty. *Ecological Applications* 17 (8): 2145–2151.

44 Nagel, L.M., Palik, B.J., Battaglia, M.A. et al. (2017). Adaptive silviculture for climate change: a national experiment in manager-scientist partnerships to apply an adaptation framework. *Journal of Forestry* 115 (3): 167–178.

17

Ecological Silviculture for Fennoscandian Scots Pine Ecosystems

Timo Kuuluvainen[1] *and Timo Pukkala*[2]

[1] *Department of Forest Sciences, University of Helsinki, Helsinki, Finland*
[2] *Forest Science and Technology Centre of Catalonia, Solsona, Spain*

17.1 Introduction

In the Eurasian continent, Scots pine (*Pinus sylvestris* L.) is one of the most important tree species ecologically and economically. It occurs over a vast geographic area, from Fennoscandia to Siberia and from Arctic tree line to Turkey [1]. In Fennoscandia, Scots pine occurs naturally in mixed species stands on several site types, but as a forest-forming species especially on nutrient-poor dry and dryish rocky or sandy soils, that are fire prone [2, 3]. Scots pine also commonly occurs as sparse monodominant stands of stunted trees on ombrotrophic nutrient-poor peatlands.

Natural Scots pine forests are often characterized by late-successional features, including spatial heterogeneity, relatively open, complex multistoried canopy structure and continuous presence of very old trees that can be 200–300 years old [4] (Figure 17.1), and abundant deadwood and its continuity [5]. Such forest structures and dynamics are typically created by recurring low- and mixed-severity fires that may kill smaller understory trees, but leave large pines alive, thus providing seed source for postfire regeneration [6]. In the absence of fire, the structural complexity of stands is further enhanced by increasing abundance of a shade-tolerant Norway spruce (*Picea abies* (L.) Karst.) understorey (Figure 17.2a–d), sometimes accompanied with deciduous trees, especially birch (*Betula* spp.) and aspen (*Populus tremula* L.) [7, 8].

Historically, Scots pine timber has been highly valued for commodity uses like construction, and smaller and dead-standing trees, termed kelo-trees [9], for firewood. Large-scale production of tar from pine for export was common in many areas of Finland, especially from sixteenth to early nineteenth century. After establishment of a sawmill industry, high-quality pine logs were targeted in harvesting, predominantly by high grading. After World War Two, intensive forest management based on clear-cut harvesting and artificial regeneration (by seeding or planting) has reduced the area of natural pine forest and the associated key ecosystem structures, such as big old living and dead trees with diverse microhabitats [8, 10]. The extent and ecological quality of dry pine forests have also diminished due to decrease in their natural disturbance, fire, and increase in atmospheric nitrogen deposition [11].

Although native tree species are mostly used in forest regeneration, intensified forestry together with other global change impacts, have profoundly impacted and simplified forest habitats

Ecological Silvicultural Systems: Exemplary Models for Sustainable Forest Management,
First Edition. Edited by Brian J. Palik and Anthony W. D'Amato.

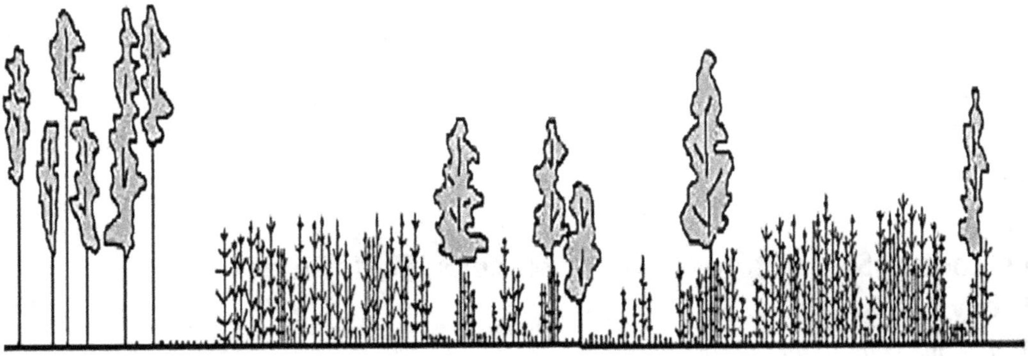

Figure 17.1 Illustration of spatially heterogeneous patchy structure of a Scots pine forest with different age cohorts of trees on sandy soil from Finnish Lapland. *Source:* Redrawn from [44] and based on [45].

Figure 17.2 Typical mature forest structures, and disturbance and regeneration dynamics in natural and managed Scots pine forests in Fennoscandia. (a, b): On dry, low-fertility sites, forest dynamics is driven by low-severity disturbances, such as partial and gap wind throws and low-severity surface fires. (c, d): On semi-fertile, dryish sites, forests are characterized by mixed-severity disturbances, including gap dynamics and fires, with longer rotation times. This is related to development of a more abundant and layered canopy structure, which under suitable weather conditions can act as *ladder fuels* increasing the probability of severe crown fire. (e, f): The monotonously applied business-as-usual even-aged management based on, thinning from below, clear-cutting with low-retention and regeneration by planting seedlings.

compared with natural ecosystem conditions [12] (Figure 17.2). This is also the main reason for the realized and anticipated future decline of many forest-dwelling species associated with Scots pine forest ecosystems [13]. For example, epixylic lichens are directly affected by changes in forest structure because they are dependent on large trees or deadwood as physical structures and growing surfaces [3]. Some species are specialized for growing on specific types of host trees, such as dead-standing trunks or scarred wood that rarely occur in managed forests [9].

As the result of extensive long-term, direct (management), and indirect (nutrient deposition) human impact, natural Scots pine forests have become rare, with the exception of some specific habitat types like cliff outcrops and unditched peatbogs. In fact, pine taiga forest is a habitat type under special protection status in the European Union. In Finland, for example, dry pine forests are recognized as a nationally threatened natural habitat type [11]. It is evident that ecosystem-based management and ecological restoration of pine forests in Fennoscandia, aimed at achieving a balance between economic profits and safeguarding biodiversity, must be based on silviculture that emulates natural dynamics and resultant forest structures [14, 15]. In this chapter, we first review the structure, dynamics, and biodiversity effects of Fennoscandian pine forests. Secondly, we provide an example of application of a novel ecological silviculture approach, termed any-aged management (AAM), and compare the performance of this approach with business-as-usual even-aged silviculture.

17.2 Structure, Dynamics, and Composition of Scots Pine Ecosystems

Scots pine is a long-lived species that in Fennoscandia commonly can reach ages up to 200–300 years [4]. In fire-prone boreal landscapes, this is related to Scots pine being distinctly a fire-adapted, pyrogenic species. Large trees with thick heat-insulating bark are fire-resistant and commonly survive low to mid-intensity fires. Another important trait is rapid pruning of lower branches, which reduces the amount of "ladder-fuels" and the risk of high-intensity crown fires [2]. Although Scots pine obviously benefits from fire compared with other major tree species, Norway spruce, and birches, it is not strictly dependent on fires and there are also non-pyrogenic pine forests [16, 17].

There are roughly two main types of natural Scots pine forest dynamics, depending on the interactions among site type, forest structure, and disturbance regime (Figure 17.3). The first type of dynamics is most common on low-productivity sites, where the scarcity of ground fuels, short disturbance cycles, and location of canopy fuels high up from the ground, limit the risk of high-severity crown fires. This results in a high-frequency low-severity disturbance regime driven by disturbance agents such as wind and, especially in the past centuries, fire of human origin (Figures 17.2a, b and 17.3). Especially high disturbance (fire) frequency plays an important role in preventing the accumulation of fuels between subsequent fires.

Such disturbance regimes typically cause only partial tree mortality (Figure 17.2), while causing a simultaneous release of growing space and resources for growth. The surviving large trees provide seed and facilitate quick regeneration of postfire sites [6]. Over time, these partial mortality dynamics result in a patchy heterogeneous mosaic of uneven-aged and relatively open-canopied pine forest, consisting of patchy cohorts of larger surviving legacy trees and those regenerated after fires (Figures 17.1 and 17.2). With this dynamics model, Scots pine can efficiently dominate a site and maintain a patchy relatively stable forest structure for long time periods [18, 19].

Figure 17.3 Kelo trees in Kalevala National Park in northwest Russia. Keloa trees are standing dead Scots pines that can remain standing for decades or even up to two centuries after tree death, slowly developing a silvery grey color. Kelo trees are a visually striking and ecologically important structural component of natural Scots pine forests in Fennoscandia. *Source:* From Refs. [9, 20].

The second type of stand dynamics (Figure 17.2c–d), which was probably more common before significant human impact, is driven by high-severity disturbances. Such disturbances usually consist of intensive stand-replacing crown fires that occur predominantly in young and dense stands or in Scots pine stands with a multilayered *Picea*-dominated understorey, often found on medium dry sites. In such cases, depending on site type and seed sources, a more linear successional sequence may occur. The early successional stages may include an abundance of deciduous trees, such as birch, rowan (*Sorbus aucuparia* L.), and aspen, but Scots pine usually establishes as well and becomes an important component in the postfire succession. Being a drought-tolerant species Scots pine can gain dominance with time, especially on dry and nutrient-poor sites. Thus, Scots pine can dominate from mid- to late-successional old stages that may last hundreds of years.

17.3 Dead Standing Kelo Trees as a Key Component of Fennoscandian Pine Forests

In unmanaged Fennoscandian Scots pine forests, large dead-standing pines, called kelos or kelo trees, are a visually distinct and ecologically important feature [9, 17] (Figures 17.2 and 17.3). Creation of kelo trees through natural forest dynamics is an extremely slow process. First, pines commonly live for up to 200–300 years of age [7]. Second, following tree death, a tree can stay standing for decades up to two centuries [20]. Over time, a standing pine tree becomes sculpture like with a characteristic silvery appearance (Figure 17.3).

There are some important stages in the development of kelo trees. First, tree death is preceded by decades of slow growth [9, 21], influenced by senescence together with various stressors [20]. One of these stressors is recurring surface fires, which damage the stem base, form fire scars, and trigger impregnation of stem wood with defensive chemicals. During this long process leading to tree death, the wood material in the stem base becomes impregnated with decay-resisting chemicals [21, 22], which makes the decay of wood very slow. Over time, the bark and fine branches fall off, and a kelo tree gets its characteristic greyish-silvery appearance (see Figure 17.2).

What makes standing dead Scots pines so important ecologically in the Fennoscandian boreal forests is that similar long-term development leading to kelo trees does not occur with the other main tree species. Hence, standing dead Scots pine dynamics are essential for the creation of specific structural features and habitats, as well as long-term carbon storage, as kelo trees last for decades to centuries. Standing dead trees are known to be important habitats for many groups of species, such as epixylic lichens, fungi, and cavity-forming and cavity-nesting birds, insects, and mammals. In particular, kelo trees provide a unique habitat for some specialist species [9, 23, 24], including lichens, fungi, and insects. When a kelo tree eventually falls, it becomes an equally unique type of deadwood habitat for many species of fungi [9].

17.4 Evolution of Ecological Silviculture of Scots Pine Forests

After the gradual end of selective cuttings by the early twentieth century, clear-cutting and even-aged silviculture became the dominant management practice in Scots pine forests. Regeneration practices included prescribed burning followed by direct seeding. Also, a seed tree method is common; with seed trees removed when regeneration is sufficient. Over the last several decades, clear-cutting followed by mechanical soil scarification and seeding or planting has become the main approach in Scots pine silviculture.

Beginning from the mid-1990s, some retention trees (see Chapter 1) are left on harvested sites [25]. Development of retention practices was a response to concerns about biodiversity loss due to clear-cutting, with the aim of maintaining some of the structure of mature native forests and their associated species [24, 25]. Retention practices are meant to reconcile the conflicting goals of timber production, conservation of biodiversity, and provisioning of other ecosystem services in forest management [26].

However, by themselves, retention practices in Fennoscandia cannot be called ecological silviculture because the prevailing even-aged management regime creates structural conditions that are far from the complex multilayered and spatially heterogeneous structure of natural pine forests (Figures 17.2 and 17.3). One obvious difference is the low level of retention generally used. For example, an average of only 1.2% of stand volume was retained as living trees during 2013–2017 in private forests in Finland, which satisfies the Finnish Programme for the Endorsement of Forest Certification standard [26]. Such retention levels are too low to maintain a continuous supply of new large-sized deadwood for obligate saproxylic species [24, 27]. Minimum retention level of 5–10% of preharvest trees is likely needed, and significantly more in many cases [27].

Another practice that is ecological in intent is *continuous cover forestry* [28]. Continuous-cover forestry refers to management without clear-cutting, with harvesting based on single-tree or group selection, or shelterwood cuttings, all of which retain a significant portion of the trees. There is evidence that continuous-cover forestry benefits species living in mature forest habitats and late-successional stages better than rotation forestry with clear-cuts [26, 28, 29].

Finally, a management planning approach, *any-aged forestry (AAF)* [30–33], has been proposed as an ecological model for Scots pine forests. With this approach, at the landscape level, no permanent choice is made between continuous cover and even-aged management, and all variants of either silvicultural approach can be used at any time. Because all types of treatment variants are permitted in AAF, it is flexibly able to create and maintain forest structures that resemble both stand-replacing and lower-intensity disturbance dynamics where tree mortality is only partial across multiple spatial scales. Also, since AAM is less constrained than either even-aged management or continuous cover forestry separately, an optimally implemented AAM approach can sustain more ecosystem benefits than the other silvicultural approaches [33].

17.5 Toward Ecological Silviculture for Scots Pine in Fennoscandia

It is clear that the current business-as-usual even-aged management model for Scots pine creates forest structures that differ drastically from those found in forests that develop under natural dynamics (Figure 17.2). The core challenge in ecological forestry is to find silvicultural approaches that reconcile between the goals of (i) production management and related forest structures and (ii) naturally emerging heterogeneous ecosystem structures that foster biodiversity, resilience, and adaptive capacity.

Important legacy structures to retain in ecological silviculture include large living, and standing dead kelo trees as well as downed logs. Important stand-level features are continuity of small-scale horizontal and vertical heterogeneity (Figure 17.2; Table 17.1). Many of these features take a long time to develop and managing their continuity requires long-term planning far beyond normal cutting cycles (Table 17.1).

Table 17.1 Comparison of some key differences between low-retention even-aged silviculture versus ecological silviculture, and benefits of an ecological approach.

Characteristic	Low-retention even-aged approach	Ecological approach	Benefits of ecological approach
Spatial scale of management	Stand- and retention patch scales; one canopy layer	Multi-scale from microsites to wildlife trees to patches to stands and landscapes. Vertically multilayered canopies	High habitat variability supporting species diversity and resilience
Temporal scale of management	Rigid, determined by clear-cut and thinning rotations	Flexible from years to decades to centuries emphasizing continuity	Continuity of key ecological structures supporting biodiversity and resilience
Large old trees	Scarcity of large and old trees	Continuity of large old trees	Continuity of structural variability, cavity microhabitat, and biodiversity
Deadwood	Small- to medium-sized deadwood; lack of continuity	All-sized deadwood including Coarse Woody Debris continuity	Continuity and variability of deadwood habitat support biodiversity
Tree species composition	Small- to medium-sized pines	Continuity of coarse trees of different species	High tree species and size variability supporting biodiversity

17.6 Reconciling Economic Profitability with Biodiversity: A Case Study Using Any-Aged Forestry

Although features beneficial for ecological silviculture are easy to list (Table 17.1), in reality in boreal Fennoscandia the starting point of management is a forest where structural variation has been grossly truncated by intensive even-aged management. Nevertheless, even limited variation can be utilized by flexibly using different silvicultural options. To demonstrate this, we present here a case study that deals with silvicultural options in a 506-ha holding of managed forest located in North Ostrobothnia, in the western part of Finland. The most common site types in this forest are mesic (240 ha), sub-xeric (167 ha), and xeric site (27 ha). The mean growing stock volume of the forest was 133.5 m^3 ha^{-1}, of which the share of pine was 105.5 m^3 ha^{-1} and the rest was mainly spruce and birch.

The case study forest was characterized by high proportion of economically mature stands (almost 50% of the area) and low percentage of young (seedling and sapling) stands (15%). In typical even-aged management, the currently recommended silvicultural approach would be to cut and regenerate the mature stands soon. The prevailing regeneration method would be clear-felling and planting on mesic sites, clear-felling and artificial seeding (sowing) on sub-xeric sites, and natural regeneration via seed trees on xeric sites. The planted tree species on mesic sites is mostly Norway spruce because Scots pine seedlings are often browsed by moose. Young stands are treated with thinning from below at about 20-year intervals after the trees reach commercial dimensions.

As an alternative to this business-as-usual even-aged silviculture [34], we provide an approach termed AAF [33], which was applied for the case study forest as one approach to implement ecological silviculture, as it enhances more diversified forest structures, compared to business-as-usual management. This approach produces and supports stand structures reflective of both stand-replacing disturbance dynamics and those emulative of partial, non-stand-replacing disturbances.

The guidelines for application of AAF consist of three rules. (i) The first rule determines whether the stand is economically "mature" and thus a candidate for harvesting. If it is, then (ii) the second rule determines whether harvesting should be final felling or partial cutting. If the determination is the latter, then (iii) the third rule determines the removal intensity in different diameter classes.

In the original formulation, the guidelines for AAF aim to maximize economic profitability and as such, do not prescribe that retention legacy trees should be left on sites. In fact, thinning treatments are often prescribed in such a way that all large trees are harvested if they are financially mature. Correspondingly, for the management plan for the even-aged approach [35], retention basal area was set to 0.2 m^2 ha^{-1}, when the stand area was 0.5 ha or more. In stands smaller than 0.5 ha, no retention trees were left. These settings roughly correspond to the current Finnish practices of leaving small amounts of retention trees [27].

To adapt the any-aged guidelines to conform the idea of ecological silviculture, the retention basal area was increased to 1 m^2 ha^{-1}, which is about 5% of stand volume and basal area, and which has been suggested as the minimum level for a positive ecological impact [25]. In addition, the maximum thinning intensity among the large trees was restricted to 70%.

The management plans for both silvicultural approaches, i.e. business-as-usual timber-focused even-aged and AAF, was developed for a period of 50 years. Net present value (NPV) calculated with a 3% discount rate was maximized in both plans. The NPV was calculated to infinity, which means that the NPV of the stands at the end of the 50-year period was taken into account. The NPV of the ending growing stock was calculated with models [33]. The stand dynamics (growth and survival of trees, and ingrowth) were simulated with recent models [36].

Figure 17.4 shows that with even-aged management, more than 50% of the area had a clear-felling prescription during the 50-year period. Most of this was clear-felling with green tree retention. The area of thinning was 176.4 ha. Small areas were treated with seed-tree cutting and subsequent harvesting of the seed trees. In contrast, in ecological silviculture most of the harvests were thinning from above with various intensities. The total thinned area over 50 years was 958 ha; so on average, the stands were thinned at approximately 25-year intervals. The area of clear-felling with green tree retention was 80.1 ha, which is 15.8% of the total area. Figure 17.5 shows examples of typical treatment schedules for the first three 10-year periods of the 50-year planning horizon and the corresponding stand development in ecological silviculture and even-aged management. Structural elements that are considered important for sustaining biodiversity of Finnish forests, and are generally scarce in forests under even-aged management, include the amount of old and large trees, the amount and diversity of deadwood, and the amount of broad-leaved tree species. In addition, tree species richness and tree size diversity are usually low in even-aged planted forest stands. Figure 17.5 shows that the development of all these structural elements is more favorable in ecological silviculture as compared to even-aged management.

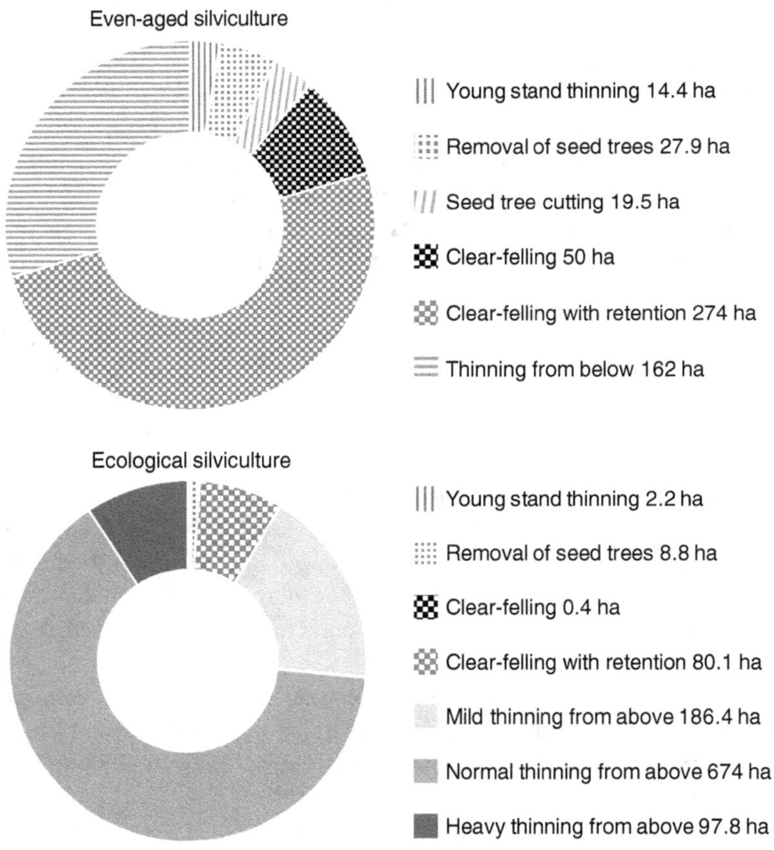

Figure 17.4 Areas of different types of management prescriptions during 50 years in business-as-usual even-aged and ecological silviculture.

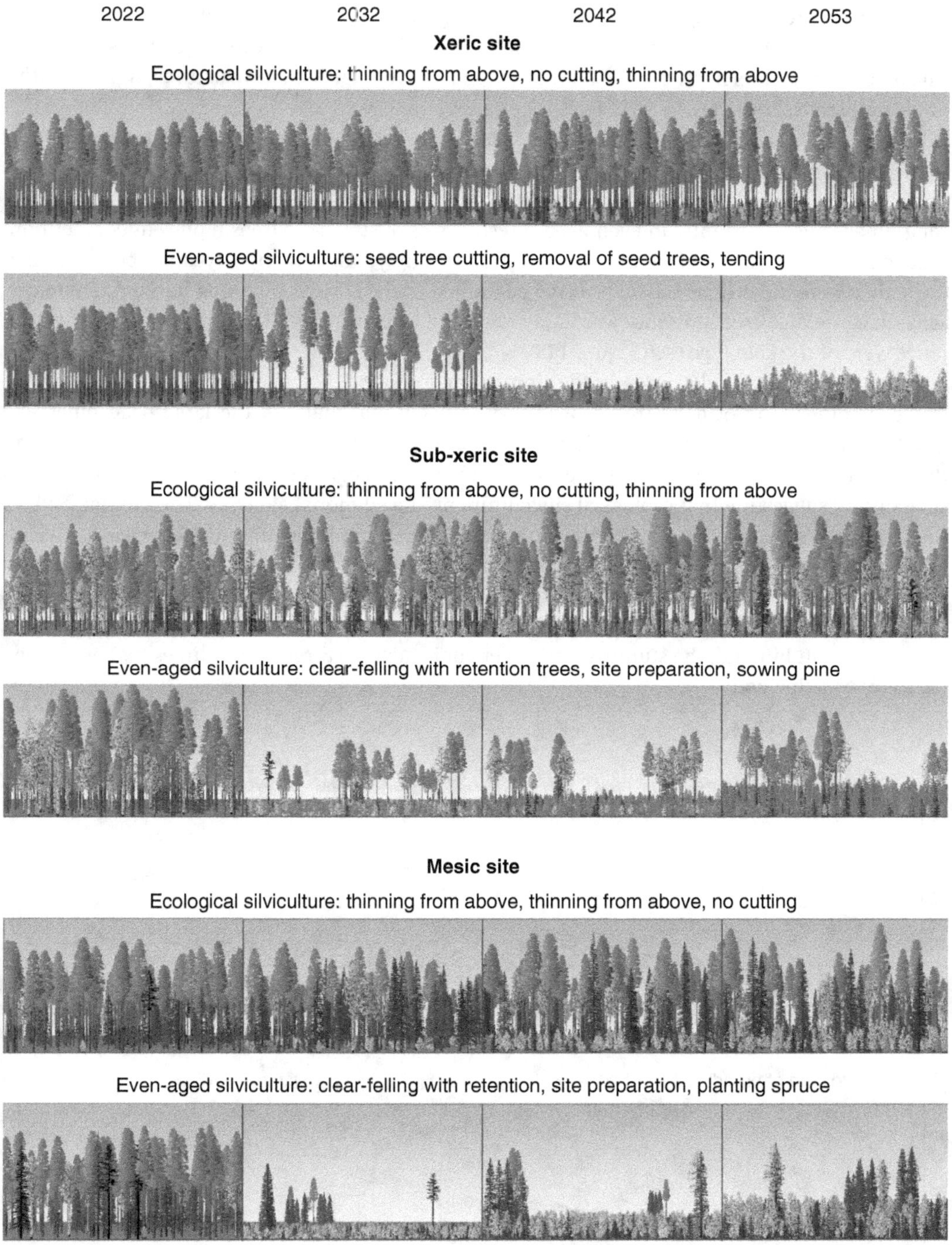

Figure 17.5 Illustration of typical treatment schedules selected for different site types in ecological silviculture and even-aged management. The treatments were simulated in the middle of the 10-year period.

However, differences in species diversity (Shannon index) and size diversity (Gini index) were rather marginal. Opposite to even-aged silviculture which maintains even-aged forests, ecological silviculture fosters more diverse both even- and uneven-aged stand structures, representative of natural landscapes [33].

In Figure 17.6, the minimum diameter of "a large tree" depended on-site productivity according to the formula suggested by [35]. As a result, trees that were classified large had larger diameters on more productive sites, and in all cases, trees that were classified large were also older than the rotation ages used currently in even-aged forestry. Most of the dead trees represent competition-induced mortality. All stands of the case study forest were younger than 160 years. About 120 ha of the stands were older than 120 years at the beginning of the 50-year planning horizon. Therefore, senescence-related tree mortality was likely low, with a consequence that only a part of the dead pines were of the kelo type (see Figure 17.2).

The tree mortality models used in the simulation do not predict whether a dying pine is a kelo tree or another kind of dead tree. However, it is likely that many of the large dead pines will develop into kelo trees. Figure 17.7 shows that the volume of large pines (dbh larger than 30 cm) that died during the 50 years was three times higher in ecological silviculture than in even-aged silviculture, suggesting that ecological silviculture produced more kelo trees. The average NPV of the treatment schedules was higher in ecological silviculture, indicating that ecological silvicultural was more profitable to the landowner than the currently recommend even-aged silviculture. The main reasons for the better profitability of ecological silviculture are lower management costs and higher share of sawlogs of harvested timber (81% in ecological silviculture, 72% in even-aged silviculture). In Finland, the stumpage price per cubic meter is three to four times higher for sawlog than for pulpwood.

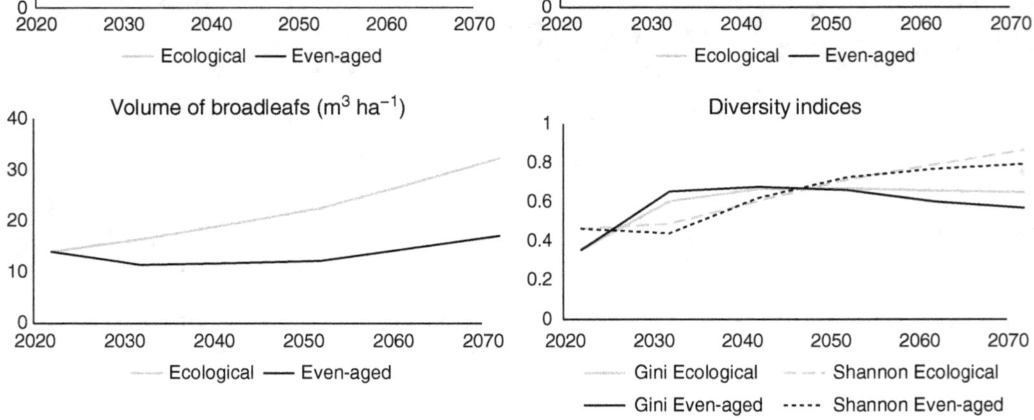

Figure 17.6 Development of large tree, deadwood, and broadleaf volumes and the indices that describe the average tree species diversity (Shannon) and tree size diversity (Gini) of the stands.

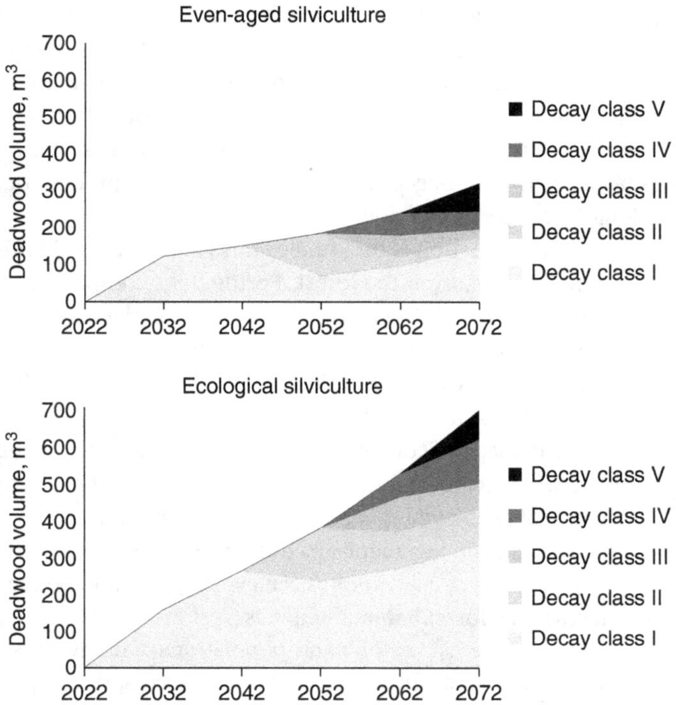

Figure 17.7 Total volume of new large-sized pine deadwood (dbh >30 cm) in the case study area in ecological and even-aged silviculture. Decay class I is recent fresh deadwood and Decay class V is the most decomposed type of deadwood.

17.7 Ecological Silviculture in Fennoscandia: Policy Context and Future Prospects

Development of ecological silviculture of Scots pine forests is based on natural models of dynamics and patterns of forest development [3, 37, 38]. The core challenge is to find management approaches that balance forestry's income and profitability goals, and those related to naturally emerging ecosystem properties, such as biodiversity, resilience, and long-term adaptive capacity [37, 39, 40]. However, reaching a multifunctional balance has not been the goal during the past decades in Fennoscandian forest management, which has mostly based on even-aged silviculture and concentrated on boosting timber production for forest industry, at the expense of other ecosystem services, such as carbon storage and sequestration, biodiversity maintenance, and recreation potential [27].

However, this is changing rapidly as it has been estimated that the regulative effects of forests, for example, carbon sequestration and groundwater management, account for 65–90% of the value of world's forests [41]. In Finland, for example, the economic value of carbon sequestration in tree biomass is already higher than the economic value of timber production if carbon price is taken from the international carbon credit markets [42]. Pricing of biodiversity maintenance is more complicated, but it can be claimed that high biodiversity is invaluable since it is the foundation of the long-term sustainability of all ecosystem services [41].

All this means that there is an urgent need to develop ecological silviculture, that would balance different ecosystem services better than current practices [40, 43] (Figure 17.2; Table 17.1). Here, the emphasis should be on maintaining key legacy structures such as deadwood, and spatial heterogeneity providing continuous but variable cover forest, emulating the outcome of natural disturbance dynamics [39]. However, it is also realistic to acknowledge, that because Fennoscandian forests have been shaped for decades by intensive even-aged management for timber production, their homogenized structure needs a lengthy restoration period.

A major reason for the dominance of timber production is that timber sales are the only significant way for the landowner to get economic benefits from the forest. Fortunately, compensation systems for the other forest benefits are gradually being developed. So far, the negative effects of timber production on the other functions have not been taxed or penalized, and restrictions concerning the avoidance of harmful effects of forestry have been minimal. As a consequence, timber has been produced with the detriment of other forest functions [26].

The case study calculations we present for managed Scots pine forests in Fennoscandia, using the all-aged forestry principles [33], show that an ecological silvicultural approach can be a cost-efficient means to reduce the negative impacts of even-aged management in Scots pine-dominated Fennoscandian forests. However, it is clear that the management schedules used in our case study may not be sufficient, e.g. for effective biodiversity conservation, because they do not produce large-enough continuous areas of late-successional forest habitat, which is a critical key habitat in unmanaged Fennoscandian forest landscapes, and necessary for viable populations of many forest-dwelling species [34]. This highlights the importance of forest protection as a necessary complement to ecological silviculture [34].

The guidelines for AAF that were used in the case study do not indicate how the partial cuttings should be used to modify the spatial distribution of trees (Table 17.1); the guidelines only tell the harvest intensity for different diameter classes. However, creating small canopy gaps and using variable-density thinning (see Chapter 1) enhance the natural regeneration of Scots pine. Therefore, a patchy, spatially heterogeneous forest that is characterized by structures and processes similar to those promoted by natural disturbance dynamics, is also beneficial for timber production [39, 40].

17.8 Conclusions

Ecological silviculture of Scots pine forests aims at improved multi-objectivity and ecological sustainability by emulating natural forest structures and dynamics [40, 43]. However, in Fennoscandia the starting point for ecological silviculture is a forest that is far from natural conditions. This is because forests have been shaped by intensive utilization for past centuries and decades, and they are void of much of the important ecological structures typical of natural forests. Long-term implementation of ecological silviculture is needed to restore more natural ecological structures to regain the associated ecosystem services.

Fennoscandian forests are to a large extent owned by hundreds of thousands of small forest owners. For them, wood production and economic profit are important goals also in ecological silviculture striving for balanced sustainability. Our example with AAF makes the point that switching to ecological forestry, which better emulates natural disturbance dynamics and more variable forest structures, can be a competitive alternative also economically. This is an encouraging result concerning the wider application and adoption of ecological silviculture approaches in Fennoscandian-managed forests.

In general, economically profitable forest management relies on growing trees with a high rate of value increment. A prerequisite for this is that stand density is quite low and that trees are harvested before they reach very large sizes and very old ages. This poses a major challenge for ecological silviculture: how to maintain important ecological "slow" legacy structures, that take a long time to develop, such as the kelo trees and coarse woody debris? It is clear that ecological silviculture of Scots pine needs to be augmented with ecologically based criteria for retaining ecological structures, such as deadwood, retention tree patches, and set-aside areas of different sizes. To conclude, ecological silviculture that maximizes the total benefit of forests, consisting of economic, biodiversity, regulation, and multiple-use benefits, calls for a combined use of AAM, retention forestry, and protection.

References

1 Richardson, D.M. (1998). *Ecology and Biogeography of Pinus*. Cambridge University Press.

2 Esseen, P.-A., Ehnström, B., Ericson, L., and Sjöberg, K. (1997). Boreal forests. *Ecological Bulletins* 46: 16–47.

3 Kuuluvainen, T. (2002). Natural variability of forests as a reference for restoring and managing biological diversity in boreal Fennoscandia. *Silva Fennica* 36: 97–125.

4 Kuuluvainen, T., Mäki, J., Karjalainen, L., and Lehtonen, H. (2002). Tree age distributions in old-growth forest sites Vienansalo wilderness, eastern Fennoscandia. *Silva Fennica* 36 (1): 169–184.

5 Karjalainen, L. and Kuuluvainen, T. (2002). Amount and diversity of coarse woody debris within a boreal landscape dominated by *Pinus sylvestris* in Vienansalo wilderness, eastern Fennoscandia. *Silva Fennica* 36 (1): 147–167.

6 Kuuluvainen, T.T. and Rouvinen, S. (2000). Understory regeneration on two sites of different fire history in a boreal *Pinus sylvestris* forest. *Journal of Vegetation Science* 11: 801–812.

7 Axelsson, L. and Östlund, L. (2000). Retrospective gap analysis in a Swedish boreal forest landscape using historical data. *Forest Ecology and Management* 147: 109–122.

8 Lilja, S. and Kuuluvainen, T. (2005). Structure of old *Pinus sylvestris* dominated forest stands along a geographic and human impact gradient in mid-boreal Fennoscandia. *Silva Fennica* 39 (3): 407–428.

9 Niemelä, T., Wallenius, T., and Kotiranta, H. (2002). The kelo tree, a vanishing substrate of specified wood-inhabiting fungi. *Polish Botanical Journal* 47 (2): 91–101.

10 Siitonen, J. (2001). Forest management, coarse woody debris and saproxylic organisms: Fennoscandian boreal forests as an example. *Ecological Bulletins* 49: 11–41.

11 Kontula, T. and Raunio, A. (ed.) (2018). *Threatened Habitat Types in Finland 2018. Red List of Habitats. Results and Basis for Assessment. The Finnish Environment 2/2019*, 254. Helsinki: Finnish Environment Institute and Ministry of the Environment.

12 Kuuluvainen, T. (2009). Forest management and biodiversity conservation based on natural ecosystem dynamics in northern Europe: the complexity challenge. *Ambio* 38: 309–315.

13 Hanski, I. (2000). Extinction debt and species credit in boreal forests: modelling the consequences of different approaches to biodiversity conservation. *Annales Zoologici Fennici* 37: 271–280.

14 Angelstam, P. (1998). Maintaining and restoring biodiversity in European boreal forests by developing natural disturbance regimes. *Journal of Vegetation Science* 9: 593–602.

15 Kuuluvainen, T., Angelstam, P., Frelich, L.E. et al. (2021). Natural disturbance based forest management: moving beyond retention and continuous-cover forestry. *Frontiers in Forests and Global Change* 4: 629020.

16 Zackrisson, O., Nilsson, M., Steijlen, I., and Hörnberg, G. (1995). Regeneration pulses and climate-vegetation interactions in nonpyrogenic boreal Scots pine stands. *Journal of Ecology* 83: 469–483.

17 Kuuluvainen, T., Aakala, T., and Várkonyi, G. (2017). Dead standing trees in a boreal forest landscape in the Kalevala National Park, northern Fennoscandia: amount, population characteristics and spatial pattern. *Forest Ecossystems* 4: 12.

18 Zackrisson, O., Nilsson, M., Steijlen, I., and Hörnberg, G. (1995). Regeneration pulses and climate-vegetation interactions in non-pyrogenic boreal Scots pine stands. *Journal of Ecology* 83: 469–483.

19 Rouvinen, S., Kuuluvainen, T., and Siitonen, J. (2002). Tree mortality in a *Pinus sylvestris* dominated forest landscape in Vienansalo wilderness, eastern Fennoscandia. *Silva Fennica* 36 (1): 127–145.

20 Venugopal, P., Julkunen-Tiitto, R., Junninen, K., and Kouki, J. (2015). Phenolic compounds in Scots pine heartwood: are kelo trees a unique woody substrate? *Canadian Journal of Forest Research* 46 (2): 225–233.

21 Venugopal, P., Junninen, K., Linnakoski, R. et al. (2016). Climate and wood quality have decayer-specific effects on fungal wood decomposition. *Forest Ecology and Management* 360: 341–351.

22 Santaniello, F., Djupström, L.B., Ranius, T. et al. (2017). Large proportion of wood dependent lichens in boreal pine forest are confined to old hard wood. *Biodiversity and Conservation* 26 (6): 1295–1310.

23 Tikkanen, O.P., Martikainen, P., Hyvärinen, E. et al. (2006). Red-listed boreal forest species of Finland: associations with forest structure, tree species, and decaying wood. *Annales Zoologici Fennici* 43 (4): 373–383.

24 Gustafsson, L., Baker, S.C., Bauhus, J. et al. (2012). Retention forestry to maintain multifunctional forests: a world perspective. *BioScience* 62: 633–645.

25 Gustafsson, L., Kouki, J., and Sverdrup-Thygeson, A. (2010). Tree retention as a conservation measure in clear-cut forests of northern Europe: a review of ecological consequences. *Scandinavian Journal of Forest Research* 25: 295–308.

26 PEFC Finland. (2009). PEFC Finland Standard. Criteria for Group Certification. Level of a Forestry Centre or a Forest Management Association. PEFC FI 1002:2009.

27 Kuuluvainen, T., Lindberg, H., Vanha-Majamaa, I. et al. (2019). Low-level retention forestry, certification, and biodiversity: case Finland. *Ecological Processes* 8: 47.

28 Peura, M., Burgas, D., Eyvindson, K. et al. (2018). Continuous cover forestry is a cost-efficient tool to increase multifunctionality of boreal production forests in Fennoscandia. *Biological Conservation* 217: 104–112. https://doi.org/10.1016/j.biocon.2017.10.01.

29 Pommerening, A. and Murphy, S.T. (2004). A review of the history, definitions and methods of continuous cover forestry with special attention to afforestation and restocking. *Forestry* 77: 27–44.

30 Pukkala, T. (2016). Which type of forest management provides most ecosystem services? *Forest Ecosystems* 3: 9. https://doi.org/10.1186/s40663-016-0068-5.

31 Haight, R.G. and Monserud, R.A. (1990a). Optimizing any-aged management of mixed-species stands. I. Performance of a coordinate-search process. *Canadian Journal of Forest Research* 20 (1): 15–25.

32 Haight, R.G. and Monserud, R.A. (1990b). Optimizing any-aged management of mixed-species stands: II. Effects of decision criteria. *Forest Science* 36 (1): 125–144.

33 Pukkala, T. (2022). Improved guidelines for any-aged forestry. *Journal of Forestry Research* https://doi.org/10.1007/s11676-022-01473-6.

34 Äijälä, O., Koistinen, A., Sved, J. et al. (2014). *Metsänhoidon Suositukset*. Publications of the Forest Development Centre Tapio, 181 pp.

35 Pukkala, T. (2022). Assessing the externalities of timber production. *Forest Policy and Economics* 135: 10.

36 Pukkala, T., Vauhkonen, J., Korhonen, K.T., and Packalen, T. (2021). Self-learning growth simulator for modeling forest stand dynamics in changing conditions. *Forestry* 94 (3): 333–346. https://doi.org/10.1093/forestry/cpab008.

37 Berglund, H. and Kuuluvainen, T. (2021). Representative boreal forest habitats in northern Europe, and a revised model for ecosystem management and biodiversity conservation. *Ambio* https://doi.org/10.1007/s13280-020-01444-3.

38 Palik, B.J., D'Amato, A.W., Franklin, J.F., and Johnson, K.N. (2021). *Ecological Silviculture.* Waveland Press.

39 Mori, A.S., Lertzman, K.P., and Gustafsson, L. (2016). Biodiversity and ecosystem services in forest ecosystems: a research agenda for applied forest ecology. *Journal of Applied Ecology* 2017 (54): 12–27. https://doi.org/10.1111/1365-2664.12669.

40 Kuuluvainen, T., Angelstam, P., Frelich, L.E. et al. (2021). Natural disturbance based forest management: moving beyond retention and continuous-cover forestry. *Frontiers in Forests and Global Change* 4: 629020.

41 Kappen, G., Kastner, E., Kurth, T., Puetz, J., Reinhardt, A., Soininen, J., 2020. The Staggering Value of Forests – and How to Save Them. Boston Consulting Group. https://www.bcg.com/publications/2020/the-staggering-value-of-forests-and-how-to-save-them (accessed 15 June 2023)

42 Pukkala, T. (2020). At what carbon price forest cutting should stop. *Journal of Forestry Research* 31: 713–727.

43 Bo Larsen, J.B., Angelstam, P., Bauhus, J. et al. (2022). *Closer-to-Nature Forest Management*, From Science to Policy 12. European Forest Institute https://doi.org/10.36333/fs12.

44 Kuuluvainen, T. and Ylläsjärvi, I. (2011). On the natural regeneration of dry heath forests in Finnish Lapland: a review of V.T. Aaltonen (1919). *Scandinavian Journal of Forest Research* 26 (Suppl 10): 34–44.

45 Aaltonen, V.T. (1919). Über die naturliche Verjungung der Heidewälder im finnischen Lappland (in Finnish with German summary). Communicationes ex Instituto Quaestionum Forestalium Finlandiae Editae 1. 319 p.

18

Silvicultural Systems in the Mountain Ash Forests of the Central Highlands of Victoria, South-eastern Australia

David B. Lindenmayer

Fenner School of Environment and Society, Australian National University, Canberra, ACT, Australia

18.1 Introduction

In Australia, community concerns about the impacts of clear-cutting date back several decades and have been, in part, at the core of major social divisions between environmentalists and advocates of forest industries [1]. There has been a particular focus on the silvicultural systems employed in the wet ash-type eucalypt forests of southeastern Australia that can be valuable for timber production but also have significant values for biodiversity, water production, and carbon storage [2]. Clear-cutting has been the conventional harvesting system in these forest types, such as those dominated by Mountain Ash (*Eucalyptus regnans* F. Muell.) in mainland southeastern Australia [3]. However, the range of negative environmental impacts of clear-cutting has increasingly been recognized [4]. Studies in Mountain Ash forests of alternative silvicultural systems to clear-cutting date back to the 1980s [5], with a designed experiment exploring aspects of variable retention harvesting (sensu (6)) having commenced in 2003 and continuing to this day [2]. The silviculture systems employed in Mountain Ash forests (and similar ash-type forests such as those dominated by Alpine Ash, *Eucalyptus delegatensis* R. T. Baker) are important in the Central Highlands region of Victoria because ~65% of all logging in that state is concentrated in this area and in these forest types [6].

In this chapter, I outline some of the key ecological characteristics of the Mountain Ash ecosystem in the Central Highlands of Victoria. I describe a variable retention harvesting experiment that has been taking place over the past ~20 years in Mountain Ash forests. I present a new conceptual model for the post-disturbance development of Mountain Ash forests based on both contrasting structural patterns that arise from variable retention harvesting versus clear-cutting and on the biological legacies (Chapter 1) that remain following stand-replacing wildfires. I conclude with commentary on the broader ecological, economic, social, and climate change domains within which silvicultural systems must sit in Victorian Mountain Ash forests. These domains will most likely overwrite efforts to transition to new kinds of silvicultural systems in Mountain Ash forests and eventually force an end to native forest logging in this ecosystem.

Ecological Silvicultural Systems: Exemplary Models for Sustainable Forest Management,
First Edition. Edited by Brian J. Palik and Anthony W. D'Amato.

18.2 Ecosystem Characteristics

Mountain Ash ecosystems are dominated by Mountain Ash overstory eucalypt trees, which are the tallest flowering plants on earth and can approach 100 m tall [7]. Mountain Ash forests occur in the Australian states of Victoria and Tasmania, and those in the Central Highlands of Victoria are the focus of this chapter (Figure 18.1). Mountain Ash forests in this region are characterized by a well-developed understorey comprised of *Acacia* spp. and temperate rainforest trees that can reach ~40+ m in height [8]. These forests also support a luxuriant shrub and tree fern layer, especially in old-growth stands exceeding 120+ years old [9]. Mountain Ash forests are not only rich in native vascular plant species richness [9] and plants in other groups such as bryophytes [10], but also support species-rich assemblages of birds and arboreal marsupials [11]. Mountain Ash forests are also among the most carbon-dense forests on earth [12] and have a critical water supply function, generating almost all of the water for the 5+ million inhabitants of the city of Melbourne [13].

Fire is the primary form of natural disturbance in Mountain Ash forests. The natural regime is infrequent, high-severity, stand-replacing wildfire that occurs on average every 75–150 years [14]. Moderate severity fires also occur in Mountain Ash forests and are partial stand-replacing events that lead to the development of multi-aged stands comprising living and dead overstorey trees [15]. The severity of wildfires in Mountain Ash forests is influenced by the age of the forest at the time it is burnt. The highest severity fires occur in young (7–36 year old) stands, whereas the lowest

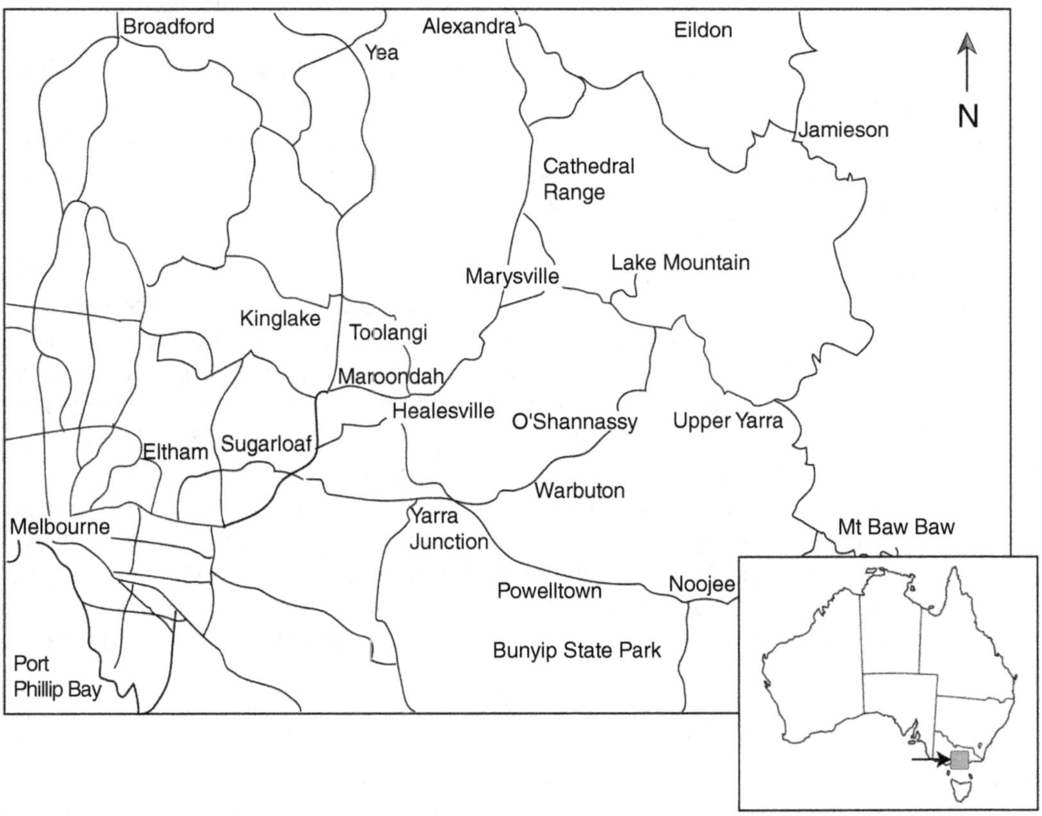

Figure 18.1 The location of the Central Highlands of Victoria, southeastern Australia.

severity fires are in stands that exceed ~100 years old [16]. Fires in older stands typically produce more biological legacies (sensu (18)) than fires in young stands [17]. The biological legacies remaining in burnt stands provide critical habitat elements for birds, mammals, vascular plants, and bryophytes [17]. Fire is essential for natural regeneration in Mountain Ash forests. However, when fire is frequent and young stands (<20–30 years old) are burnt, then a paucity of viable seed can result in forests being replaced by *Acacia* spp. woodland [18] – sometimes termed the "interval squeeze" problem.

18.3 Prevailing Silvicultural Systems in Mountain Ash Forests

18.3.1 Clear-Cutting

Clear-cutting is the conventional silvicultural system in Mountain Ash forests and has been employed for ~50 years [3]. Mountain Ash forests are predominately on State Government-owned land, and Timber Release Plans specify those areas where cutblocks will be located, the size of cutblocks, and where logging exclusion areas occur [19]. The clear-cutting approach is relatively simple and entails: (i) all merchantable trees being removed off-site for subsequent processing as sawn timber or for pulpwood, (ii) logging slash being left on the ground for one or more years to dry, (iii) a high-intensity regeneration burn being applied to consume the dried and cured logging slash and create a bed of ashes in which Mountain Ash seed is dropped on the cutblock to instigate stand regeneration [3] (Figure 18.2). The typical size for cutblocks ranges from 15 to 40 ha, depending on topography and other factors [20].

Clear-cutting is relatively efficient in producing large quantities of timber. However, it also has a number of negative environment impacts. These are: (i) eroding biodiversity such as reducing populations of arboreal marsupials and birds as well as some resprouting native plants like tree ferns [21]; (ii) simplifying stand structures such as through promoting the losses of large old trees which are key habitat attributes for cavity-dependent fauna [22]; (iii) generating carbon emissions, including creating smoke pollution as part of the regeneration burning component of the silvicultural system [23]; (iv) depleting key soil nutrients and degrading soil structure [24]; and (v) reducing water yields from watersheds [13]. Some of these effects are at the stand level such as changes in soil environments [24], whereas others, like impacts on biodiversity, are at both stand and landscape levels [21]. An additional major environmental problem associated with logging in Mountain Ash forests is that harvested and then regenerated stands are at increased risk of elevated fire severity relative to old-growth stands [16]. That is, young logged forests are at risk of reburning at high severity thereby precluding their maturation to an older (less fire-prone) stage [18].

The dense regenerating stands of Mountain Ash forest that develop following clear-cutting are sometimes targeted for thinning 15–30 years after stand initiation [3] and this has been proposed as a way to reduce the risk of high-severity fire [25]. However, empirical studies following the 2009 wildfires in the Central Highlands of Victoria have indicated that thinning either makes little difference to fire severity or in some cases serves to increase the risk of higher fire severity [26].

18.3.2 Variable Retention Harvesting Systems

In 2004, the Victorian Government, in partnership with researchers from The Australian National University, established a blocked and replicated experiment that compared the plant and animal biota on cutblocks subject to variable retention harvesting, clear-cutting, and uncut forest

Figure 18.2 Stages of harvesting in clear-cutting in Mountain Ash forests. Part A shows the initial stages where merchantable trees are removed and logging debris remains. Part B shows the high-intensity burn lit to consume logging debris and prepare a bed of ashes into which seeds are dropped to initiate a new stand of trees. Part C shows the recently burnt logged stand prior to the germination of new stand. *Source:* Photos by David Lindenmayer.

(Figure 18.3). The overarching aim of the experiment was to help develop new approaches to Mountain Ash silviculture that would result in greater levels of stand structural complexity in logged areas. This, in turn, aimed to promote biodiversity persistence in, and biodiversity recolonization of, areas subject to timber harvesting. To quantify the potential for such responses, data on a range of response variables have been gathered as part of the experiment, including the occurrence of forest birds [27], the abundance of small mammals (since abandoned) [28], vascular plants, and rates of tree deterioration and collapse.

Figure 18.3 Retention islands in a cutblocks subject to variable retention harvesting in the Mountain Ash forests of the Central Highlands of Victoria.

The results of studies to date have shown that variable retention harvesting systems contribute to on-site persistence of both small mammals and forest birds, particularly relative to clear-cut sites, although (as expected) they support fewer bird species and a generally lower abundance of small mammals than uncut control sites [28, 29]. Retention patch size (0.5 vs 1.5 ha) appears to make limited difference to mammal and bird persistence in otherwise logged areas. Data remain unpublished for plants, but preliminary analyses suggest that retained patches in areas subject to variable retention harvesting systems support a wide range of vascular plant species.

The Government of Victoria has indicated an intention to employ variable retention harvesting extensively across areas proposed for logging from 2019 onwards [19, 30]. However, there are some key issues associated with current applications of this new silvicultural approach in Mountain Ash forests. First, high-intensity regeneration burns continue to be applied in all logged areas, including those subject to variable retention harvesting. These fires often result in retained islands being badly damaged (see Figure 18.4). This can defeat the purpose of retention and preclude the development of multi-aged stands of Mountain Ash forest, which as outlined above, was one of the key overarching aims of developing new silvicultural systems in these ecosystems. Part of the reason for such high-intensity regeneration burns is the large amount of debris that remains in the forest after harvesting operations – approaching 450 tonnes ha^{-1} [23].

A second issue with the deployment of variable retention harvesting in Mountain Ash forests has been that almost all of the retention has been at the margins of cutblocks and not within the boundaries of logged areas (Figure 18.5). This is inconsistent with the definition of variable retention harvesting in which retained structures need to be a part of, not apart from, the cutblock area [31].

Figure 18.4 High-intensity regeneration burns damaging a retention island within a cutblock subject to variable retention harvesting in the Mountain Ash forests of the Central Highlands of Victoria. *Source:* Photo sequence by the late David Blair.

Figure 18.5 A cutblock in the Central Highlands of Victoria that was originally registered as an area that had been subject to variable retention harvesting, but which highlights the lack of retained forest on-site. *Source:* Photo by Chris Taylor.

A third issue is that under variable retention harvesting, less wood will be generated from cutblocks (because more forest will be retained). More cutblocks will be needed to meet legislated timber yields from Mountain Ash forests. This demands more roading and more cutblocks within an ecosystem, which is already highly fragmented [6]. Indeed, the average distance from an intact area in wood production forest to a disturbance boundary in Mountain Ash forest is currently just 71 m [6].

Finally, Mountain Ash ecosystems are dominated by extensive areas of young, highly flammable forest, with more than 98.8% of the ecosystem currently ~82 years or less [32]. The predominance of young logged and regenerated forest in this ecosystem means it is vulnerable to reburning in the near future [33]. Ongoing logging will add to this already skewed age class structure and further increase the fire proneness of the ecosystem (see Figure 18.2). Partial or even full replacement of clear-cutting with the variable retention harvest system will likely not reduce this major fire risk problem, given that other methods such as stand thinning have had limited impact on fire risk, and in some cases may elevate it [26].

18.4 Natural Development Model and Silviculture

As mentioned, wildfires are the most common form of natural disturbance in Mountain Ash forests. The severity of wildfires can be variable depending on a range of factors including fire weather at the time of a conflagration as well as the age of a forest when it is burnt; old-growth forests burn at significantly lower severity than young logged and regenerated stands [16] (Figure 18.3). Wildfires, especially those in old-growth stands, tend to leave large amounts of biological legacies [34] including large old fire-scarred living and dead trees, resprouting understorey trees and shrubs, seeds, and living animals. The presence of some of these types of legacies can result in the development of multi-aged stands [15] in which the living overstorey trees surviving previous fires then persist in the recovering stand and create structurally complex forests comprised several (up to three or sometimes more) age cohorts [35]. Indeed, old-growth Mountain Ash forests are rarely a single age, but rather they support trees for two and sometimes more age cohorts [35]. Moreover, multi-aged stands are where species richness of groups such as arboreal marsupials is highest [15]. Such levels of stand complexity contrast markedly with clear-cutting where the majority of these legacies are either highly depleted or lost entirely, producing structurally simple forests often lacking a diversity of overstorey tree sizes, particularly large old trees [36]. Disturbance regimes associated with variable retention harvesting, in which retention islands result in the persistence of mature trees within cutblocks (see Figure 18.2) are therefore more akin to those natural disturbances regimes in terms of the biological legacies and hence the stand structural complexity that is produced (Figure 18.6). Variable retention harvesting systems therefore better emulate this natural model of canopy disturbance than does clear-cutting.

It is possible that the variable retention harvesting system could provide the backbone for future silvicultural treatments to recreate the natural kinds of complex stand structural complexity characteristic of old-growth Mountain Ash forests (but see the following section for key riders on this premise). Application of the variable retention harvesting system in Mountain Ash forests would entail leaving multiple "retention islands" within the boundaries of cutblocks and ensuring appropriate slash treatments reduce the risk of damage to those islands – if regeneration burns continue to be applied. A key issue will be to plan the location of islands within cutblocks well in advance and over several cutting cycles so that they continue to support multiple age cohorts in relatively close proximity and thereby better emulate the kinds of multi-aged forests that can be so important for the biota inhabiting Mountain Ash forest. In planning the location of islands within cutblock boundaries, there would be value in anchoring islands around key natural assets that might otherwise be vulnerable to the effects of harvesting operations such as large old trees and resprouting

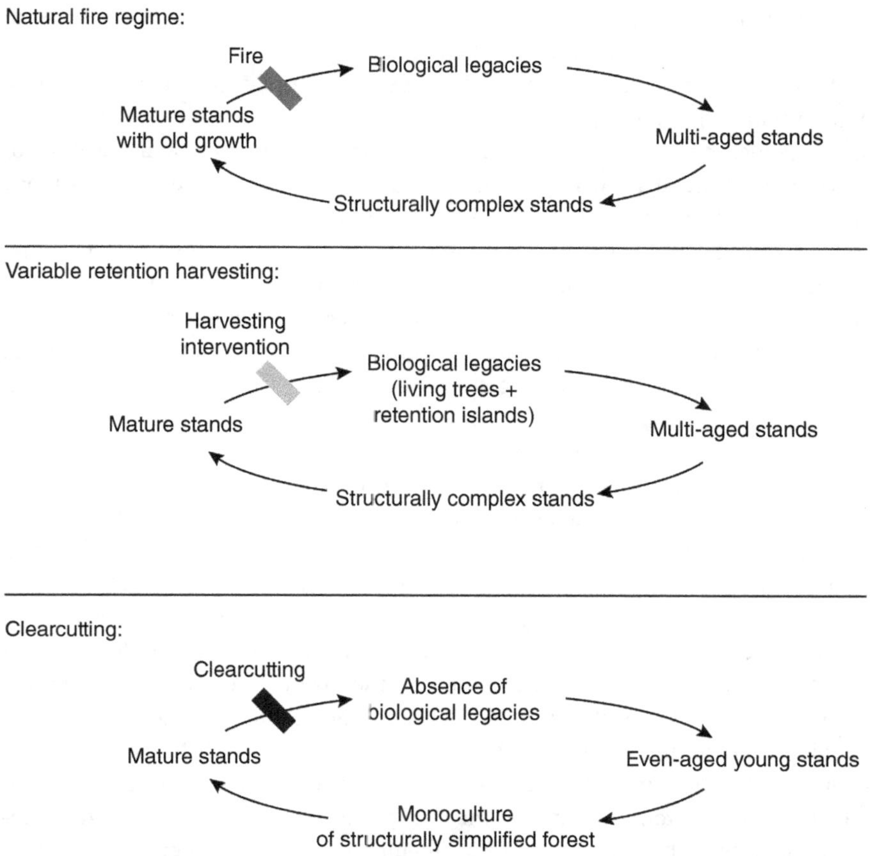

Figure 18.6 Conceptual model showing the differences in stand structural complexity associated with natural disturbance, variable retention harvesting, and clear-cutting. The model highlights how variable retention harvesting better emulates post-disturbance stand dynamics relative to clear-cutting.

understorey trees and shrubs [37]. Actions like small-scale ecological thinning also may be useful in growing larger diameter trees faster, although this does not necessarily accelerate the development of key structural attributes like cavities (as these are typically a function of tree age and tree physiology rather than tree size *per se*, see [38].). A further consideration is that activities like thinning may elevate the risks of high-severity wildfires in Mountain Ash forests [26] and simplify understorey structure (which is a key component of the habitat requirements of many species in Mountain Ash forests [39]. As discussed elsewhere in this chapter, there are, however, a number of key considerations associated with the application of new silvicultural systems in Mountain Ash forests.

18.5 The Challenges for Mountain Ash Silviculture: Climate Change and Other Drivers

Empirical evidence from the experiment in the Central Highlands of Victoria indicates that the variable retention harvest system has distinct environmental benefits relative to clear-cutting. On this basis, the Victorian Government has indicated that it plans to roll out variable retention harvest system on a proportion of cutblocks in Mountain Ash forest [30]. Whilst this change in harvest policy is welcome, there are some substantial environmental, economic, and social issues in native forests in Victoria that will likely overwrite the widespread application of variable retention harvesting in Mountain Ash forests.

First, the native forest logging industry in Victoria has become uneconomic, with the costs of operations far exceeding financial returns [40]. Annual reports for the government agency responsible for native forest logging (VicForests) indicate that native forest logging has rarely been profitable since the mid-2000s, with the extent of losses increasing over time. Environmental and economic accounting have clearly indicated that other natural assets and ecosystem services like water production, carbon storage, and nature-based tourism are of far greater value to regional GDP than native forest logging [41]. Indeed, the Victorian Government has set in a train an exit from native forest harvesting statewide by 2030 (https://www.vic.gov.au/timber-harvesting).

Second, there is a rapidly declining wood supply in Mountain Ash forests as a result of past widespread overcutting, recurrent wildfire, and the interaction of the two (driven by the slow responses to reduce sustained yield following resource losses due to major wildfires such as those in 2009). Repeated high-severity wildfires are a key issue in this regard. Current evidence indicates that fire frequency is increasing and that the return interval for successive fires will prevent trees growing to an age where they are suitable for sawlog production [14]. This will create enormous resource uncertainty. This problem is overlaid on Mountain Ash landscapes that are already heavily disturbed by past fire and logging and that are now dominated by young, fire-prone forests [6], little of which is at a sufficient level of maturity to produce sawlogs [14]. Climate change is also contributing to increasing wildfire in Mountain Ash forest (as well as other ecosystems in southeastern Australia; [42]. Repeated wildfires, coupled with ongoing logging and climate change, could eventually trigger the collapse of the entire Mountain Ash ecosystem and with it industries such as the native forest logging industry [43]. Repeated fire also will create problems with water yields – which are highest from intact watersheds dominated by old-growth forest [44].

A third issue is that because almost all native forest in Victoria is publicly owned, continued wood production from that forest requires a strong social license to continue to operate. Despite the potential for variable retention harvesting to reduce environmental impacts relative to clear-cutting, social surveys indicate that the majority of Victorians do not want native logging to continue in State-owned native forests, even in rural communities living adjacent to native forests used for wood production (https://www.google.com/search?q=DELWP+social+licence+to+keep+logging+in+Victoria&rlz=1C1GGRV_enAU795AU795&oq=DELWP+social+licence+to+keep+logging+in+Vioctoria&aqs=chrome..69i57j33i10i160.10785j0j1&sourceid=chrome&ie=UTF-8). This may be another reason why the Victorian Government has a policy to cease native forest logging within the next decade.

There are other reasons why the economic, social, and environmental context for native forest logging in Mountain Ash forests remains problematic, including: (i) areas targeted for future harvesting under current Timber Release Plans overlap with areas of high conservation value for biodiversity [45]; (ii) logging contributes to flammability and fire risk in Mountain Ash forests (see Figure 18.2); and (iii) that VicForests has repeatedly failed in its attempts to obtain Forest Stewardship Council certification for its harvesting operations. Collectively, these issues mean that although variable retention harvesting would be an improvement in clear-cutting, other factors suggest that large-scale native forest logging will likely eventually cease in Mountain Ash forests.

18.6 A New Silvicultural Model for Mountain Ash Forests

A key question given the issues outlined above with both clear-cutting and the variable retention harvesting system is: *Is there an appropriate model for ecologically sustainable silviculture in Mountain Ash forests?* Finding such a model will be a major challenge, especially given the marginal economics, high levels of resource uncertainty, and low levels of community-based social license that currently exist in the native forest timber industry in Victoria. I suggest that a more sustainable silvicultural system in Mountain Ash forests could be developed and it would be based on the variable retention harvest system. However, it would have to overcome some key hurdles outlined above and below.

First, ongoing logging operations would have to be focused away from forests of high conservation value [45]. This is a major challenge given that there is very limited remaining forest in Victoria that has not been logged, burnt, or both in the past few decades [32], with biodiversity values in currently relatively undisturbed forests being particularly pronounced [45].

Second, the level of overall harvesting in Victorian forests (including in Mountain Ash forest) would need to be reduced dramatically otherwise more areas of forest would need to be cut by variable retention to meet high wood quotas. The reduction in timber yield would need to be marked and enduring, with the level of cut (of all timber – both sawlogs and pulpwood) potentially being as little as 10% of what it is currently. In addition, current distortions in the wood market would need to be removed in which almost 85% of all cut timber ends up in the pulpwood and woodchip stream [41], but timber yields are set by unreasonably high extraction rates for dwindling sawlog resources [46]. Therefore, variable retention harvesting systems in Mountain Ash forests could not be implemented without major industry restructure, downsizing, and comprehensive policy and management reform.

Third, the agency responsible for logging in Victoria (VicForests) must adopt harvest approaches in which retained patches are located within (and not at the edge of) the boundary areas of cutblocks (cf. Figure 18.5). This is critical because current attempts to conduct variable retention harvesting do not conform to how such a system is defined or is typically practiced in other forest types globally. Retained patches within cutblocks under such an appropriately implemented variable retention harvesting system must then be far better protected from damage by regeneration fires. This will demand better dispersal of logging slash away from retained islands to reduce the intensity of burns lit to facilitate the regeneration of cutblocks.

Fourth, better planning and implementation of harvesting operations will be critical to ensure that populations of key structural features of Mountain Ash forests like large old trees are maintained (which are in rapid decline across the ecosystem [43] and then recruited *en masse*). All individuals of such trees need to be protected by buffers of uncut forests, especially as they can take centuries to develop but are easily lost by poor management practices. Existing large old trees could become anchor points around which to establish retention islands in variable retention harvesting operations with the trees in the protective buffers then being recruited to eventually be added to the population of large old trees in the future. Special provisions will be needed to ensure the long-term protection and retention of these forest attributes, and therefore maintain their persistence through multiple stand entries and successive harvest rotations.

Fifth, ongoing monitoring, including the current variable retention harvesting experiment will be critical to quantify the extent to which such silvicultural systems contribute to the medium and long-term maintenance of key elements of stand structural complexity and associated forest biodiversity in Mountain Ash forests.

Prescriptions to ensure that variable retention harvesting systems are effective will likely add considerably to the costs of logging operations in Mountain Ash forests and therefore further increase already not inconsiderable losses in the native forest logging industry in Victoria [40]. It remains unclear whether governments, or the general public, will maintain an appetite to continue to support such a loss-making industry, nor maintain a government agency responsible for logging on public land that currently has limited social license to operate.

Acknowledgments

I thank Brian Palik and Tony D'Amato for asking me to write this chapter. Lachlan McBurney and the late Dave Blair have made major contributions to the data collection aspects of the work reported here. Tabitha Boyer assisted with many aspects of manuscript preparation.

References

1 Dargarvel, J. (2018). Views and perspectives: why does Australia have forest wars? *International Journal of Environmental History* 4: 33–51.
2 Lindenmayer, D.B., Blair, D., and McBurney, L. (2019). Variable retention harvesting in Victoria's Mountain Ash (Eucalyptus regnans) forests (southeastern Australia). *Ecological Processes* 8: Art. 2.
3 Flint, A. and Fagg, P. (2007). *Mountain Ash in Victoria's State Forests*. Melbourne: Department of Sustainability and Environment.
4 Lefoe, M., Rendall, A.R., McKinnon, F., and Whisson, D.A. (2021). Logging and wildfire limit the distribution of a vulnerable arboreal mammal. *Forest Ecology and Management* 503: 119773.
5 Squire, R.O., Dexter, B., Smith, R. et al. (1987). *Evaluation of Alternative Systems for Victoria's Commerically Important Mountain Forests – Project Brief.* Melbourne: Lands and Forests Division, Department of Conservation, Forests and Lands.
6 Taylor, C. and Lindenmayer, D.B. (2020). Temporal fragmentation of a critically endangered ecosystem. *Austral Ecology* 45: 340–354.
7 Ashton, D.H. (1975). The root and shoot development of *Eucalyptus regnans* F. Muell. *Australian Journal of Botany* 23: 867–887.
8 Lindenmayer, D.B., Blanchard, W., Bowd, E. et al. (2021). What factors influence the extent and abundance of midstorey Acacia in Mountain Ash forests? *Austral Ecology*.
9 Bowd, E.J., Lindenmayer, D.B., Banks, S.C., and Blair, D.P. (2018). Logging and fire regimes alter plant communities. *Ecological Applications* 28: 826–841.
10 Pharo, E.J., Meagher, D.A., and Lindenmayer, D.B. (2013). Bryophyte persistence following major fire in eucalypt forest of southern Australia. *Forest Ecology and Management* 296: 24–32.
11 Lindenmayer, D.B., Blair, D.P., McBurney, L., and Banks, S.C. (2015). *Mountain Ash: Fire, Logging and the Future of Victoria's Giant Forests*. Melbourne: CSIRO Publishing.
12 Keith, H., Mackey, B.G., and Lindenmayer, D.B. (2009). Re-evaluation of forest biomass carbon stocks and lessons from the world's most carbon-dense forests. *Proceedings of the National Academy of Sciences* 106 (28): 11635–11640.
13 Taylor, C., Blair, D., Keith, H., and Lindenmayer, D.B. (2019). Modelling water yields in response to logging and representative climate futures. *Science of The Total Environment* 688: 890–902.

14 Cary, G., Blanchard, W., Foster, C.N., and Lindenmayer, D.B. (2021). Effects of altered fire regimes on critical timber production and conservation rotations. *International Journal of Wildland Fire* 30: 322–328.

15 Lindenmayer, D.B. and McCarthy, M. (1998). Multi-aged mountain ash forest, wildlife conservation and timber harvesting. *Forest Ecology and Management* 104 (1–3): 43–56.

16 Taylor, C., McCarthy, M.A., and Lindenmayer, D.B. (2014). Non-linear effects of stand age on fire severity. *Conservation Letters* 7: 355–370.

17 Lindenmayer, D.B., Westgate, M.J., Scheele, B.C. et al. (2019). Key perspectives on early successional forests subject to stand-replacing disturbances. *Forest Ecology and Management* 454: 117656.

18 Lindenmayer, D.B., Hobbs, R.J., Likens, G.E. et al. (2011). Newly discovered landscape traps produce regime shifts in wet forests. *Proceedings of the National Academy of Sciences of the United States of America* 108: 15887–15891.

19 VicForests (2019). Timber release plan 2019: spatial data Melbourne. http://www.vicforests.com.au/planning-1/spatial-data (accessed 5 May 2019).

20 Department of Natural Resources and Environment (1996). *Code of Practice. Code of Forest Practice for timber production. Revision No. 2, November 1996*. Melbourne: Department of Natural Resources and Environment.

21 Lindenmayer, D.B., Blanchard, W., Blair, D. et al. (2020). The response of arboreal marsupials to long-term changes in forest disturbance. *Animal Conservation* 24 (2): 246–258.

22 Lindenmayer, D.B., Blanchard, W., Blair, D. et al. (2018). Empirical relationships between tree fall and landscape-level amounts of logging and fire. *PLoS One* 13 (2): e0193132.

23 Lindenmayer, D.B. and Taylor, C. (2018). Where there is fire, there is smoke. *Science* 361: 341.

24 Bowd, E.J., Banks, S.C., Strong, C.L., and Lindenmayer, D.B. (2019). Long-term impacts of wildfire and logging on forest soils. *Nature Geoscience* 12: 113–118.

25 Volkova, L., Bia, H., Hilton, J., and Weston, C.J. (2017). Impact of mechanical thinning on forest carbon, fuel hazard and simulated fire behaviour in Eucalyptus delegatensis forest of south-eastern Australia. *Forest Ecology and Management* 405: 92–100.

26 Taylor, C., Blanchard, W., and Lindenmayer, D.B. (2020). Does forest thinning reduce fire severity in Australian eucalypt forests? *Conservation Letters* 14 (2): e12766.

27 Lindenmayer, D.B., Wood, J., McBurney, L. et al. (2015). Single large versus several small: the SLOSS debate in the context of bird responses to a variable retention logging experiment. *Forest Ecology and Management* 339: 1–10.

28 Lindenmayer, D.B., Knight, E., McBurney, L. et al. (2010). Small mammals and retention islands: an experimental study of animal response to alternative logging practices. *Forest Ecology and Management* 260: 2070–2078.

29 Lindenmayer, D.B., McBurney, L., Blair, D. et al. (2018). From unburnt to salvage logged: quantifying bird responses to different levels of disturbance severity. *Journal of Applied Ecology* 55: 1626–1636.

30 VicForests (2019). *Harvesting and Regeneration Systems*. Melbourne, Australia: VicForests.

31 Franklin, J.F., Berg, D.E., Thornburgh, D.A., and Tappeiner, J.C. (1997). Alternative silvicultural approaches to timber harvest: variable retention harvest systems. In: *Creating a Forestry for the 21st Century* (ed. K.A. Kohm and J.F. Franklin), 111–139. Covelo, CA: Island Press.

32 Lindenmayer, D.B. and Taylor, C. (2020). Extensive recent wildfires demand more stringent protection of critical old growth forest. *Pacific Conservation Biology* https://doi.org/10.1071/PC20037.

33 Lindenmayer, D.B., Zylstra, P., Kooyman, R. et al. (2022). Logging effects outweighed fire weather effects to promote crown fire in the 2019–20 Australian forest fires. *Nature Ecology and Evolution* 6: 533–535.

34 Franklin, J.F., Lindenmayer, D.B., MacMahon, J.A. et al. (2000). Threads of continuity. *Conservation Practices* 1 (1): 8–17.

35 Lindenmayer, D.B., Cunningham, R.B., Donnelly, C.F., and Franklin, J.F. (2000). Structural features of old growth Australian montane ash forests. *Forest Ecology and Management* 134 (1–3): 189–204.

36 Lindenmayer, D.B., Blanchard, W., Blair, D. et al. (2016). Environmental and human drivers of large old tree abundance in Australian wet forests. *Forest Ecology and Management* 372: 266–235.

37 Lindenmayer, D.B. (2017). Conserving large old trees as small natural features. *Biological Conservation* 211: 51–59.

38 Gibbons, P. and Lindenmayer, D.B. (2002). *Tree Hollows and Wildlife Conservation in Australia*. Melbourne: CSIRO Publishing.

39 Bowd, E.J., Banks, S.C., Bissett, A. et al. (2021). Direct and indirect disturbance impacts in forests. *Ecology Letters* 24: 1225–1236.

40 Parliamentary Budget Office (2020). *End native forest logging in Victoria*. Melvourne, Victoria: Parliamentary Budget Office.

41 Keith, H., Vardon, M., Stein, J.A.R. et al. (2017). Ecosystem accounts define explicit and spatial trade-offs for managing natural resources. *Nature Ecology and Evolution* 1: 1683–1692.

42 van Oldenborgh, G.-J., Krikken, F., Lewis, S. et al. (2021). Attribution of the Australian bushfire risk to anthropogenic climate change. *Natural Hazards and Earth System Sciences* 21: 941–960.

43 Lindenmayer, D.B. and Sato, C. (2018). Hidden collapse is driven by fire and logging in a socioecological forest ecosystem. *Proceedings of the National Academy of Sciences* 115: 5181–5186.

44 Vertessy, R.A., Watson, F.G.R., and O'Sullivan, S.K. (2001). Factors determining relations between stand age and catchment water balance in mountain ash forests. *Forest Ecology and Management* 143: 13–26.

45 Taylor, C. and Lindenmayer, D.B. (2019). The adequacy of Victoria's protected areas for conserving its forest-dependent fauna. *Austral Ecology* 44: 1076–1090.

46 Lindenmayer, D.B. (2017). Halting natural resource depletion: engaging with economic and political power. *The Economic and Labour Relations Review* 28: 41–56.

19

Ecological Silviculture for European Beech-Dominated Forest Ecosystems

Thomas A. Nagel[1], Miroslav Svoboda[2], Lucie Vítková[2], and Dušan Roženbergar[1]

[1] *Department of Forestry and Renewable Forest Resources, University of Ljubljana, Ljubljana, Slovenia*
[2] *Department of Forest Ecology, Faculty of Forestry and Wood Sciences, Czech University of Life Sciences Prague, Prague, Czech Republic*

19.1 Introduction

This chapter focuses on European beech (*Fagus sylvatica* L.) forest ecosystems across the temperate zone of Europe. Forest communities dominated by European beech (hereafter beech) or mixtures of beech and other broadleaf and conifer species are widespread across the temperate zone of Europe [1], including large areas in France, Germany, and mountain regions across south and eastern Europe [2] (Carpathians, Dinaric and Balkan mountains, Apennines, southern Alps, and Pyrenees) (Figure 19.1). Consequently, they play an essential role in European society by provisioning ecosystem services and contributing to the forest economy. Beech timber, for example, has a number of important uses, such as furniture, flooring, and plywood, and is a main source of firewood in the region. Moreover, there is increasing interest in growing admixed high-value "noble" broadleaf species, including species of maple (*Acer pseudoplatanus* L., *A. platanoides* L.), ash (*Fraxinus excelsior* L.), elm (*Ulmus glabra* Huds.), lime (*Tilia cordata* Mill.) and cherry (*Prunus avium* L.), for both economic and ecological functions.

19.2 Characteristics and Natural Dynamics of European Beech-Dominated Ecosystems

19.2.1 Composition

Although beech is often the dominant tree species within forest stands across much of its range, in some cases forming monodominant stands, beech communities are often comprised of a mixture of broadleaf and conifer species depending on site conditions [3–7]. At lower elevations or warm slope positions, where the climate is prone to drought, sessile oak (*Quercus petraea* (Matt.) Liebl.) and other thermophilic species are present, while at higher elevations, where winters are long and cold, Norway spruce (*Picea abies* (L.) Karst.) gains dominance [1]. Between these elevation zones, a number of other species often coexist with beech, particularly fir (*Abies alba* Mill.) and a mixture of broadleaf species (Table 19.1). Local species mixtures are largely a result of unique disturbance

Ecological Silvicultural Systems: Exemplary Models for Sustainable Forest Management,
First Edition. Edited by Brian J. Palik and Anthony W. D'Amato.

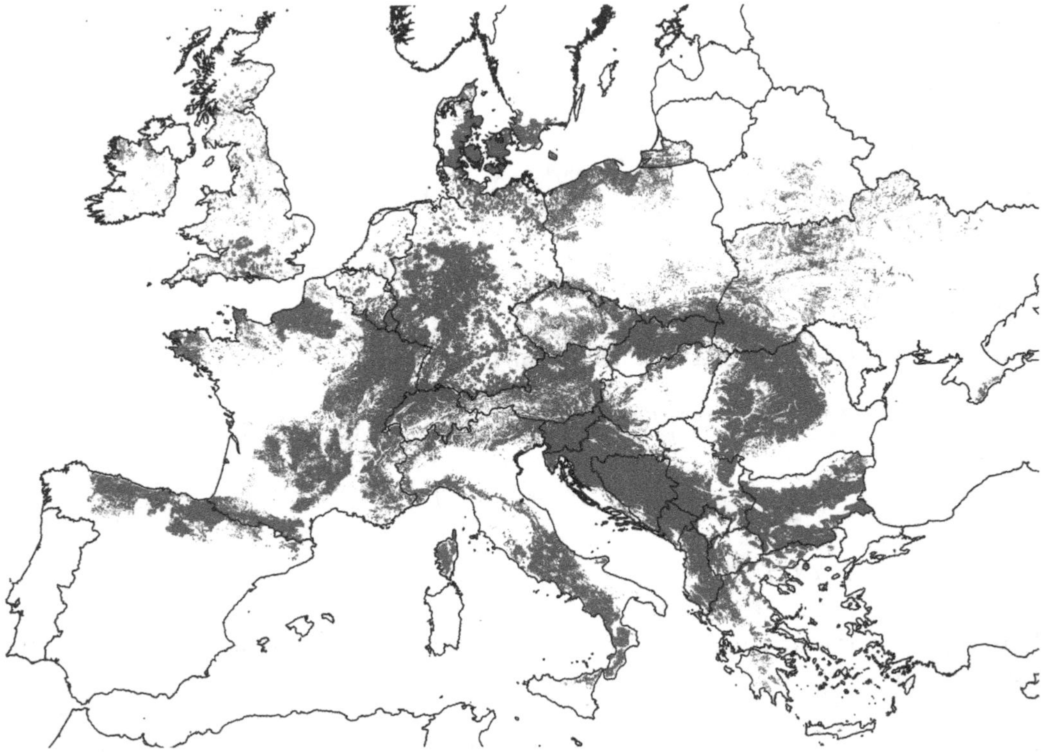

Figure 19.1 Beech-dominated ecosystems are widespread across the temperate region of Europe, as indicated by the distribution of European beech. *Source:* Adapted from Brus et al. [2].

Table 19.1 Key traits of selected tree species found in beech dominated forest ecosystems.

Species	Shade tolerance	Drought tolerance	Lifespan	Palatability
Fagus sylvatica	Tolerant	Intolerant	Very long	Intermediate
Abies alba	Tolerant	Mid-tolerant	Long	High
Picea abies	Mid-tolerant	Intolerant	Long	Low
Acer pseudoplatanus	Mid-tolerant	Mid-tolerant	Long	High
Acer platanoides	Mid-tolerant	Mid-tolerant	Long	High
Fraxinus excelsior	Mid-tolerant	Mid-tolerant	Long	High
Ulmus glabra	Mid-tolerant	Mid-tolerant	Long	High
Tilia cordata	Mid-tolerant	Tolerant	Long	Intermediate
Quercus petraea	Intolerant	Tolerant	Very long	High
Prunus avium	Intolerant	Mid-tolerant	Short	High
Betula pendula	Intolerant	Mid-tolerant	Short	Low
Populus tremula	Intolerant	Mid-tolerant	Short	Low

Source: Refs. [5–7].

histories, coupled with species life history traits [3]. For example, shade-tolerant beech and fir are often present as saplings or pole-sized trees in the understory and access the canopy following formation of small gaps, while larger openings (>500 m^2) created by disturbances foster recruitment of mid-tolerant species (e.g. maple, ash, elm). Pioneer species, such as aspen (*Populus tremula* L.), birch (*Betula pendula* Roth), and willow (*Salix* spp.), may establish in larger disturbance patches [4].

19.2.2 Disturbance Regime

The natural disturbance regime in beech-dominated ecosystems of Europe is generally characterized by relatively continuous formation of small canopy gaps, punctuated by periodic mesoscale disturbances that cause intermediate severity damage to stands. The gap disturbance regime has been quantified across many old-growth stands in the Dinaric and Carpathian mountains (e.g. [8, 9]). Gap fractions generally range from 5% to 15%, and gap sizes typically exhibit a negative exponential distribution, whereby most gaps are less than 100 m^2. These small openings often result from standing death or "background" mortality of one or several canopy trees; rates of background mortality average about 1% annually [10]. Periodic intermediate severity disturbances are caused by a variety of agents, including strong wind from thunderstorms, ice storms, and spring or autumn snowstorms, and create much larger canopy openings (e.g. >1000 m^2). Such events may remove between 30% and 50% of the canopy at stand scales (i.e. tens of hectares), but damage is often heterogenous across a variety of scales [8, 11]. In rare cases, patch-scale severities of >80% canopy removal have also been observed [8, 11]. Rotation periods for events that remove 30–50% of the canopy are approximately 200–500 years [3], within the range of the lifespan of canopy trees.

19.2.3 Developmental Model

The natural developmental model for European beech ecosystems begins with disturbance, and progresses with young, mature, and old forest stages (OFSs) (see Chapter 1). It is important to note that homogeneous patches of forest in a given developmental stage are unusual in natural beech dominated forests because severe, canopy replacing disturbances are exceptionally rare, with recurrence intervals of several thousand years. Consequently, the developmental stages below describe how forests develop at the scale of a large forest gap or patch created by wind disturbance, while the OFS describes a stand-scale condition that includes a range of different age classes maintained by low to intermediate-severity disturbance. Nevertheless, these stages serve as a useful framework to introduce various silvicultural treatments because many managed beech-dominated forests in Europe are evenaged at stand scales.

19.2.3.1 Disturbance and Legacy Creation

Intermediate severity disturbances in beech-dominated ecosystems, such as a microburst from a local thunderstorm, may disturb stands over tens of hectares, and cause heterogeneous damage patterns, ranging from scattered small gaps to extensive thinning of canopy trees to larger blowdown patches (e.g. 0.1–1 ha) [8]. These larger disturbance patches are often "messy" with complex shapes, and include both snapped and uprooted canopy trees, as well as "windfirm" live canopy tree legacies (Figure 19.2).

19.2.3.2 Preforest Stage

The Preforest stage is rare in beech-dominated ecosystems, primarily because advanced regeneration of beech is often well developed in forest understories, and leads to rapid site capture

Figure 19.2 Typical "messy" damage patterns from intermediate severity disturbances in beech-dominated ecosystems.

Figure 19.3 Advance regeneration of beech in oldgrowth (left) and managed (right) stands.

following disturbance [12] (Figure 19.3). However, in some cases following more severe disturbance, there may be a protracted period of herb or shrub dominance, such as from *Rubus* spp., particularly when regeneration is heavily browsed [13].

19.2.3.3 Young Forest Stage

In disturbance patches where advanced beech regeneration is well developed, the Young Forest Stage (YFS) is dominated by beech, or in some cases mixtures with fir, such that the composition of the YFS is similar to the pre-disturbance stand. In cases when the pre-disturbance stand had a closed canopy and a less developed layer of advanced beech regeneration, a number of mid-tolerant species, including maple, ash, elm, oak, lime, and spruce, can make up a prominent component of the YFS composition due to the more rapid height growth of these species compared to beech [3]. On some sites with larger disturbance patches, short-lived, pioneer species with highly dispersed seeds, such as willow, aspen, and birch, can make up a large component of the YFS community [4, 13]. Most individuals in this pioneer cohort senesce by the end of the YFS and are replaced by beech or other mid-tolerant species. Likewise, most of the downed beech logs from the initial disturbance event completely decompose by about 50 years [14].

19.2.3.4 Mature Forest Stage

The Mature Forest Stage (MFS), often referred to as the "optimum phase" in European literature, follows a period of intense density-dependent thinning in the YFS, yet some canopy thinning from competition continues in the MFS stage. Forest development in this stage is mostly structural, with less change in composition. Mortality of canopy trees from competition or disturbance may introduce light into an otherwise shaded understory, which can increase structural complexity by promoting recruitment of understory stems. However, a prominent feature of the MFS stage in stands dominated by beech is that structural complexity is often very low due to the high crown plasticity of beech – small gaps formed during this stage are rapidly filled by lateral crown expansion, giving rise to structurally simple, homogenous stands [15]. These "vault-like" stands are well documented in developmental phase literature in European beech forests.

19.2.3.5 Old Forest Stage

In regions with primary forest, the natural disturbance regime in beech-dominated ecosystems gives rise to a forested landscape that is dominated by the Old Forest State (OFS), often referred to as old-growth, with smaller portions of the landscape recovering from more severe disturbances that remove most of the canopy layer [11]. The structure and composition of the OFS are maintained by relatively continuous gap-scale disturbance and periodic intermediate severity events (Figure 19.4). The combination of these processes promotes structurally complex stands [15, 16], dominated by beech, or mixtures of beech and fir, and includes other species with intermediate shade tolerance (e.g. spruce, maple, ash, elm, lime), depending on site conditions and disturbance history. Advance regeneration of shade-tolerant beech and fir is often abundant, with saplings and

Figure 19.4 Typical structural features of beech-dominated ecosystems in an old-growth stage, including complex structures, large-old trees, habitat trees with cavities, snags, and downed dead wood.

pole-sized trees that recruit in patches with sufficient understory light. Age structure in the OFS is typically irregular, with cohorts associated with past intermediate severity disturbances [3]. Size distributions in old-growth stands, however, are poor correlates with age structure and often converge to a rotated sigmoid structure at scales approaching 1 ha [16, 17]. A prominent feature of the OFS includes an abundance of large, old habitat trees and deadwood [16, 18]. Deadwood legacies include large snags, snapped boles, and uprooted trees (i.e. tip and mounds) in multiple stages of decay. Snags and habitat trees with canopy deadwood are important sources of tree microhabitats, which are abundant in the OFS [19].

19.3 Conventional Silvicultural Approach

Most beech-dominated forest ecosystems in Europe are currently managed with rotational forest management, mainly using uniform shelterwood systems (67%) with natural regeneration and subsequent release through overstory removal (Figure 19.5), and some with clear-cut (5%) and coppice (7%) systems. Continuous cover approaches using uneven-aged silvicultural systems are also applied (20%) [20], mainly using irregular shelterwood, as well as group and single tree selection [21]. The timing and pattern of harvest with most of these conventional silvicultural systems are outside the range of variability with regard to natural disturbance frequency and severity in beech-dominated ecosystems [20]. These silvicultural systems have led to a lack of structural features associated with the OFS, as well as a loss of early successional stages across forest landscapes. For example, the amount of deadwood and large-old habitat trees is depauperate in most beech-dominated ecosystems. Old-growth stands often have in excess of 150 m^3 ha^{-1} of high-quality deadwood (large logs in multiple stages of decay), while managed forests typically have tenfold less deadwood volume, often made up of smaller sized logs and stumps [18]. Given the diverse

Figure 19.5 Typical beech-dominated forest landscape managed with a conventional even-aged shelterwood system.

assemblage of taxa that require deadwood for food or habitat in beech forests (e.g. fungi, lichens, bryophytes, insects, birds, and bats), many species have experienced substantial declines over much of their range [22].

19.4 Ecological Silviculture for European Beech-Dominated Ecosystems

The silvicultural system described in this section attempts to mimic the gap-scale and intermediate severity disturbances, including the ecological legacies associated with these processes, that are typical of beech-dominated ecosystems. The goal of this system is to simultaneously provide habitat conditions for native biodiversity associated with the OFS and to supply high-quality timber – balancing these two functions should be economically profitable for forest owners. We also emphasize that this type of management should be placed in a broader context of zoning biodiversity and timber production functions across large regions (see discussion of *Context* in Chapter 1). Satisfying the growing demand for timber at least cost to native biodiversity, while also facilitating forest adaptation to climate change, will likely require careful optimization of forest area managed under strict protection, ecological silviculture, and intensive timber production.

It is important to note that this section describes silvicultural treatments for mountain forests typically dominated by beech, but often including a significant component of fir and spruce, as well as mid-tolerant broadleaf trees. This system would require some adaptation depending on the specific forest community type and site conditions, such as at lower elevations, where oak and other species mix with beech. Application of specific silvicultural treatments is carried out within the same developmental stages described in the previous section (Figure 19.6). Because many managed beech forest ecosystems in Europe are currently in a late YFS or MFS, the silvicultural treatments described in this section attempt to accelerate the transition toward the structure and composition found in the OFS.

19.4.1 Disturbance and Legacy Creation

One of the primary goals of this stage is to initiate regeneration across a range of gap sizes that emulate natural disturbances, thereby promoting structural and compositional complexity in stands that are often in the YFS and MFS. This includes harvesting single and small groups of trees in a closed canopy matrix to mimic gap-scale disturbances (e.g. 0.01–0.1 ha) (similar to variable-density thinning (VDT) described in Chapter 1), and irregular shelterwood and variable retention harvesting (VRH, see Chapter 1) to emulate intermediate severity events (Figures 19.7 and 19.8). To mimic the size and damage heterogeneity associated with intermediate severity disturbances, VRH treatments should be carried out at stand scales that cover several hectares with varying levels and patterns of retention across a treatment. A key concept is to create variability over space, which may include retention as dispersed trees, aggregated retention with or without some dispersed trees, as well as larger openings (e.g. 0.1–0.5 ha) with or without dispersed trees. Regeneration harvests should both release advanced regeneration of shade-tolerant beech and fir but also provide regeneration opportunities for high-value mid-tolerant species. For the latter, some larger openings created with VRH should therefore be placed in areas with little to no advanced regeneration of beech. Care should also be taken to ensure that adult seed-producing trees of target species are retained in the stand.

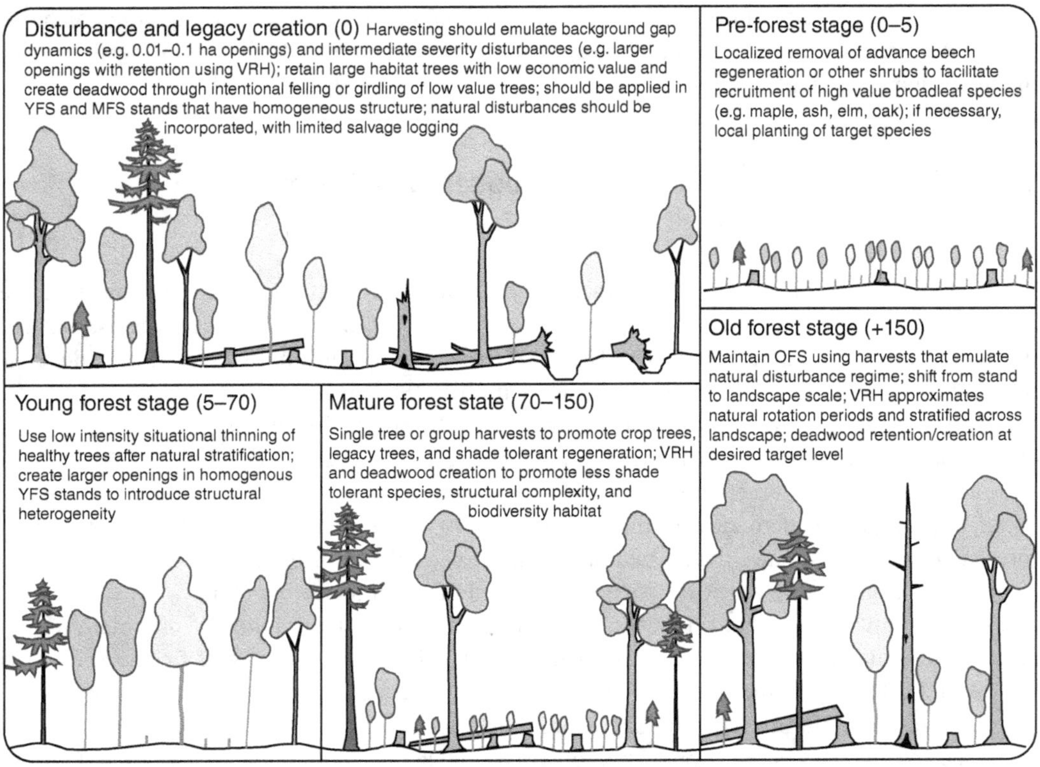

Figure 19.6 An ecological silvicultural system for beech-dominated forest ecosystems showing key treatments across development stages, including the approximate time span of each stage (years).

Figure 19.7 Structural complexity resulting from irregular shelterwood harvests that emulate intermediate severity disturbances.

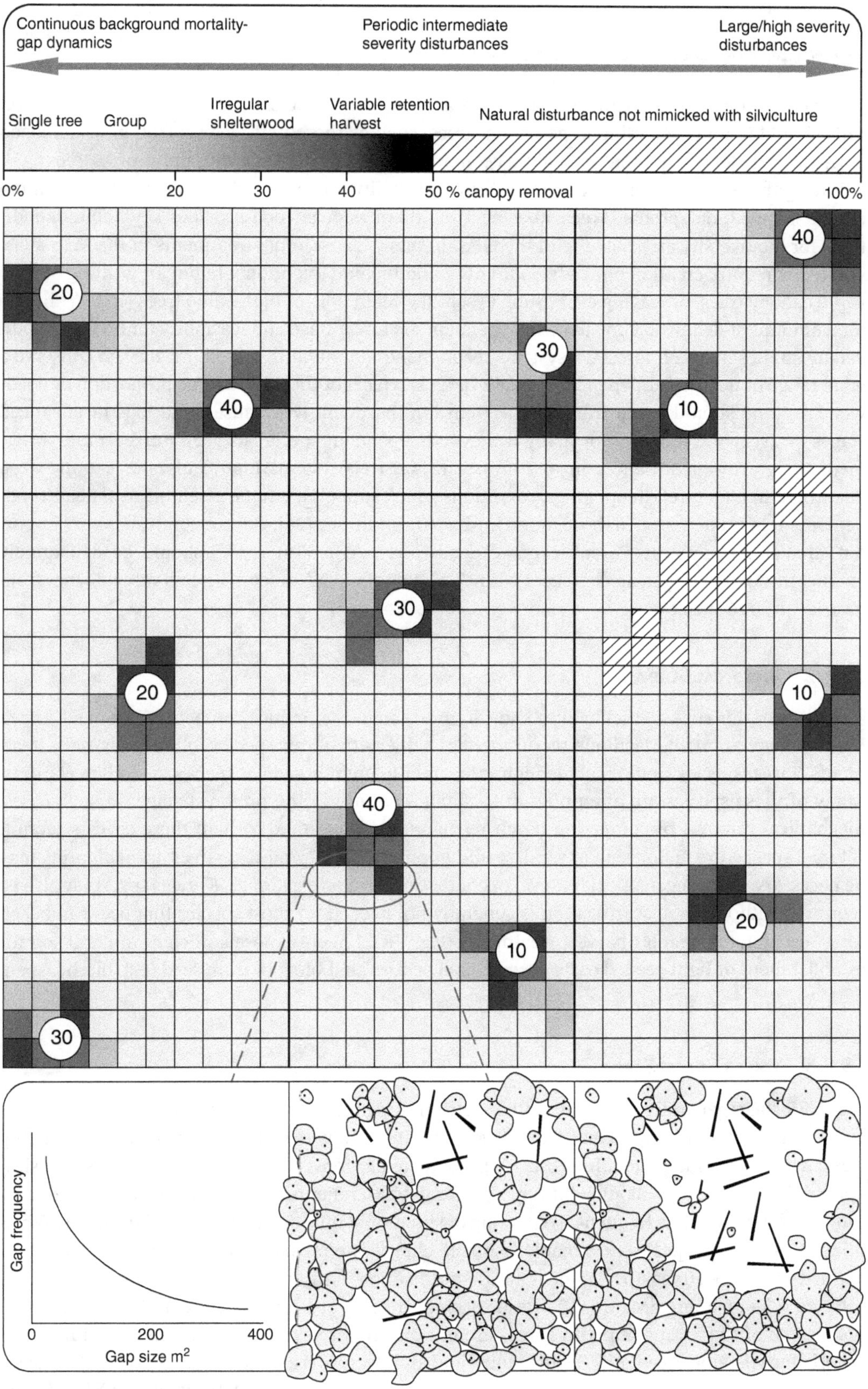

Figure 19.8 Hypothetical 900 ha landscape depicting spatial and temporal harvest prescriptions during the Old Forest Stage (OFS) based on the natural disturbance regime in beech-dominated ecosystems. The numbers indicate the timing (years) of treatments that emulate intermediate-severity disturbances.

Another primary goal of this stage is to retain or create legacies associated with disturbance. This should include retaining large habitat trees, such as those with defects (rotten or damaged boles, forked stems, etc.) that have little economic value, aiming for a minimum of 10 trees ha^{-1}. Allowing these trees to die from various natural mortality processes, which can result in standing death, snapped, or uprooted trees, and retaining them as deadwood provides key habitat conditions for biodiversity associated with the OFS. In many cases, initial treatments in the MFS stage will likely require creation of deadwood. This could include intentional felling or girdling of low-value canopy trees, or leaving everything except the main bole of high-value trees [23]. While the total amount of deadwood created will depend on the forest stage and feasibility for owners, biodiversity studies in beech forest ecosystems recommend maintaining >20–50 m^3 ha^{-1} to conserve a wide range of deadwood-dependent taxa [24]. As is typical of the OFS, a large proportion of deadwood volume should come from logs >50 cm dbh, particularly because large logs in advanced stages of decay are required for many deadwood-dependent specialists [25]. Furthermore, intermediate severity disturbances in natural beech forest ecosystems input pulses of canopy-sized deadwood in large, well-lit openings, conditions which support a unique assemblage of deadwood-dependent taxa compared to those found in shaded conditions [26]. Consequently, it is very important that larger amounts of sun-exposed deadwood (e.g. >50 m^3 ha^{-1}) should be deliberately created through felling in areas treated with VRH; this could again focus on less valuable logs from a variety of tree species.

19.4.2 Preforest Stage

The focus of this stage, which will be short in most cases, is to reduce competition from advanced beech regeneration and facilitate recruitment of a diversity of tree species following harvest treatments. Target species would include high value, mid-tolerant broadleaf species, as well as conifers. Many of these species are often present as small seedlings in the preforest stage; in such cases, mechanical removal of competing beech regeneration is required to help these species recruit. If target species are not present as small seedlings or adult seed-producing trees, local planting may be necessary. Local mechanical removal of competing shrubs (e.g. *Rubus, Corylus* spp.) may also be required. Although soil scarification is generally not needed for most species that occur in beech forest ecosystems, regular harvest treatments (e.g. VRH) likely expose sufficient mineral soil for establishment of light-seeded early successional species that blow in from seed (e.g. birch, aspen, willow).

19.4.3 Young Forest Stage

The YFS stage generally relies on natural self-thinning and resulting stratification but includes some *situational* thinning to guide the development of a low density of high-quality future crop trees, which also reduces tending costs [27]. Compared to conventional thinning at this stage, situational thinning selects about eight times less crop trees, relies on natural stratification, and thinning treatments are less frequent. The timing of situational thinning depends on the species. When mid-tolerant target species (e.g. maple, ash elm, cherry) are mixed with beech, tending should be applied after a 5–10 year period of natural stratification. Crop trees of mid-tolerant species are selected with an approximate 8–10 m grid spacing, yielding about 50–120 trees ha^{-1}, in which case one or two of the most competitive trees around the crop tree are removed. The most important criteria for selection are vitality/health status and architecture (i.e. clear "winner" trees after stratification). For beech, no intervention is needed for at least the first 30–50 years of the YFS, during which time there is significant stratification due to strong intraspecific competition. Situational tending begins when about 6–8 m of the lower bole is branch-free, again selecting crop trees with an 8–10 m grid spacing (unless a mid-tolerant tree was already selected) and removal of one or two stronger competitors. Crop trees can be released several times to accelerate the development of large trees. The remainder of the cohort is left to develop naturally, which also yields potential crop trees through natural stratification, and provides an input of small-diameter deadwood from mortality. In addition to situational thinning, entire stands in the YFS may require creation of larger openings (e.g. 100–400 m^2) using VDT (see Chapter 1) to introduce structural heterogeneity at this stage.

19.4.4 Mature Forest Stage

The main focus of this stage is to accelerate the transition to the OFS. This includes promotion of new age classes and large trees, creation and retention of live and dead tree legacies, and increasing the proportion of mid-tolerant tree species. Initial entries at the MFS include harvest treatments outlined in the Disturbance and Legacy creation stage. Care should be taken to not over-disturb the stand during initial entries with VRH treatments, which will depend on the size of the stand and initial structure. In cases when the starting condition has homogeneous structure, typical of many beech forest ecosystems in the MFS, initial entry might use VRH on about 20–30% of the stand area. In the example shown here (Figure 19.7), VRH emulates the damage patterns that are characteristic of periodic windthrow events from thunderstorms. Treatments could be limited to 10 ha patches, but adjacent 1 ha compartments could vary from 20% to 50% canopy removal, reflecting the variability that is typical of storm damage. At the other end of the severity gradient, harvesting single trees could be used to accelerate the development of large crop trees (for both future harvest or for retention of habitat trees), while harvesting tree groups (e.g. gaps of 0.04–0.1 ha) could be used to promote recruitment of advance regeneration of shade-tolerant beech and fir or mid-tolerant species present in the stand (such treatments are similar to VDT described in Chapter 1). Stand reentry might occur on 10-year intervals and would use gap harvests to further accelerate large tree growth, or to expand existing openings and release regeneration.

19.4.5 Old Forest Stage

Due to the rarity of the OFS across the temperate region of Europe, any stand in this stage should ideally be protected in a strict reserve where wood extraction is not permitted. Here we focus on regularly managed forests, where the ultimate goal of ecological silviculture is to create forest stands that approach the structural and compositional conditions typical of the OFS, while also producing high-quality timber. Given sufficient time, a cohort of canopy trees, such as those selected for habitat trees, will eventually die from natural mortality processes and deadwood will begin to accumulate. These processes indicate a transition to the OFS, and the silvicultural system should focus on maintaining a mosaic of harvests that mimic gap-scale and intermediate severity disturbances. Ideally, planning of treatments should shift from stand to landscape scale. For example, in the 900 ha landscape in Figure 19.8, three 10 ha intermediate severity treatments could be prescribed each decade, giving a 300-year rotation period for the entire landscape, emulating the timing of these disturbances in the OFS. Treatments could be stratified across the landscape to help ensure a relatively equal distribution of early seral habitat across the landscape, which may aid the dispersal of specialist species, such as those dependent on larger amounts of sun-exposed deadwood. Finally, any natural disturbances that damage the forest should be incorporated into the planning. This might involve limited salvage logging of high-value boles, but effort should be made to avoid damaging the complex biological legacies created by natural disturbance.

19.5 Climate Change Considerations

In addition to warming temperatures, climate change is predicted to include an increase in global-change-type hot droughts in Europe. Temperature increases and hot drought have already led to marked declines in growth and increased mortality of beech trees across large parts of its range, trends that are predicted to increase in the future [28, 29]. If ecological silviculture in beech-dominated ecosystems is to succeed in the future, immediate treatments that consider climate change will need to be implemented.

Several aspects of the ecological silviculture outlined in this chapter should inherently promote some degree of resilience to climate change, such as by increasing species diversity and frequent establishment of regeneration to facilitate gradual adaptation of the local gene pool. However, we advocate a more active approach that accelerates the transition toward climate-adapted forest compositions. One approach is to use silvicultural treatments to increase the proportion of heat and drought-tolerant species already present in a stand. In mixed mountain forests, this might include increasing the proportion of fir and mid-tolerant broad-leaf species [30]. At lower elevations, oak (*Q. petraea*) should be promoted, a species that is both valuable for timber production and drought tolerant [31], but requires silvicultural tending to reduce competition with beech [32].

To keep pace with climate change, we also advocate assisted movement of drought-adapted species or provenances that are not present in current stand compositions. For example, this may include planting oak at higher elevations in beech-dominated forests, planting fir in beech monocultures, or adding drought-tolerant provenances of beech. This does not mean that climate-sensitive species should be excluded; keeping spruce or larch (*Larix decidua* Mill.) in the mixture may remain an important way to maintain forest productivity and for owners to ensure profitable forestry operations.

Finally, because recruitment of healthy regeneration is a basic requirement of ecological silviculture and climate change adaptation, maintaining ungulate populations at densities that allow recruitment is of utmost importance. Valuable mid-tolerant species and species advocated under climate change are palatable species that are susceptible to regeneration failures due to deer browsing. Constructive dialogue among foresters, hunters, and the public will help to ensure sustainable management of forests and ungulate populations in the future.

19.6 Summary

An ecological silviculture system for beech-dominated forest communities emulates key processes and legacies of the natural disturbance regime, with the goal of increasing biodiversity habitat while maintaining an economically viable level of wood production. Important differences from conventional silviculture in beech-dominated forests include a focus on stand and landscape scale heterogeneity, creation of early seral habitat with pioneer and mid-tolerant tree species, retention of large habitat trees, intentional creation of deadwood, and limited salvage logging. A key silvicultural tool to establish or release regeneration includes VRH, which emulates the complex damage patterns and light conditions created by intermediate severity disturbances. Successful implementation of ecological silviculture in beech-dominated forests should include production of high-value timber, active adaptation to a warming climate by altering species composition, and management of ungulate populations to allow recruitment of desired tree species.

References

1 Leuschner, C. and Ellenberg, H. (2017). *Ecology of Central European Forests: Vegetation Ecology of Central Europe*. Springer International Pubishing.

2 Brus, D.J., Hengeveld, G.M., Walvoort, D.J.J. et al. (2012). Statistical mapping of tree species over Europe. *European Journal of Forest Research* 131: 145–157.

3 Nagel, T.A., Svoboda, M., and Kobal, M. (2014). Disturbance, life history traits, and dynamics in an old-growth forest landscape of southeastern Europe. *Ecological Applications* 24: 663–679.

4 Jaloviar, P., Saniga, M., Kucbel, S. et al. (2017). Seven decades of change in a European old-growth forest following a stand-replacing wind disturbance: a long-term case study. *Forest Ecology and Management* 399: 197–205.

5 Leuschner, C. and Meier, I.C. (2018). The ecology of Central European tree species: trait spectra, functional trade-offs, and ecological classification of adult trees. *Perspectives in Plant Ecology, Evolution, and Systematics* 33: 89–103.

6 Gill, R.M.A. (1992). A review of damage by mammals in north temperate forests: 1. deer. *Forestry* 65: 145–169.

7 Pavlin, J., Nagel, T.A., Svitok, M. et al. (2021). Disturbance history is a key driver of tree life span in temperate primary forests. *Journal of Vegetation Science* 32 (5): https://doi.org/10.1111/JVS.13069.

8 Nagel, T.A., Mikac, S., Dolinar, M. et al. (2017). The natural disturbance regime in forests of the Dinaric Mountains: a synthesis of evidence. *Forest Ecology and Management* 388: 29–42.

9 Kucbel, S., Jaloviar, P., Saniga, M. et al. (2010). Canopy gaps in an old-growth fir-beech forest remnant of Western Carpathians. *European Journal of Forest Research* 129: 249–259.

10 Nagel, T.A., Firm, D., and Rozman, A. (2021). Intermediate disturbances are a key driver of long-term tree demography across old-growth temperate forests. *Ecology and Evolution* 11: 16862–16873.

11 Frankovic, M., Janda, P., Mikolas, M. et al. (2021). Natural dynamics of temperate mountain beech-dominated primary forests in Central Europe. *Forest Ecology and Management* 479: https://doi.org/10.1016/j.foreco.2020.118522.

12 Nagel, T.A., Svoboda, M., and Diaci, J. (2006). Regeneration patterns after intermediate wind disturbance in an old-growth *Fagus-Abies* forest in southeastern Slovenia. *Forest Ecology and Management* 226: 268–278.

13 Fidej, G., Rozman, A., and Diaci, J. (2018). Drivers of regeneration dynamics following salvage logging and different silvicultural treatments in windthrow areas in Slovenia. *Forest Ecology and Management* 409: 378–389.

14 Privetivy, T., Janik, D., Unar, P. et al. (2016). How do environmental conditions affect the deadwood decomposition of European beech (*Fagus sylvatica* L.)? *Forest Ecology and Management* 381: 177–187.

15 Stiers, M., Willim, K., Seidel, D. et al. (2018). A quantitative comparison of the structural complexity of managed, lately unmanaged and primary European beech (*Fagus sylvatica* L.) forests. *Forest Ecology and Management* 15: 357–365.

16 Kucbel, S., Saniga, M., Jaloviar, P., and Vencurik, J. (2012). Stand structure and temporal variability in old-growth beech-dominated forests of the northwestern Carpathians: a 40-year perspective. *Forest Ecology and Management* 264: 125–133.

17 Alessandrini, A., Biondi, F., Di Filippo, A. et al. (2011). Tree size distribution at increasing spatial scales converges to the rotated sigmoid curve in two old-growth beech stands of the Italian Apennines. *Forest Ecology and Management* 262: 1950–1962.

18 Nagel, T.A., Firm, D., Pisek, R. et al. (2017). Evaluating the influence of integrative forest management on old-growth habitat structures in a temperate forest region. *Biological Conservation* 216: 101–107.

19 Kozak, D., Mikolas, M., Svitok, M. et al. (2018). Profile of tree-related microhabitats in European primary beech-dominated forests. *Forest Ecology and Management* 429: 363–374.

20 Aszalós, R., Thom, D., Aakala, T. et al. (2022). Natural disturbance regimes as a guide for sustainable forest management in Europe. *Ecological Applications* 32: e2596.

21 Mason, W.L., Diaci, J., Carvalho, J., and Valkonen, S. (2021). Continuous cover forestry in Europe: usage and the knowledge gaps and challenges to wider adoption. *Forestry: An International Journal of Forest Research* 95: 1–12.

22 Brunet, J., Fritz, O., and Richnau, G. (2010). Biodiversity in European beech forests-a review with recommendations for sustainable forest management. *Ecological Bulletins* 53: 77–94.

23 Doerfler, I., Mueller, J., Gossner, M.M. et al. (2017). Success of a deadwood enrichment strategy in production forests depends on stand type and management intensity. *Forest Ecology and Management* 400: 607–620.

24 Muller, J. and Butler, R. (2010). A review of habitat thresholds for dead wood: a baseline for management recommendations in European forests. *European Journal of Forest Research* 129: 981–992.

25 Gossner, M.M., Lachat, T., Brunet, J. et al. (2013). Current near-to-nature forest management effects on functional trait composition of saproxylic beetles in beech forests: functional diversity of beetles. *Conservation Biology* 27): 605–614.

26 Seibold, S., Bässler, C., Brandl, R. et al. (2016). Microclimate and habitat heterogeneity as the major drivers of beetle diversity in dead wood. *Journal of Applied Ecology* 53: 934–943.

27 Ammann, P. (2012). Jungwaldpflegekonzepte mit biologischer Rationalisierung. *Zurcher Wald* 2: 12–15.

28 Schuldt, B., Buras, A., Arend, M. et al. (2020). A first assessment of the impact of the extreme 2018 summer drought on Central European forests. *Basic and Applied Ecology* 45: 86–103.

29 del Castillo, E.M., Zang, C.S., Buras, A. et al. (2022). Climate-change-driven growth decline of European beech forests. *Communications Biology* 5 (1): 163.

30 Vitali, V., Buntgen, U., and Bauhus, J. (2017). Silver fir and Douglas fir are more tolerant to extreme droughts than Norway spruce in south-western Germany. *Global Change Biology* 23: 5108–5119.

31 Mette, T., Dolos, K., Meinardus, C. et al. (2013). Climatic turning point for beech and oak under climate change in Central Europe. *Ecosphere* 4 (12): 1–19.

32 Stimm, K., Heym, M., Uhl, E. et al. (2021). Height growth-related competitiveness of oak (*Quercus petraea* (Matt.) Liebl. and *Quercus robur* L.) under climate change in Central Europe. Is silvicultural assistance still required in mixed-species stands? *Forest Ecology and Management* 482: 118780.

20

Ecological Silviculture for Chilean Temperate Rainforests

Pablo J. Donoso[1] *and Daniel P. Soto*[2]

[1] *Facultad de Ciencias Forestales y Recursos Naturales, Instituto de Bosques y Sociedad, Universidad Austral de Chile, Valdivia, Chile*
[2] *Departamento de Ciencias Naturales y Tecnología, Universidad de Aysén, Coyhaique, Chile*

20.1 Introduction

Chilean temperate rainforests range from 37 to 56° S latitude [1], including the Valdivian Temperate Rainforests (37–42° S), the North Patagonian Rainforests (42–47° S), and the Magellanic Rainforests (47–56° S) [2, 3]. Growing within the region of the Valdivian and North Patagonian Rainforests, the Evergreen Forest Type (EFT; [4]) is the largest in South Central Chile (3.5 million ha; Figure 20.1) [5]. The northern distribution of the EFT contains a greater diversity of species compared to its southern distribution and experiences a greater human pressure due to land-use change and high demand for the goods and services these forests provide. Consequently, the landscape now includes many secondary forests and partially harvested forests in the most accessible areas, plus undisturbed old-growth forests, especially in private or public forest reserves and less accessible areas. Most current secondary forests are 60–100 years old, so they are either in a stem exclusion or understory regeneration phase of development (*sensu* [6]), i.e. they are young or mature. Regenerating forests following clear-cuts or partial cuts are dominated by a mixture of pioneer tree species of seed origin combined with species with vigorous regeneration from vegetative origin [7]. Studies in young and old-growth forests have generated relevant knowledge on their ecology and dynamics [8] and constitute a valuable framework to develop recommendations and approaches for ecological silviculture in forests at different successional stages.

We provide a developmental model for forests now common on the landscape, from initial stages following large-scale disturbances to old-growth forests, and we then address silvicultural approaches for these stages that are aimed at meeting the principles of ecological silviculture [9, 10], i.e. silviculture aimed at following a natural model of disturbance and development.

The implementation of some ecological silviculture in Chile has occurred mostly as experiments, so its operational application remains a challenge, especially in public forests or among private landowners interested in bringing their forests to more natural conditions given the lack of on-the-ground examples. Our approaches for ecological silviculture are based on recent experiments implemented in young, mature, and old forests, and predictions stemming from knowledge about the silvics and stand dynamics of these forests.

Ecological Silvicultural Systems: Exemplary Models for Sustainable Forest Management,
First Edition. Edited by Brian J. Palik and Anthony W. D'Amato.

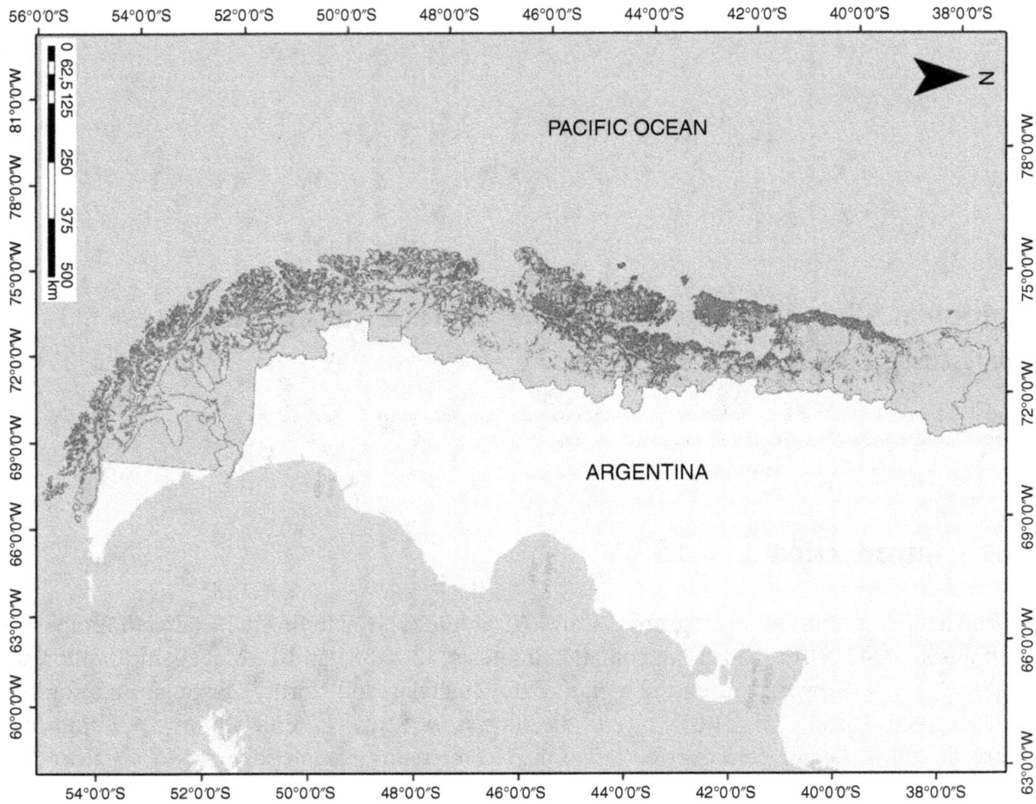

Figure 20.1 Distribution of the EFT in Chile, showing its dominance in the Coastal range and Archipelago, and partially in the Andes range (western slope) where it grows to approximately 600 m a.s.l., above which other forest types become dominant.

20.2 Characteristics of the Evergreen Forest Type (EFT)

20.2.1 Composition, Structure, and Growth

Evergreen rainforests in Chile include several tree species that may be dominant at some point through succession, mostly in response to their shade tolerance and reaction to different types of disturbance (Table 20.1). Many of these species can reach great heights and diameters, especially those that are pioneer species and long lived, such as Coihue (*Nothofagus dombeyi* (Mirbel) Oersted) and Ulmo (*Eucryphia cordifolia* (Cav.)). These forests in their northern distribution are among the richest in tree species composition in Chile, with nearly 30 tree and arborescent species [16, 17], and when old they are comprised by three or more well-differentiated tree strata, depending largely on the existence of pioneer species that reach emergent positions [11, 12, 18, 19].

Growth rates of these forests depend upon location, forest type, species, and successional stage, but they are generally high owing to a mild climate, abundant rainfall, and deep and nutrient-rich soils of volcanic origin, especially in the Andes and Central Valley. Growth rates have been estimated from 4 to 10 m^3 ha^{-1} $year^{-1}$ for recently managed uneven-aged forests [20] and up to 20 m^3 ha^{-1} $year^{-1}$ in 20–30-year-old secondary forests [21, 22].

Table 20.1 Some silvical characteristics of the main tree species in the hardwood-dominated evergreen forest type in South Central Chile.

Species		Shade tolerance[a]				Regeneration mode[b]				
Common name	**Latin name**	**VI**	**I**	**M**	**T**	**Pio**	**Gap**	**Con**	**Max. *h* (m)**	**Max. *d* (cm)**
Coihue	*Nothofagus dombeyi*	x	X			X			53	420
Coihue de Chiloé	*Nothofagus nitida*		x	X		X	X		40	200
Tineo	*Weinmannia trichosperma*		X			X			40	200
Ulmo	*Eucryphia cordifolia*			X		x	X		51	230
Tepa	*Laureliopsis philippiana*				X			X	35	150
Olivillo	*Aextoxicon punctatum*				X			X	35	100
Lingue	*Persea lingue*			X	x		X	X	35	100
Mañío Hembra	*Saxegothaea conspicua*				X		X	X	35	200
Mañío de Hojas Largas	*Podocarpus salignus*			X			X	x	40	100
Mañío Macho	*Podocarpus nubigena*				X			X	25	100
Avellano	*Gevuina avellana*		x	X		x	X		25	60
Notro	*Embothrium coccineum*	X	x			X			20	60
Canelo	*Drimys winteri*		x	X		X	X	x	30	100
Trevo	*Dassyphyllum diacanthoides*			X			X		35	100
Laurel	*Laurelia sempervirens*			X			X	x	50	200
Luma	*Amomyrtus luma*				X			X	20	60
Meli	*Amomyrtus meli*				X			X	20	80
Patagua	*Myrceugenia planipes*				X			X	15	30

[a] Shade tolerance: VI, very intolerant; I, intolerant; M, mid-tolerant; T, tolerant.
[b] Regeneration mode: Pio, pioneer; Gap, gap; Con, continuous. Main category marked with X, second best with x.
Sources: Refs. [7, 8, 11–15].

20.2.2 Forest Disturbance and Dynamics

Natural forest dynamics of the EFT is dominated by small-scale endogenous disturbances from windstorms that cause the fall of large trees that usually create openings up to 500 m^2 [18, 23, 24], favoring the regeneration of shade-tolerant and mid-tolerant tree species. Larger canopy openings (up to >1600 m^2) may occur in forests with very large emergent trees ([23], and references therein), but rarely allow the regeneration of pioneer shade-intolerant species (Coihue, Tineo [*Weinmannia trichosperma* Cav.]). As a consequence, old-growth forests are tall, complex in their structure and composition, and contain large amounts of biomass (average 539 ton ha^{-1} [23]), especially in very large trees (Figure 20.2).

Both continuous and gap regeneration modes are common in the EFT (Table 20.1). Continuous tree regeneration includes the shade-tolerant upper canopy hardwoods Tepa (*Laureliopsis philippiana* (Looser) Schodde), Olivillo (*Aextoxicon punctatum* Ruiz & Pav), several lower canopy species

Figure 20.2 Interior of an old-growth (left) and a secondary mixed-species forest (right) of the EFT, illustrating their differences in the vertical profiles and structural components.

of the Mirtaceae family, such as Luma (*Amomyrtus luma* (Mol) Legr et Kaus), Meli (*Amomyrtus meli* (Phil.) Legrand & Kausel), and Patagua (*Myrceugenia planipes* (Hook. & Arn.) Berg), and the Podocarpaceae Mañío Hembra (*Saxegothaea conspicua* Lindl.) and Mañío Macho (*Podocarpus nubigenus Lindl.*), the latter especially in poorly drained or low-quality soils. Gap regeneration favors mostly mid-tolerant species, which have low tree densities and usually reach greater sizes. These species include mainly Ulmo in larger openings, Canelo (*Drimys winteri* (J. R. Forster & G. Forster)) and Coihue de Chiloé (*Nothofagus nitida (Phil.) Krasser*) in wet or poorly drained soils, and Trevo (*Dassyphyllum diacanthoides* (Less.) Cabrera) in some locations. At lower elevations in the northern distribution of the EFT (200–500 m a.s.l.) other common canopy species may include the hardwoods Lingue (*Persea lingue* Nees) and Laurel (*Laurelia* sempervirens (Ruiz & Pav.) Tul.), and the Podocarpaceae Mañío de Hojas Largas (*Podocarpus salignus* D. Don.). Many small tree or arborescent species are also components of these forests.

Natural large-scale disturbances are infrequent in the EFT since this forest type occupies low elevations (mostly <600 m) in both the Andean and Coastal ranges [17], i.e. they are generally far away from the effects of volcanic eruptions or large landslides that are frequent in the Andes (in the lifetime of these hardwood forests [Table 20.1]). However, these events occasionally occur in some regions where low- to medium elevation Evergreen forests are close to volcanic cordons, generating secondary forests dominated by Coihue, Canelo, Ulmo, or other pioneer species (including relatively short-lived Proteaceae species). Early research on forest dynamics in Chilean rainforests by Veblen [24] referred to stand-devastating windstorms as a common disturbance in these

Figure 20.3 A landscape of standing dead trees and other legacies, plus abundant regeneration of shrubs, bamboos (Chusquea spp.), ferns, and some tree species, from seeds and sprouts, following theeruption of the Chaitén volcano (42° S Latitude) in 2008 in a region dominated by the Evergreen forest type (a). *Source:* Photo credit: Alvaro Promis. A Coihue tree with its characteristic buttress growing on a fallen log in a stand with no evidence of fire, therefore suggesting that the original stand-devastating disturbance was a windstorm (b). *Source:* Tomás Riquelme-Buitano.

forests based on the fact that old-growth forests had emergent trees of pioneer species (mostly Coihue and Tineo in the Andes, and Ulmo in the Coast). Nevertheless, these events were not documented in the EFT by this early research or subsequent studies, suggesting their rarity in the landscape (but see Figure 20.3).

Gutiérrez and Huth [23] described the different successional trajectories for these forests and highlighted that increasing basal area accumulation of shade-tolerant species becomes important and is stabilized after 500 years in terms of structure and composition. In absence of stand disrupting (large- and mid-severity) disturbances, the EFT could reach a stand condition similar to the *climax condition* dominated by shade-tolerant tree species, and with abundant understory vegetation. These conditions can be found mostly in the Coastal range, which has remained largely free of large-scale stand-disrupting disturbances (*sensu* [8, 24]).

20.2.3 Secondary Forest Development Following Human Disturbances

Fire and logging are the main disturbances of human origin in the EFT [25], and are followed by different initial successional pathways [17]: (i) areas containing abundant understory vegetation (*Chusquea* spp., ferns, etc.) that develop into a forest dominated by more shade-tolerant tree species, (ii) areas experiencing mineral soil exposure that develop into forests dominated by pioneer species, and (iii) areas dominated by stump sprouting and root suckering from different tree species across a range of shade tolerances and regeneration strategies, including those of seed origin, which develop into mixed forests (Figure 20.2). Fire or heavy logging may result in pathway (i), while fire generates the conditions in (ii), and partial logging the conditions in (iii). All of these successional pathways end with different composition and structure over time but may converge into similar forests with similar attributes following the death of pioneer species after several centuries. However, despite these different successional pathways, the common trend is that initial density and biomass or basal area accumulation is high in these forests [21, 22, 26, 27], especially in stands dominated by shade-intolerant Coihue in areas with medium- to well-drained soils (e.g. [22]) or by mid-tolerant Canelo, sometimes combined with Coihue de Chiloé, in poorly drained soils [21, 26].

20.3 Natural Developmental Model

The natural development model for EFT is based upon developmental patterns observed for younger stages following some large-scale disturbances, the known silvics of the species, and the reported development of secondary forests. For the older stages, we build from multiple studies of forest dynamics in old-growth forests.

20.3.1 Disturbance and Legacy Creation

The main disturbance in old-growth forests corresponds to gaps (of larger sizes when created following the death or fall of emergent trees) with legacy creation and cohort development happening at this scale. In contrast, secondary forests developing following forest fires (as those caused by volcanic eruptions) are abundant in dead trees and downed wood, plus exposed mineral soil across larger areas. As mentioned earlier, following these stand-replacing disturbances there are different possible trajectories, one of which may include the dominance of bamboos (*Chusquea* spp.) and ferns, creating a preforest stage that may last for decades if the abundance of *Chusquea* is such that it can arrest succession [28]. Otherwise, forest development proceeds quite rapidly ([21, 22]; Figure 20.4).

20.3.2 Young Forest Stage

Fire-origin secondary forests are quite pure in tree species composition, with one or two species densely packed in one vertical stratum, allowing little light to reach the forest floor. As a consequence, there is little understory layer until mortality of some canopy trees occurs and partially opens the canopy, which in Canelo forests occurs when trees reach around 15 cm in dbh [26, 29], and is likely similar in Coihue forests [22]. Usually in these forests, some relatively short-lived Proteaceae species (especially Avellano (*Gevuina avellana* Molina; [30]) and Notro (*Embothrium coccineum* J. R. Forst. & G. Forst.; [27]) are important component species. Secondary forests that develop after windstorms are rare, as mentioned earlier; they can also develop into high-density forests, but likely with more downed wood than those developed after fires.

20.3.3 Mature Forest Stage

Pioneer and long-lived shade-intolerant or mid-tolerant tree species have reached near 20–25 (Canelo) or 30–35 (Coihue) m in height by the Mature Forest Stage, and small canopy gaps create light conditions that trigger more abundant regeneration and the recruitment of established trees into the mid-story, resulting in a two-storied forest. These transitional forests are rich in tree species richness, which is similar to or greater than in old-growth forests [16, 30]. All these forests have low levels of coarse dead woody material present.

Table 20.2 A developmental model for evergreen forests in Chile, starting with a stand-replacing disturbance. Secondary forests developed from these disturbances are 60–100 years old (young to mature) so that the schemes for the mature stage and especially the old stage are based on studies of natural or seminatural forests since there is limited understanding of how current secondary forests transition to old-growth forests.

Developmental stage	Attributes	Cause and consequence	Schematic representation (successional stage)
Disturbance and legacy creation	Most of the canopy destroyed; large snags and downed wood	Establishment of hardwoods from seeds (shade-intolerant) or roots sprouts (more shade-tolerant species), mixed with *Chusquea* spp., ferns, etc.	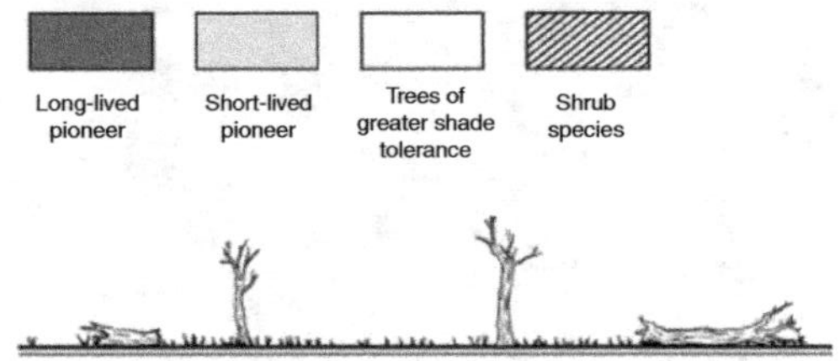
Young	Pure or mixed-species secondary forest dominated by hardwoods, likely even- or two-aged forests.	Results from major disturbance that exposes mineral soil or has stand-replacing effects.	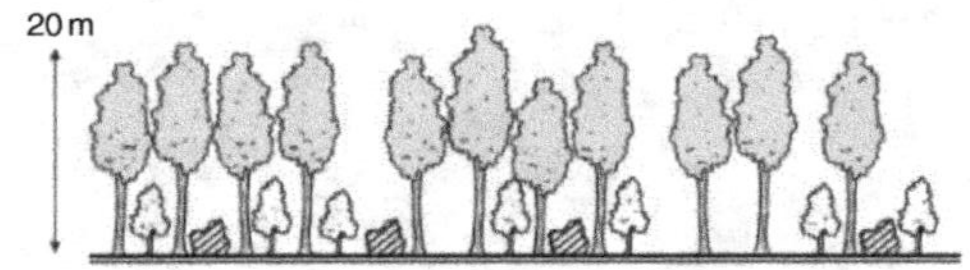
Mature	Low density of fast-growing pioneer species (intolerant or mid-tolerant to shade) with a well-defined mid-story of shade-tolerant species.	Results from continuous growth and mortality of the long-lived pioneer species, creating a scattered open canopy.	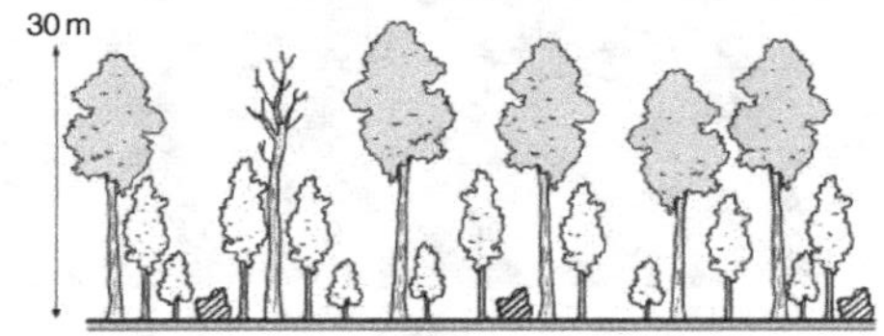
Old	Emergent scattered pioneer hardwood trees (except *D. winteri*), abundant canopy gaps, epiphytes, lianas, snags, and downed wood. Peak of biomass.	Highly variable light environment caused by the presence of some emergent trees, gaps of variable sizes following tree death, and a forest canopy dominated by shade-tolerant species.	

Figure 20.4 Variable-density thinning in secondary forests of the evergreen forest type in the Llancahue experimental forest in South Central Chile. (a) Mixed-species forest two to three years after a cut showing a 300 m^2 patch with two dead *Drimys winteri* trees in the background after being girdled. (b) Interior of the same thinned forest with a reserve (no cut) area in the background. (c) *Nothofagus dombeyi*-dominated secondary forest seven years after a cut with a 300 m^2 patch in the foreground, released *E. cordifolia* trees in the background and a girdled and dead snag of *N. dombeyi* in the right of the photo. (d) A closer look at a 300 m^2 patch, which has been covered by a suit of species, including short-lived pioneer arborescent species (pillo-pillo (*Ovidia pillopillo* (Gay) Meisn.), the tallest), and mostly mid-tolerant long-lived species.

20.3.4 Old Forest Stage

Old-growth forests have large basal areas, are multi-cohort and dominated by late-successional shade-tolerant species, and are complex in structure (vertical and horizontal) and composition [31]. Old-growth forests for the EFT differ basically in only one aspect, which is that those with *N. dombeyi* have large (>200 cm in dbh) and tall (>45 m) emergent trees, which also results in more basal area (additive basal area phenomenon; [32, 33]), and larger gaps. In the old-growth forests of the EFT, the upper canopy is 25–35 m tall, and have two- to three distinctive tree layers [11, 12]: they also include large snags, a rich understory, and abundant downed wood, plus multiple lianas and epiphytes. Although often dominated by shade-tolerant species, Ulmo is a mid-tolerant species that can opportunistically occupy canopy gaps especially due to its great root sprouting capacity and is able to persist as dominant or emergent in old-growth forests [7, 8, 16, 18, 23]. Emergent Ulmo trees in these old-growth forests are more common than emergent Coihue or Tineo trees, which are very shade-intolerant species.

20.4 An Ecological Silvicultural System for the Chilean Hardwood-Dominated Evergreen Forest Type

20.4.1 Disturbance and Legacy Creation

In the region where the EFT occurs, large-scale disturbances mostly occurred after human impacts during the last century, and their natural occurrence is quite rare. Managed forests where landowners aim to maintain the dominance of valuable and rapid-growth shade-intolerant Nothofagus species should maintain legacy trees, downed wood, and some horizontal variability throughout, in addition to enhancing tree diversity that may be common even in early stages due to the successful vegetative regeneration of many species. Under these conditions and also in the conditions associated with heavily high-graded forests with abundant downed wood and understory vegetation but a scarcity of regeneration (arrested succession; [28]), ecological silviculture should focus on release cuts, creating some snags if necessary, controlling dense thickets of understory vegetation, and conducting some improvement cuttings, especially focusing on releasing late-successional tree species. However, in following the natural developmental model for EFT, the disturbance and legacy creation stage is more appropriately applied and conceptualized at the gap or mesoscale given the dominant natural gap-phase disturbance regime for these systems.

20.4.2 Young Forest Stage

Variable-density thinning is an approach that can be used in the young forest stage to increase structural heterogeneity (vertical and horizontal) in dense forests by creating gaps or patches, no-cut areas, and thinning between these end points. Gaps or patches should follow the range of sizes close to those most commonly found in old-growth forests (<500 m^2, see above), with some also emulating less common, larger-sized gaps that occasionally naturally occur. Thinning may be heavy, down to 50% residual density, to increase light reaching the understory so as to promote regeneration of late-successional species. In these thinnings, the residual trees, including pioneer species like Coihue, Canelo, Avellano, and Notro, can have different fates: some could remain as crop trees to reach large sizes as early possible, some could be girdled to become snags, and others could be felled to become downed wood. Trial implementation of a VDT approach in a mixed forest and in a forest dominated by Coihue, both with simple structure (Figure 20.4), seems promising for accelerating old-growth attributes [34, 35], although bird species fitness did not show significant changes compared to unmanaged forests four years after VDT implementation [36]. The approach may also be appropriate for secondary forests dominated by Canelo, Ulmo, or a mixture of species.

Figure 20.5 An irregular shelterwood in a mature forest dominated by Coihue can release mid-tolerant and shade-tolerant tree species in the midstory (as some Canelo and Ulmo trees seen here), and promote understory growth to accelerate a greater complexity inthe managed forest.

20.4.3 Mature Forest Stage

Long-lived pioneer tree species are either dominant in the forest canopy or start to emerge above the main canopy at this stage, with mortality of some of these trees contributing to snags or downed wood, although not with the abundance found in old-growth forests. A heavy thinning focused on diminishing the density of these trees to increase light for the least dominant, mid canopy, and lower-canopy late-successional species, plus the understory, would accelerate the growth of the existing cohorts and the development of new cohorts. This can be approached as an irregular shelterwood cut (Figure 20.5; [37]) or even as a selection cut (Figure 20.6) if the forest has developed enough cohorts and relatively large-diameter trees (e.g. 50 cm).

20.4.4 Old Forest Stage

Forests in the old stage have a multilayered vertical structure with trees of different ages and a tree species composition dominated by shade-tolerant and mid-tolerant species so that the implementation of the single-tree selection system is recommended (Figure 20.6). Selection systems applied to these forests based on residual basal areas of 40–60 $m^2\ ha^{-1}$ did not result in significant changes in tree species richness and diversity but did significantly reduce diameter and height variability compared to old-growth forests, which naturally reach basal areas near 100 $m^2\ ha^{-1}$ ([38]; Figure 20.6). As a result, it is suggested that this approach retain a certain number of large-sized and dominant or emergent trees as permanent legacy trees to better emulate natural models of development [38]. Good growth rates and abundant regeneration of both mid-tolerant and shade-tolerant tree species have been reported in these forests in transition to fully managed forests under the selection system after only one cut [1, 20].

Figure 20.6 A stand with a selection cut immediately after its implementation (left) and 10 years after (right), which shows trees across all diameter classes and abundant tree and understory regeneration. Leaving standing old and very large trees, above the maximum residual diameter defined for the cut is essential to maintain the vertical and horizontal complexity of old-growth forests in managed settings.

20.5 Summary: Ecological Silviculture for Chilean Temperate Forests

An important element of sustainable forest management in landscapes historically managed with even-aged stands (as in South Central Chile) is to promote managed forests with tree mixtures and enhanced structural complexity to increase forest resilience and provisioning of ecosystem services [29, 39, 40]. While the natural developmental model of the forests of the EFT is mostly conducive to complex old-growth forests with a gap-phase dynamics that promotes the development of shade-tolerant and mid-tolerant tree species, there are several species of high timber value that require larger disturbances for regeneration and development and are desired by landowners. Therefore, silvicultural approaches are needed to meet the complexity of natural forests at different stages of development in managed forests. Several silvicultural approaches have been developed in Chile aimed at managing forest ecosystems to provide a wide range of goods and services while increasing (or maintaining) their structural complexity [1, 38, 41]. These have included restoration and variable-density thinning in secondary forests to increase their old-growth attributes [34, 35], irregular shelterwood where forest cover of mature trees is retained during extended periods of time (*sensu* [41]) to accommodate special management objectives [27, 38], and uneven-aged silviculture aimed at maintaining or increasing structural complexity and species diversity common in old-growth or mature forests [23, 37, 38]. Early results from all these approaches illustrate the feasibility of implementing them and their ability to provide timber or firewood (depending upon forest development, species, and the conservation status of the forest) for landowners.

While native forests in the landscape of South Central Chile should include even- and uneven-age management in response to the diversity of dominant human and natural disturbances [42], these forests should also contribute to mitigating climate change by storing carbon and should be diverse and structurally complex to be adaptive to climate change and disturbances. These features can be more easily met through multi-aged management systems [43]. As such, the conversion of some secondary forests to complex multi-aged forests should be promoted, especially considering the array of late-successional tree species of high timber value that could render the management of these mixed forests as profitable as even-aged pure forests.

Among forests that are already multi-aged, an irregular shelterwood may be convenient when the purpose includes the maintenance of more shade-intolerant species as part of a productive forest, i.e. species needing large canopy openings and variable degrees of soil disturbance. Otherwise, if more shade-tolerant species are favored, reflecting those species most common in old-growth forests, a selection system seems appropriate, varying residual basal area and cutting cycle length as needed, and maintaining legacy trees. Recent experiences with ecological silviculture in the EFT show that these approaches are possible to implement, demonstrating the potential of their application in other temperate native forests of Chile and Argentina [1, 40].

Acknowledgments

The authors thank FONDECYT grant N°1210147 from the Agencia Nacional de Investigación y Desarrollo (ANID), and Ignacio Gómez for the forest sketches.

References

1 Donoso, P.J., Promis, A., Loguercio, G.A. et al. (2022). Silviculture of South American temperate native forests. *New Zealand Journal of Forestry Science* 52: 2.

2 Veblen, T.T., Schlegel, F.M., Oltremari, J.V. 1983. Temperate broad-leaved evergreen forests of South America. En: Ovington JD (Ed.), Temperate Broad-Leaved Evergreen Forests. Amsterdam. Elsevier Science Publishers. pp. 5–31.

3 Gutiérrez, A.G., Armesto, J.J., Aravena, J.C. et al. (2009). Structural and environmental characterization of old-growth temperate rainforests of northern Chiloé Island, Chile: regional and global relevance. *Forest Ecology and Management* 258: 376–388.

4 Donoso, C. (1981). *Tipos forestales de los bosques nativos de Chile. Investigación y desarrollo forestal (CONAF/PNUD/FAO)*, Documento de Trabajo N° 38. Santiago, Chile: Publicación FAO. 83 p.

5 CONAF (Corporación Nacional Forestal) (2021). Cifras oficiales catastros usos de la tierra y recursos vegetacionales. https://sit.conaf.cl/exp/ficha.php (accessed August 2021).

6 Oliver, C.D. and Larson, B.C. (1996). *Forest Stand Dynamics*. Update edition. New York: Wiley. 520 p.

7 Donoso, C. (2006). *Las especies arbóreas de los bosques templados de Chile y Argentina. Autoecología*. Valdivia, Chile: Marisa Cuneo Ediciones. 688 p.

8 Donoso, C. (2015). *Estructura y dinámica de los bosques del Cono Sur de América*. Santiago, Chile: Editorial Universidad Mayor. 405 p.

9 Palik, B.J., D'Amato, A.W., Franklin, J.F., and Johnson, K.N. (2021). *Ecological Silviculture. Foundations and Applications*. Illinois: Waveland Press, Inc. 343 p.

10 D'Amato, A.W. and Palik, B.J. (2021). Building on the last "new" thing: exploring the compatibility of ecological and adaptation silviculture. *Canadian Journal of Forest Research* 51 (2): 172–180.

11 Oyarzún, A., Donoso, P.J., and Gutiérrez, A.G. (2019). Patrones de distribución de alturas de bosques antiguos siempreverde del centro-sur de Chile. *Bosque* 40 (3): 355–364.

12 Parada, T., Jara, C., and Lusk, C. (2003). Distribución de alturas máximas de especies en rodales antiguos de selva Valdiviana, Parque Nacional Puyehue. *Bosque* 24 (2): 63–67.

13 Veblen, T.T. 1992. Regeneration dynamics. En Glenn-Lewin C, RK Peet, TT Veblen (Eds.). Plant Succession: Theory and Prediction. London, Chapman and Hall. pp. 153–187.

14 Gutiérrez, A.G. (2016). Árboles monumentales: un patrimonio natural no reconocido en Chile. *Bosque* 37 (3): 445–449.

15 Lusk, C.H. (1996). Stand dynamics of the shade-tolerant conifers *Podocarpus nubigena* and *Saxegothaea conspicua* in Chilean temperate rain forest. *Journal of Vegetation Science* 7 (4): 549–558.

16 Donoso, P.J., Ponce, D., Pinto, J.B., and Triviño, I.l. (2018). Cobertura y regeneración arbórea en bosques siempreverdes en diferentes estados sucesionales en el sitio experimental de Llancahue, Cordillera de la Costa de Valdivia, centro sur de Chile. *Gayana Botánica* 75 (2): 657–662.

17 Donoso, C. (1989). Antecedentes básicos para la silvicultura del tipo forestal Siempreverde. *Bosque* 10 (1): 37–53.

18 Gutiérrez, A.G., Aravena, J.C., Carrasco-Farías, N.V. et al. (2008). Gap-phase dynamics and coexistence of a long-lived pioneer and shade-tolerant tree species in the canopy of an old-growth coastal temperate rain forest of Chiloé Island, Chile. *Journal of Biogeography* 35 (9): 1674–1687.

19 Donoso, C., Escobar, B., and Urrutia, J. (1985a). Estructura y estrategias regenerativas de un bosque virgen de Ulmo (*Eucryphia cordifolia* Cav.) - Tepa (*Laurelia philippiana* Phil.) Looser en Chiloé, Chile. *Revista Chilena de Historia Natural* 58: 171–186.

20 Donoso, P.J., Ojeda, P.F., Schnabel, F., and Nyland, R.D. (2020). Initial responses in growth, production, and regeneration following selection cuttings in hardwood-dominated temperate rainforests in Chile. *Forests* 11 (4): 412.

21 Navarro, C., Donoso, C., Sandoval, V. 1999. Los bosques secundarios de Canelo. En Donoso C, A Lara (Eds.). Silvicultura de los bosques nativos de Chile. Santiago, Chile. Editorial Universitaria. pp. 341–380.

22 Donoso, P.J., Cabezas, C., Lavanderos, A., and Donoso, C. (1999). Estudio comparativo de la estructura y crecimiento de renovales de Coihue (*Nothofagus dombeyi*) en la precordillera de la Costa y de los Andes de la provincia de Valdivia. *Bosque* 20 (2): 9–23.

23 Gutiérrez, A.G. and Huth, A. (2012). Successional stages of primary temperate rainforests of Chiloé Island, Chile. *Perspectives in Plant Ecology, Evolution and Systematics* 14 (4): 243–256.

24 Veblen, T.T. (1985). Forest development in tree-fall gaps in the temperate rain forests of Chile. *National Geographic Research* 1 (2): 162–183.

25 González, M.E., Szejner, P., Donoso, P.J., and Salas, C. (2015). Fire, logging and establishment patterns of second-growth forests in south-central Chile: implications for their management and restoration. *Ciencia e Investigación Agraria* 42 (3): 427–441.

26 Donoso, P.J., Soto, D.P., and Bertín, R.A. (2007). Size-density relationships in *Drimys winteri* secondary forests of the Chiloe Island, Chile: effects of physiography and species composition. *Forest Ecology and Management* 239: 120–127.

27 Donoso, P.J., Schlegel, B.C., Salas-Eljatib, C., and Donoso, C. (2019). Development of 26-year-old mixed forests following different regeneration cutting treatments in Andean temperate rainforests of south-central Chile. *Forest Ecology and Management* 432: 686–693.

28 Soto, D.P. and Puettmann, K.J. (2020). Merging multiple equilibrium models and adaptive cycle theory in forest ecosystems: implications for managing succession. *Current Forestry Reports* 6: 282–293.

29 Soto, D.P. and Donoso, P.J. (2006). Patrones de regeneración en renovales de *Drimys winteri* en el centro-norte de la Isla de Chiloé: cambios de acuerdo al tamaño y la densidad relativa. *Bosque* 27 (3): 241–249.

30 Donoso, P.J., Ponce, D., Salas, C. 2018b. Opciones de manejo para bosques secundarios de acuerdo a objetivos de largo plazo y su aplicación en bosques templados del centro-sur de Chile. En: Donoso P, A Promis, D Soto (Eds.), Silvicultura en bosques nativos. Experiencias en silvicultura y restauración en Chile, Argentina y el este de los Estados Unidos. Estudios en Silvicultura de Bosques Nativos. Volumen 2. Valdivia, Chile. OSU College of Forestry, Corvallis. pp. 93–116.

31 Ponce, D.B., Donoso, P.J., and Salas, C.E. (2017). Differentiating structural and compositional attributes across successional stages in Chilean temperate rainforests. *Forests* 8 (9): 329.

32 Donoso, P.J. and Lusk, C. (2007). Differential effects of emergent *Nothofagus dombeyi* on growth and basal area of canopy species in an old-growth temperate rainforest. *Journal of Vegetation Science* 18 (5): 675–684.

33 Donoso, P.J. and Soto, D.P. (2016). Does site quality affect the additive basal area phenomenon? Results from Chilean old-growth temperate rainforests. *Canadian Journal of Forest Research* 46 (11): 1330–1336.

34 Donoso, P.J., Puettmann, K.J., D'Amato, A.W. et al. (2020b). Short-term effects of variable-density thinning on regeneration in hardwood-dominated temperate rainforests. *Forest Ecology and Management* 464: 118058.

35 Salas-Eljatib, C., Riquelme, J., Donoso, P.J. et al. (2022). Analysing changes in spatial point patterns: a proposal using data from a forest thinning experiment. *Forest Ecosystems* 9: 100081.

36 Biscarra, G., McFadden, T.N., Donoso, P.J. et al. (2021). Short-term avian responses to experimental variable-density thinning: initial stand attributes outweigh effects of management in Austral temperate rainforests. *Forest Ecology and Management* 502: 119711.

37 Donoso, P.J., Soto, D.P., Salas-Eljatib, C., and Riquelme-Buitano, T.R. (2022b). Present and prospects for multi-aged silviculture in Chilean temperate forests: targeting secondary forests in transition and partially harvested old-growth forests. *Frontiers Forest and Global Change* 5: 965599. https://doi.org/10.3389/ffgc.2022.965599.

38 Schnabel, F., Donoso, P.J., and Winter, C. (2017). Short-term effects of single-tree selection cutting on stand structure and tree species composition in Valdivian rainforests of Chile. *New Zealand Journal of Forest Science* 47: 21.

39 Messier, C., Puettmann, K., and Coates, D. (2013). *Managing Forests as Complex Adaptive Systems.* New York, Chile: Routledge. 353 p.

40 Loguercio, G.A., Donoso, P.J., Müller-Using, S, Dezzotti, A., Urretavizcaya, M.F., Navarro, C.O., Martin, M., Schlegel, B., Müller-Using, B., Mujica, R., González Peñalba, M., Attis Beltrán, H., Caselli, M. 2017. Silviculture of temperate mixed forests from South America. In: pg. 271–317, A. Bravo-Oviedo, Hans Pretzsch, Miren del Río (eds.), Dynamics, Silviculture and Management of Mixed Forests, Managing Forest Ecosystems 31, Springer International Publishing AG.

41 Raymond, P., Bédard, S., Roy, V. et al. (2009). The irregular shelterwood system: review, classification, and potential application to forests affected by partial disturbances. *Journal of Forestry* 107: 405–413.

42 Drever, C.R., Peterson, G., Messier, C. et al. (2006). Can forest management based on natural disturbances maintain ecological resilience? *Canadian Journal of Forest Research* 36 (9): 2285–2299.

43 D'Amato, A.W., Bradford, J.B., Fraver, S., and Palik, B.J. (2011). Forest management for mitigation and adaptation to climate change: insights from long-term silviculture experiments. *Forest Ecology and Management* 262 (5): 803–816.

21

The Place of Ecological Silviculture, Now and in the Future

Anthony W. D'Amato[1] *and Brian J. Palik*[2]

[1] *Rubenstein School of Environment and Natural Resources, University of Vermont, Burlington, VT, USA*
[2] *USDA Forest Service, Northern Research Station, Grand Rapids, MN, USA*

21.1 Introduction

Elements of ecological silviculture have been in existence for over a century [1], yet many of the principles and applications did not emerge broadly until the 1990s (e.g. [2]), with particular emphasis on Douglas-fir-dominated forests of the Pacific Northwest [3]. The chapters in this book highlight how far this silvicultural model has evolved over the past several decades, both in geographic extent and the diversity of forest ecosystems and management contexts for which natural models have been applied. In contrast to generic mischaracterizations of ecological silviculture as solely being concerned with postharvest retention, the chapters in this volume have clearly demonstrated the art and creativity associated with practicing ecological silviculture and the resultant diversity of silvicultural approaches now being applied across various regions of the globe. In this chapter, we review the ecological silvicultural systems presented in preceding chapters, looking at commonalities and differences that reflect varying types of forests, societal concerns and objectives, and economic environments and what the growth opportunities are moving forward.

21.2 A Diversity of Approaches for a Diversity of Forests

In developing this book, we hoped to illustrate a range of ecosystems and regions where natural models of development were being used to inform ecological silviculture. In total, we showcased 19 ecological silvicultural systems spanning 19 distinctly different forests across six countries and four continents (Figure 21.1). Our coverage was heavily skewed toward North American forests with 14 contributions coming from either the United States, Canada, or both. The heavy contribution from North America reflects both our professional experience, as well as the state of the development in the field as early applications of ecological silviculture in North America often served as a catalyst for later exploration and adoption in other countries [4]. Nevertheless, similar ideas have long been pursued elsewhere, (e.g. close to nature forestry in Europe; [5]), but have a distinctly

Ecological Silvicultural Systems: Exemplary Models for Sustainable Forest Management,
First Edition. Edited by Brian J. Palik and Anthony W. D'Amato.

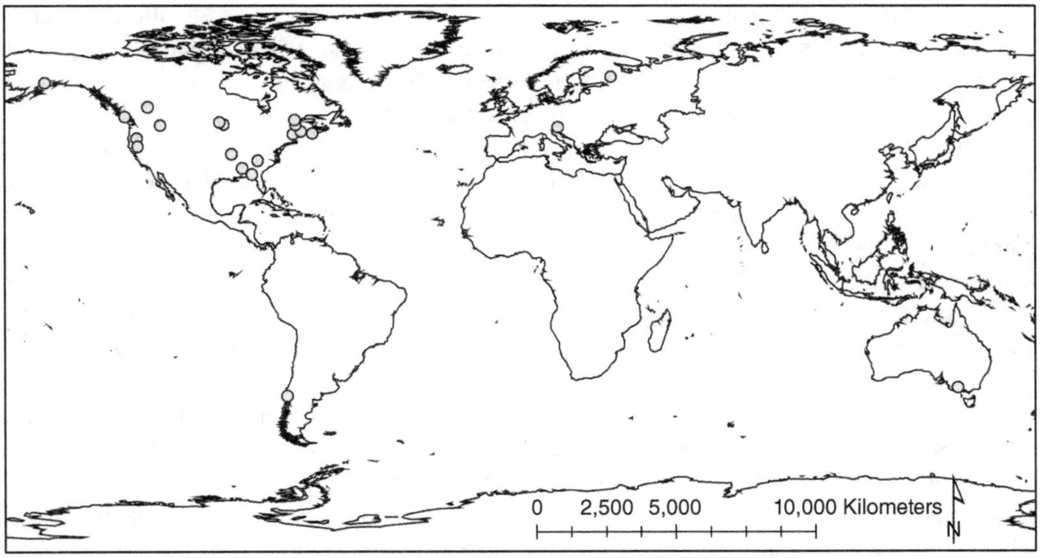

Figure 21.1 General locations of forest systems and associated ecological silvicultural systems highlighted in this text.

different approach than one based on emulation of a natural model. Regardless, the growth in application of ecological silviculture within North America and other countries reflects increasing recognition of the potential shortcomings of traditional silviculture for sustaining the full range of ecosystem services. Additionally, on a positive note, this growth reflects both increasing operational experience with implementing these strategies, as well the consistency of outcomes of ecological silviculture with societal goals around biodiversity conservation and ecosystem resilience and associated forest policies and certification schemes. As such, we would expect future compilations of ecological silviculture in practice to extend far beyond the geography and bioclimatic domains covered in this initial volume.

21.2.1 Commonalities Among Ecological Silvicultural Approaches

Despite the diversity of forest ecosystems covered by this book, a common element is for application of ecological silviculture in the context of forests simplified by historical, intensive land use and timber-focused silviculture. This includes secondary forests developing following widespread harvesting, or recovering of land cleared for agriculture, as well as restoration of industrial plantations. The desire to restore complexity in composition, structure, and dynamics to these forests represents a broad recognition of the evolution of management goals that are inclusive of sustainability of ecosystem services broadly, maintenance of native biota, and adaptation to changing global conditions. Although traditional silvicultural models exist for many of the forests and regions examined, often focusing on sustainability of commodity outputs, the collection of chapters in this text illustrates these are not always sufficient for meeting these other diverse goals.

Importantly, the silvicultural systems we present are inherently more ecological by design, rather than simple modifications to traditional timber-focused approaches. Often these approaches are based on a deeper understanding and emulation of natural models of development, including the natural disturbance dynamics for an ecosystem of interest and the corresponding patterns in structural and compositional development. Of note, in each included forest attention is given to regeneration methods and intermediate treatments necessary to sustain economically important tree species in these forests, underscoring the necessity to continue sustaining wood production from these forests, but with a shift toward ecological models to do so.

From a purely silvicultural sense, all of the chapters demonstrate the growing application of multi-aged silvicultural methods to achieve diverse management objectives (Figure 21.2). This is not surprising given the greater consistency between multi-aged approaches and natural developmental dynamics for most, if not all, global forest ecosystems; however, it stands in stark contrast with the silvicultural methods currently providing almost half of our global wood needs (i.e. even-aged plantation systems; [6]). As has been pointed out elsewhere, there is a continued need for areas of intensive production to minimize the landbase required to meet wood demands, as this in turn increases area available for ecological reserves and the practice of ecological silviculture [7]. Yet, the incongruity between ecological conditions created by intensive models and the values and services society increasingly expects argues for an expansion of ecological systems to a larger portion of the landscape, in fact, the majority of the global forest might best be managed using an ecological approach. Our goal with this book has been to demonstrate the many domains across which this expansion is possible, with many more likely not captured by this work.

In examining the regeneration methods used in the various chapters, there is general consistency in methods chosen based on prevailing natural disturbance regime (Figure 21.2). For example, variable retention harvest (VRH) were most commonly used in forests characterized by infrequent severe disturbance, such as Douglas-fir-western hemlock forests in the Pacific Northwest and mountain ash forests in Australia. In contrast, these methods were rarely used in forests dominated by frequent gap-scale disturbances (e.g. Chilean temperate forests, Central European mixed

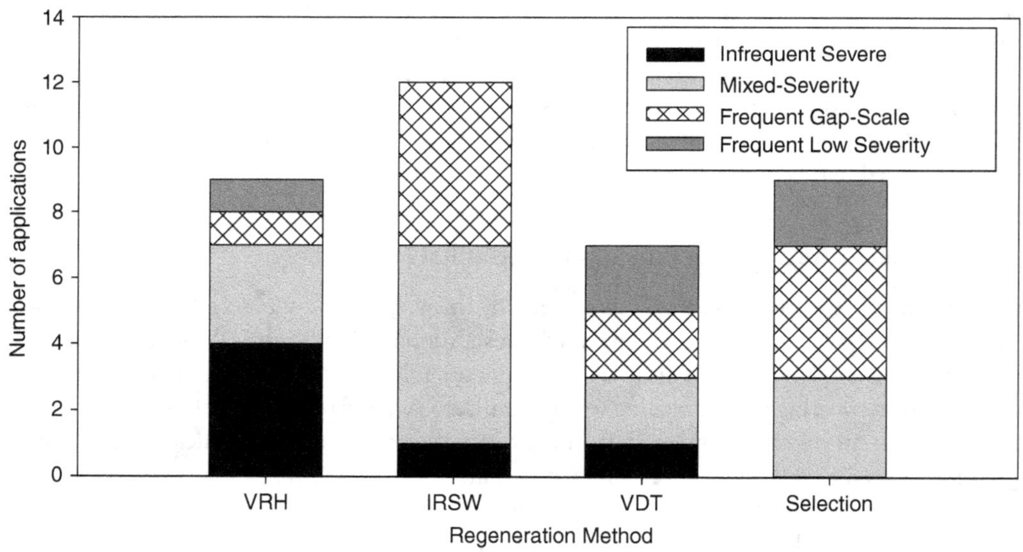

Figure 21.2 Frequency of regeneration methods used in ecological silvicultural systems presented in this text in relation to prevailing natural disturbance regimes for a given forest type. Most ecological silvicultural systems presented included multiple potential regeneration methods as options. VRH, variable retention harvest methods; IRSW, irregular shelterwood methods; VDT, variable density thinning methods; Selection, selection methods.

species forests) or frequent low-severity fire regimes (e.g. longleaf pine in the southeastern United States). For these ecosystems, selection systems, variable density thinning, and irregular shelterwood methods were often commonly applied to approximate the fine-scale mosaics generated by natural disturbance over time in these forests. Not surprisingly, forests characterized by mixed-severity disturbance regimes included examples from each multi-aged regeneration method presented in Figure 21.2, reflecting the variation in disturbance scale, severity, and frequency associated with these forests. Overall, irregular shelterwoods were the most commonly employed method across the ecosystems presented, due to their flexibility in accommodating diverse forest compositions, structures, and disturbance regimes and the consistency of silvicultural outcomes associated with these approaches with those following natural dynamics for many forest ecosystems [8].

21.2.2 Differences Among Ecological Silvicultural Approaches

There is no one-size-fits-all approach to ecological silviculture, as the underlying guide is applying the natural development model associated with a given ecosystem, in combination with creativity of the forester. As such, it is not surprising that differences existed among contributors regarding the ecological silvicultural systems developed, as there was wide variation in interpretations of natural models of development as a basis for a silvicultural system. Consequently, not all contributors present their silvicultural systems using the type of developmental model we outlined in Chapter 1, although the similarities largely outweigh the differences.

Where differences did exist in a silvicultural system compared to our generalized model (Chapter 1), they were often related to (i) differences in scale of emphasis (e.g. approaches to landscape allocation of harvesting strategies versus stand-level prescriptions); (ii) reliance on historical silvicultural models and frameworks for presenting silvicultural strategies; or (iii) limited examples of ecological silviculture in practice for a region. These differences point to the nascent stage of ecological silviculture relative to traditional models in some regions and the opportunities for growth where operational translation of natural models into on-the-ground practice has not yet occurred.

Additionally, the differences among contributions also reflect varying opinions among authors on the usefulness of historical silvicultural terminology and concepts for describing ecological silvicultural prescriptions. Although aspects of historical silvicultural systems are useful starting points for developing the regeneration methods and other silvicultural treatments contained within an ecological silvicultural system, the timber-focused intent of these strategies results in key differences between the desired outcomes of ecological silviculture and timber-focused models. As pointed out in Chapter 1, and also reflected in several of the chapters, it is imperative that terminology and silvicultural approaches evolve to encompass the methods being used (or desired) by stakeholders to meet diverse objectives beyond the regions presented in this book.

Finally, there are some notable differences in tone across chapters regarding the ability of an ecological silvicultural approach to be widely adopted and make a meaningful difference in how forests are managed. This very much speaks to the socioeconomic and management contexts within which natural models are being applied, with some regions facing significant hurdles to operationalize ecological silviculture given heavy current reliance on traditional models for supplying wood. For example, the prevalence of widespread, industrial clear-cutting in Australia (Chapter 18) has generated significant barriers to implementation of ecological silviculture, including mismatches with industry expectations of short-term forest yields, limited manager experience with alternative silvicultural systems, and reduced social trust in forest management given a legacy of poor ecological outcomes. Similarly, areas such as the Pacific Northwest or southeastern United States in which high-yield plantations are the dominant management model represent a challenging environment to promote ecological silviculture, given the emphasis on short-term commodity production. However, even in these regions, increased public demands for diverse outputs from forests have motivated policies and certification approaches that encourage application of silviculture following natural models. This, coupled with the growing application or expansion potential of ecological silviculture in many regions addressed in this book, underscores the value of continued experimentation and evolution of local approaches building from natural models, even if current contexts are less conducive to widespread application.

21.3 Conclusions

The last decade has been a time of significant change in regard to social expectations and values tied to forests, as well as the environmental stressors and challenges facing the long-term sustainability of the services they provide. Changing values have included greater emphasis on enhancing carbon sequestration and storage benefits, conserving native biodiversity, restoring Indigenous access and rights, adapting to changing climate, disturbances, and invasive pests, and increasing opportunities for wellness and renewal.

During this same time, there have been growing calls to eliminate forest management as a tool, particularly from large areas of native forest, due to public reactions to clear-cutting and elevated focus on carbon storage benefits of older, unmanaged forests. This growing sentiment, particularly in more affluent regions [9], starkly juxtaposes against the continued increases in the proportion of global wood provided by industrial plantations and suggests we may be further careening into the bifurcated global forest estate proposed by Bennett [10]. Such a bifurcation, in which there is no middle ground between protected areas and intensive plantations, represents an unacceptable solution from both a global equity perspective, as well as in regard to the future of the forestry profession. If we cannot be entrusted to manage native forests for the multiple benefits and objectives society expects, then we have failed at achieving what has been a central tenet of forestry and silviculture for almost a century; managing forests to benefit people.

We argue that ecological silviculture, as evidenced by the diversity of contributions to this text, is uniquely poised to provide a fruitful pathway forward for our profession by growing the middle ground between the two endpoints of the forest management bifurcation using a natural models approach. These contributions showcase the potential for silvicultural strategies across diverse geographies that address the challenge of sustaining biodiversity, and all other aspects of ecosystems, with potential application over the bulk of the forest estate in ways that satisfy the public's desire for a wide range of ecosystem services. Forests arguably have never been more highly valued by society, yet they also have never been so threatened by changing environmental dynamics, necessitating the need for creativity, balance, and an adaptive mindset shared among ecological foresters. We believe, as do many of the contributors to this book, that growth of the middle ground and wider application of ecological silviculture provides the pathway needed to sustain and elevate the role of foresters as trusted professionals for meeting the challenges described above.

References

1 D'Amato, A.W., Palik, B.J., Franklin, J.F., and Foster, D.R. (2017). Exploring the origins of ecological forestry in North America. *Journal of Forestry* 115 (2): 126–127.

2 Seymour, R.S. and Hunter, M.L. (1992). *New Forestry in Eastern Spruce-Fir Forests: Principles and Applications to Maine*. Orono, ME: College of Forest Resources, University of Maine.

3 Franklin, J.F., Berg, D.R., Thornburgh, D.A., and Tappeiner, J.C. (1997). Alternative silvicultural approaches to timber harvesting: variable retention harvest systems. In: *Creating a Forestry for the 21st Century* (ed. K.A. Kohm and J.F. Franklin), 111–140. Washington, DC: Island Press.

4 Gustafsson, L., Baker, S.C., Bauhus, J. et al. (2012). Retention forestry to maintain multifunctional forests: a world perspective. *Bioscience* 62 (7): 633–645.

5 Schutz, J.P. (1999). Close-to-nature silviculture: is this concept compatible with species diversity? *Forestry* 72 (4): 359–366.

6 Payn, T., Carnus, J.-M., Freer-Smith, P. et al. (2015). Changes in planted forests and future global implications. *Forest Ecology and Management* 352: 57–67.

7 Himes, A., Betts, M., Messier, C., and Seymour, R. (2022). Perspectives: thirty years of triad forestry, a critical clarification of theory and recommendations for implementation and testing. *Forest Ecology and Management* 510: 120103.

8 Raymond, P., Bédard, S., Roy, V. et al. (2009). The irregular shelterwood system: review, classification, and potential application to forests affected by partial disturbances. *Journal of Forestry* 107 (8): 405–413.

9 Moomaw, W.R., Masino, S.A., and Faison, E.K. (2019). Intact Forests in the United States: Proforestation Mitigates Climate Change and Serves the Greatest Good. *Frontiers in Forests and Global Change* 2: https://doi.org/10.3389/ffgc.2019.00027.

10 Bennett, B.M. (2015). *Plantations and Protected Areas: A Global History of Forest Management*. The MIT Press.

Index

Ecological Silvicultural Systems: Exemplary Models for Sustainable Forest Management, First Edition. Edited by Brian J. Palik and Anthony W. D'Amato.

b

C

d

f

n

O

S

y

www.ingramcontent.com/pod-product-compliance
Lightning Source LLC
LaVergne TN
LVHW080526210926
848481LV00022B/3098